Concepts in Viral Pathogenesis III

Concepts in
Viral Pathogenesis III

Edited by
Abner Louis Notkins
Michael B.A. Oldstone

With 60 Illustrations, Including 5 Color Illustrations

Springer-Verlag
New York Berlin Heidelberg
London Paris Tokyo

ABNER LOUIS NOTKINS, M.D.
Laboratory of Oral Medicine, National Institute of Dental Research,
NIH, Bethesda, Maryland 20892, U.S.A.

MICHAEL B.A. OLDSTONE, M.D.
Department of Immunology, Scripps Clinic and Research
Foundation, La Jolla, California 92037, U.S.A.

Cover Photograph: Computer graphic representation of the structure of the human class I histocompatibility antigen, HLA-A2 (blue), with a deep groove (red) representing the antigen recognition site. The structure is from 3.5 Å resolution of electron density maps as reported by Bjorkman et al. (*Nature 329:*506-518, 1987). The peptide (presumed to be the density in the groove) is restricted by class I MHC and recognized by virus-specific cytotoxic thymus-derived lymphocytes. Photomicrograph supplied by Don C. Wiley, Harvard University. Reprinted with permission from *Nature,* Copyright © Macmillan Magazine Limited.

Library of Congress Cataloging-in-Publication Data
Concepts in viral pathogenesis III/edited by Abner Louis Notkins,
  Michael B.A. Oldstone.
    p. cm.
    Includes bibliographies and index.
    1. Virus diseases—Pathogenesis. I. Notkins, Abner Louis.
  II. Oldstone, Michael B.A. III. Title: Concepts in viral pathogenesis 3.
    [DNLM: 1. Virus Diseases. 2. Viruses—pathogenicity. QW 160 C7442]
  QR201.V55C66 1989
  616'.0194—dc19
  DNLM/DLC   89-5940

Printed on acid-free paper

Typeset by TCSystems, Inc., Shippensburg, Pennsylvania.
Printed and bound by R.R. Donnelley & Sons, Harrisonberg, Virginia.
Printed in the United States of America.

9 8 7 6 5 4 3 2 1

ISBN 0-387-96974-8 Springer-Verlag New York Berlin Heidelberg
ISBN 3-540-96974-8 Springer-Verlag Berlin Heidelberg New York

Preface

Volume III of *Concepts in Viral Pathogenesis* follows the format established in Volumes I and II. It consists of 43 mini-reviews/editorials, about 1,500 words in length, on research that is at the cutting edge of virology. The purpose of this series is to allow busy investigators, teachers, and students to have available, in one place, a core of up-to-date information in brief and easily readable form.

The content of Volume III is entirely different from that of Volumes I and II, covering different areas and written by different authors. The book is divided into seven sections. It begins with chapters on viral structure/function relationships and gene regulation, and then turns to the cellular biology of viral infections. The next section deals with expression of viral genes in transgenic mice, retroviruses, and AIDS. The book ends with chapters on evolving concepts in viral infections and the control of viral disease.

Together, Volumes I, II, and III of *Concepts in Viral Pathogenesis* consist of 145 chapters written by leading investigators in the field.

Abner Louis Notkins, M.D.
Bethesda, Maryland

Michael B.A. Oldstone, M.D.
La Jolla, California

Contents

Cell Biology of Viral Infections

Transgenic Mice: Expression of Viral Genes

Retroviruses

Evolving Concepts in Viral Diseases

Control of Viral Diseases

Contributors

RAFI AHMED, Department of Microbiology and Immunology, UCLA School of Medicine, Los Angeles, CA 90024-1747, U.S.A.

SUSAN ALPERT, Cold Spring Harbor Laboratory, P.O. Box 100, Cold Spring Harbor, NY 11724, U.S.A.

MINAS ARSENAKIS, Department of Genetics, Development and Molecular Biology, Aristotelian University of Thessaloniki, Thessaloniki 54006, Greece

VICTORIA L. BAUTCH, Cold Spring Harbor Laboratory, P.O. Box 100, Cold Spring Harbor, NY 11724, U.S.A.

THOMAS L. BENJAMIN, Department of Pathology, Harvard Medical School, Boston, MA 02115, U.S.A.

DAVID H.L. BISHOP, NERC Institute of Virology, Mansfield Road, Oxford OX1 3SR, England

WILLIAM A. BLATTNER, Viral Epidemiology Section, National Cancer Institute, NIH, Bethesda, MD 20892, U.S.A.

SAMUEL BRODER, Director, National Cancer Institute, NIH, Bethesda, MD 20892, U.S.A.

PAOLO CASALI, Laboratory of Oral Medicine, National Institute of Dental Research, NIH, Building 30, Room 121, Bethesda, MD 20892, U.S.A.

FRANCIS V. CHISARI, Research Institute of Scripps Clinic, 10666 N. Torrey Pines Road, La Jolla, CA 92037, U.S.A.

FRANCOIS CLAVEL, Laboratory of Molecular Microbiology, National Institute of Allergy and Infectious Diseases, NIH, Bethesda, MD 20892, U.S.A.

TOD CRITCHLOW, Department of Molecular Biology, Research Institute of Scripps Clinic, 10666 N. Torrey Pines Road, La Jolla, CA 92037, U.S.A.

CLYDE S. CRUMPACKER, Division of Infectious Diseases, Beth Israel Hospital, Boston, MA 02115, U.S.A.

CLYDE J. DAWE, Department of Pathology, Harvard Medical School, Boston, MA 02115, U.S.A.

ROBERT W. DOMS, Department of Cell Biology, Yale University School of Medicine, New Haven, CT 06510, U.S.A.

THOMAS W. DUBENSKY, Department of Pathology, Harvard Medical School, Boston, MA 02115, U.S.A.

FRANK J. DUTKO, Sterling-Winthrop Research Institute, Rensselaer, NY 12144, U.S.A.

DAVID J. FILMAN, Department of Molecular Biology, Research Institute of Scripps Clinic, 10666 N. Torrey Pines Road, La Jolla, CA 92037, U.S.A.

ROBERT FREUND, Department of Pathology, Harvard Medical School, Boston, MA 02115, U.S.A.

D. CARLETON GAJDUSEK, Laboratory of Central Nervous System Studies, National Institute of Neurological Disorders and Stroke, NIH, Bethesda, MD 20892, U.S.A.

HAROLD S. GINSBERG, Departments of Medicine and Microbiology, College of Physicians and Surgeons, Columbia University, New York, NY 10032, U.S.A.

JOHN W. GNANN, JR., University of Alabama at Birmingham, 229 Tinsley Harrison Tower, University Station, Birmingham, AL 35294, U.S.A.

ERICA GOLEMIS, Biology Department and Center for Cancer Research,

Massachusetts Institute of Technology, 77 Massachusetts Avenue, Cambridge, MA 02139, U.S.A.

DOUGLAS HANAHAN, Cold Spring Harbor Laboratory, P.O. Box 100, Cold Spring Harbor, NY 11724, U.S.A.

STEPHEN C. HARRISON, Howard Hughes Medical Institute and Department of Biochemistry and Molecular Biology, Harvard University, 7 Divinity Avenue, Cambridge, MA 02138, U.S.A.

ALAN J. HAY, Division of Virology, National Institute for Medical Research, Mill Hill, London NW7 1AA, England

ARI HELENIUS, Department of Cell Biology, Yale University School of Medicine, New Haven, CT 06510, U.S.A.

OKIO HINO, Department of Pathology, Cancer Institute, Kami-Ikebukuro, Toshima-ku, Tokyo 170, Japan

PETER HANS HOFSCHNEIDER, Max-Planck-Institut für Biochemie, D-8033 Martinsried, F.R.G.

JAMES M. HOGLE, Department of Molecular Biology, Research Institute of Scripps Clinic, 10666 N. Torrey Pines Road, La Jolla, CA 92037, U.S.A.

KATHRYN V. HOLMES, Department of Pathology, Uniformed Services University of the Health Sciences, Bethesda, MD 20814, U.S.A.

NANCY HOPKINS, Biology Department and Center for Cancer Research, Massachusetts Institute of Technology, 77 Massachusetts Avenue, Cambridge, MA 02139, U.S.A.

COLIN R. HOWARD, London School of Hygiene and Tropical Medicine, Keppel Street, London WC1E 7HT, England

PETER M. HOWLEY, Laboratory of Tumor Virus Biology, National Cancer Institute, NIH, Bethesda, MD 20892, U.S.A.

JAMES A. HOXIE, Department of Medicine, University of Pennsylvania School of Medicine, Philadelphia, PA 19104, U.S.A.

DAVID HUSO, Division of Comparative Medicine, The Johns Hopkins University School of Medicine, Meyer Building, Room 6-181, 600 N. Wolfe Street, Baltimore, MD 21205, U.S.A.

DAVID JACOBSON, Department of Molecular Biology, Research Institute of Scripps Clinic, 10666 N. Torrey Pines Road, La Jolla, CA 92037, U.S.A.

OSWALD JARRETT, University of Glasgow Veterinary School, Bearsden, Glasgow G61 1QH, Scotland

GILBERT JAY, NIH, Building 41, Room A101, Bethesda, MD 20892, U.S.A.

PAULINE JOLLY, Division of Comparative Medicine and Department of Neurology, The Johns Hopkins University School of Medicine, Meyer Building, Room 6-181, 600 N. Wolfe Street, Baltimore, MD 21205, U.S.A.

REINHARD KANDOLF, Max-Planck-Institut für Biochemie, D-8033 Martinsried; and Medizinische Klinik I, Klinikum Großhadern der Universität München, D-8000 München 70, F.R.G.

ALEXANDER KRÄMER, Viral Epidemiology Section, National Cancer Institute, NIH, Bethesda, MD 20892, U.S.A.

PAUL F. LAMBERT, Laboratory of Tumor Virus Biology, National Cancer Institute, NIH, Bethesda, MD 20892, U.S.A.

ARNOLD J. LEVINE, Department of Molecular Biology, Princeton University, Princeton, NJ 08544, U.S.A.

JAY A. LEVY, Department of Medicine, University of California, School of Medicine, San Francisco, CA 94143, U.S.A.

DOUGLAS R. LOWY, Laboratory of Cellular Oncology, National Cancer Institute, NIH, Building 37, Room 1B-26, Bethesda, MD 20892, U.S.A.

PRESTON A. MARX, California Primate Research Center, University of California, Davis, CA 95616, U.S.A.

ALISON A. MCBRIDE, Laboratory of Tumor Virus Biology, National Cancer Institute, NIH, Bethesda, MD 20892, U.S.A.

MARK A. MCKINLAY, Sterling-Winthrop Research Institute, Rensselaer, NY 12144, U.S.A.

VOLKER TER MEULEN, Institut fur Virologie und Immunobiologie, Universitat Wurzburg, Versbacher Strasse 7, 8700 Wurzburg, F.R.G.

LUC MONTAGNIER, Unite'd'oncologie Virale, Institut Pasteur, Paris, 75724 Cedex 15, France

OPENDRA NARAYAN, Division of Comparative Medicine and Department of Neurology, The Johns Hopkins University School of Medicine, Meyer Building, Room 6-181, 600 N. Wolfe Street, Baltimore, MD 21205, U.S.A.

MICHAEL I. NERENBERG, Department of Immunology, Scripps Clinic and Research Foundation, 10666 N. Torrey Pines Road, La Jolla, CA 92037, U.S.A.

ABNER LOUIS NOTKINS, Laboratory of Oral Medicine, National Institute of Dental Research, NIH, Building 30, Room 121, Bethesda, MD 20892, U.S.A.

MICHAEL B.A. OLDSTONE, Department of Immunology, Scripps Clinic and Research Foundation, 10666 N. Torrey Pines Road, La Jolla, CA 92037, U.S.A.

PER A. PETERSON, Department of Immunology, Scripps Clinic and Research Foundation, 10666 N. Torrey Pines Road, La Jolla, CA 92037, U.S.A.

JOHN L. PORTIS, Laboratory of Persistent Viral Diseases, National Institute of Allergy and Infectious Diseases, NIH, Rocky Mountain Laboratories, Hamilton, MT 59840, U.S.A.

GREGORY PRINCE, Laboratory of Infectious Diseases, National Institute of Allergy and Infectious Diseases, NIH, Bethesda, MD 20892, U.S.A.

VINCENT R. RACANIELLO, Department of Microbiology, Columbia University College of Physicians and Surgeons, New York, NY 10032, U.S.A.

JAY RAPPAPORT, Laboratory of Tumor Cell Biology, National Cancer Institute, NIH, Building 37, Room 6A09, Bethesda, MD 20892, U.S.A.

CHARLES E. ROGLER, Marion Bessin Liver Research Center, Departments of Medicine and Microbiology and Immunology, Albert Einstein College of Medicine, Bronx, NY 10461, U.S.A.

BERNARD ROIZMAN, The Marjorie B. Kovler Viral Oncology Laboratories, The University of Chicago, Chicago, IL 60637, U.S.A.

MICHAEL G. ROSSMANN, Purdue University, West Lafayette, IN 47907, U.S.A.

JOHN T. SCHILLER, Laboratory of Cellular Oncology, National Cancer Institute, NIH, Building 37, Room 1B-26, Bethesda, MD 20892, U.S.A.

DAVID A. SHAFRITZ, Marion Bessin Liver Research Center, Departments of Medicine and Cell Biology, Albert Einstein College of Medicine, Bronx, NY 10461, U.S.A.

THOMAS SHENK, Department of Biology, Princeton University, Princeton, NJ 08544, U.S.A.

MAJA A. SOMMERFELT, Institute of Cancer Research, Chester Beatty Laboratories, Fulham Road, London SW3 6JB, England

BARBARA A. SPALHOLZ, Laboratory of Tumor Virus Biology, National Cancer Institute, NIH, Bethesda, MD 20892, U.S.A.

NANCY SPECK, Biology Department and Center for Cancer Research, Massachusetts Institute of Technology, 77 Massachusetts Avenue, Cambridge, MA 02139, U.S.A.

TOON STEGMANN, Department of Cell Biology, Yale University School of Medicine, New Haven, CT 06510, U.S.A.

CHARLES B. STEPHENSEN, Department of Pathology, Uniformed Services University of the Health Sciences, Bethesda, MD 20814, U.S.A.

MAX D. SUMMERS, Texas Agricultural Experiment Station, Department of Entomology, Texas A & M University, College Station, TX 77843, U.S.A.

RASHID SYED, Department of Molecular Biology, Research Institute of Scripps Clinic, 10666 N. Torrey Pines Road, La Jolla, CA 92037, U.S.A.

DAVID A. TALMAGE, Department of Pathology, Harvard Medical School, Boston, MA 02115, U.S.A.

HOWARD M. TEMIN, McArdle Laboratory, University of Wisconsin, Madison, WI 53706, U.S.A.

ROBERT R. WAGNER, University of Virginia, Charlottesville, VA 22903, U.S.A.

ROBIN A. WEISS, Institute of Cancer Research, Chester Beatty Laboratories, Fulham Road, London SW3 6JB, England

J. LINDSAY WHITTON, Department of Immunology, Scripps Clinic and Research Foundation, 10666 N. Torrey Pines Road, La Jolla, CA 92037, U.S.A.

RICHARD K. WILLIAMS, Department of Pathology, Uniformed Services University of the Health Sciences, Bethesda, MD 20814, U.S.A.

FLOSSIE WONG-STAAL, Laboratory of Tumor Cell Biology, National Cancer Institute, NIH, Building 37, Room 6A09, Bethesda, MD 20892, U.S.A.

RICHARD YANAGIHARA, Laboratory of Central Nervous System Studies, National Institute of Neurological Disorders and Stroke, NIH, Bethesda, MD 20892, U.S.A.

TODD O. YEATES, Department of Molecular Biology, Research Institute of Scripps Clinic, 10666 N. Torrey Pines Road, La Jolla, CA 92037, U.S.A.

Structure–Function Relationship

CHAPTER 1
Common Features in the Design of Small RNA Viruses

STEPHEN C. HARRISON

The determination of high-resolution, three-dimensional structures for a number of plant, vertebrate, and insect RNA viruses has led to a remarkable discovery—all the small, icosahedral, positive-strand RNA viruses examined so far are constructed through variations on a common design. This similarity probably shows a uniform solution to uniform functional problems—the requirements for packaging and uncoating of message-sense RNA. This chapter summarizes what is known about the design of icosahedral, positive-strand RNA viruses, and discusses possible functional reasons for the common structural features.

Common Elements of Design

The diagram in Figure 1.1**A** can be taken as a schematic representation of nearly all the structures to be described. These structures represent the classes of positive-strand RNA viruses listed in the first part of Table 1.1 [1–11]. The coats of all these viruses are constructed of 180 major "building blocks." The blocks are protein domains based on an all β-sheet organization, known variously as a "jelly-roll β barrel," a "Swiss-roll β barrel," or an "eight-stranded, antiparallel β barrel." Since it clearly represents a standard folded design in RNA Viral Capsids, we will refer to this structure for simplicity as the RVC-fold or RVC-domain. Because of the 60-fold character of icosahedral symmetry, the 180 RVC-domains in a viral shell must consist of 60 identical sets. The three domains in one set may belong to identical and independent polypeptide chains, as in many plant and insect viruses; to independent chains cleaved from a common precursor, as in the picornaviruses; or even to the same chain, as in the comoviruses (in which two of the three are covalently linked). When the chains are different, they

Table 1.1. RNA virus structures determined by X-ray diffraction.

Virus	Composition	References
Structures with 180 RVC-domains		
Mammalian viruses		
Poliovirus	VP1,VP2 (+4), VP3	1
Human rhinovirus 14	VP1, VP2 (+4), VP3	2
Mengovirus	VP1,VP2 (+4), VP3	3
Foot-and-mouth disease virus	VP1, VP2 (+4),VP3	(Stuart et al., in preparation)
Insect viruses		
Black beetle virus	T = 3	4
Plant viruses		
Tomato bushy stunt virus	T = 3	5
Southern bean mosaic virus	T = 3	6
Turnip crinkle virus	T = 3	7
Cowpea mosaic virus	Small (VP1 equivalent) and large (VP2 + VP3 equivalent) subunits	8
Bean-pod mottle virus	Small (VP1 equivalent) and large (VP2 + VP3 equivalent) subunits	(Johnson et al., in preparation)
Structures with 60 RVC-domains		
Satellite of tobacco necrosis virus	T = 1	9
Alfalfa mosaic virus	T = 1	10
and southern bean mosiac virus		11

turn out to have similar folds. The commas in Figure 1.1 correspond to VP1 (blue), VP2 (red), and VP3 (green) in the picornaviruses: The central part of each of these viral proteins is an RVC-domain. When the chains are identical, the three in a set are conformationally distinct, but the arrangement in Figure 1.1 has the property that the packing of their RVC-domains is actually quite similar. This property is known as "quasi-equivalence" [12]. It means that similar interactions occur, for example, at all three kinds of the "neck-to-neck" interactions between commas in Figure 1.1 (red/blue, blue/green, and green/red). The conventional designation of conformers in such structures is A (blue), B (green), and C (red). The designation of the overall structure is T = 3, referring to the presence of three conformers in one-sixtieth of the particle (one icosahedral unit).

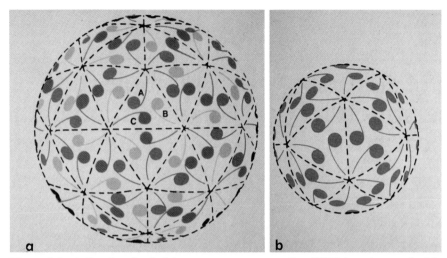

Figure 1.1. **A** One hundred eighty commas, representing RVC-domains, packed in the surface of an icosahedrally symmetrical structure. The commas are necessarily of three types (A, blue; B, green; and C, red), since icosahedral symmetry involves a 60-fold replication of a unique structure. **B** Sixty commas, representing RVC-domains, packed in the surface of an icosahedrally symmetrical structure. All commas are identical.

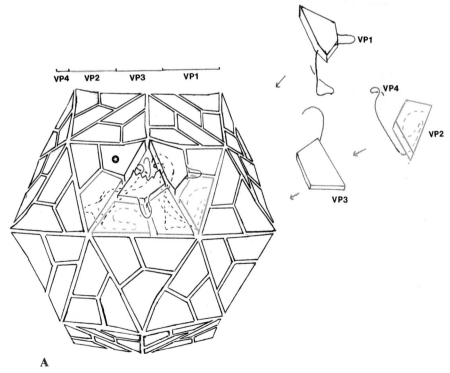

Figure 1.3 A. Caption on next page.

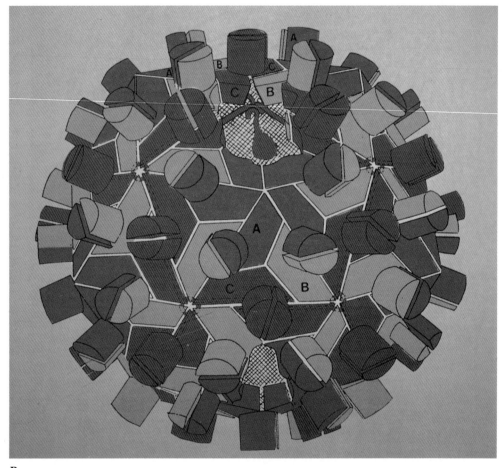

B

Figure 1.3. Packing of subunits in a picornavirus (polio[1]) and in a T = 3 plant virus (TBSV[5]). Each "block" is an RVC-domain. The N-terminal extensions form an interdigitating internal framework. In the case of the picornaviruses, all three subunits contribute to this framework; in the case of the T = 3 plant viruses, only the C-subunits contribute. **A** The proteins of the poliovirus capsid are cleaved from a precursor, as shown. VP1–3 are represented by wedge-shaped blocks (the RVC-domains) with N- and C-terminal extensions. The N-terminal extensions interdigitate to form an internal framework. The GH loop of VP1 is particularly prominent and lies across VP2 and VP3. This loop projects to create a cavity that runs from VP1 to VP1 around the fivefold axis. This cavity is called the canyon in HRV14[2]. An "exploded" view of one protomer is shown on the right. **B** In TBSV the subunit can adopt three conformations (A, B, and C). A (blue) and B (green) are very similar, with disordered N-terminal arms; C (red) has a folded arm that runs along the base of the RVC-domain.

Some alternative packings for RVC-domains have been found. In one class, represented by satellite of tobacco necrosis virus (STNV) and by the aberrant assemblies of turnip crinkle virus (TCV) and southern bean mosaic virus (SBMV) coat proteins, exactly 60 subunits form an icosahedrally symmetrical structure (Figure 1.1**B**). In these so-called T = 1 structures, all 60 subunits interact equivalently and in a way that is similar to their packing in the T = 3 shell. The alfalfa mosaic virus (AMV) coat protein is also believed to contain an RVC-fold. It can form a T = 1 icosahedral assembly [10], but in virions it forms a set of elongated structures for packaging each of the three viral RNA segments. Details of interactions in these nonicosohedral variants have yet to be worked out [13,14].

Important similarities among the RNA viruses are found at four levels: subunit architecture, subunit packing, coordination of interactions in assembly by "arms," and transition to an expanded virion. These similarities will be discussed in turn.

Subunit Architecture
The viral subunits are modular. The central module is an RVC-domain. Its design is based on a framework of two sandwiched β sheets, each with four antiparallel strands. The strands are laid out as shown in Figure 1.2. They are denoted by capital letters B through I. [A is missing, because it was originally used to refer to an additional strand contributed by the N-terminal arm of the C-subunit in tomato bushy stunt virus (TBSV) and SBMV.] Strands B, I, D, G make up one sheet, which twists strongly and thereby forms both the inner surface and one "side wall" of the wedge-shaped framework. Strands C, H, E, F make up the second, smaller sheet, which forms the other side wall of the wedge. These framework elements are relatively invariant in geometry, but their amino-acid sequences are not particularly conserved. The wedgelike aspect of the RVC-fold allows the domains to pack tightly around symmetry axes: See Figure 1.3.

Though the framework geometries of RVC-domains in each of the picornaviral subunits VP1-3 and in the plant and insect viruses are remarkably constant, the loops and N- and C-terminal extensions are not. Indeed it is the variation in the loops and extensions that gives each subunit and each virus a distinctive character. Two of the loops (CD and EF), which are generally quite long, often contain α-helices important for subunit contacts. Loop CD forms the outer surface of the wedge and, together with loop HI, one part of the "rear wall." Loop EF forms the other part of the rear wall.

Figure 1.4 shows examples of RVC-domains from picornaviral subunits VP1-3 and the subunit of a plant virus, tomato bush stunt virus (TBSV). Despite conservation of the framework geometry in the RVC-fold, it is difficult to detect homology by examining the sequences. Before the first picornavirus structures were determined, there was no suspicion of the

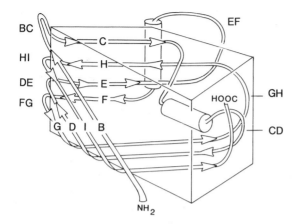

Figure 1.2. The polypeptide chain fold in an RVC-domain. Framework β strands are denoted B–I. Loops are BC, CD, etc. The diagram is oriented so that the top part of the domain faces the outside of the virion.

similarities among VP1, VP2, and VP3. Similarly structural alignment of TBSV, TCV, SBMV, and Carnation mottle virus shows no particular tendency for conserved residues to cluster in the framework elements [15]. In contrast to immunoglobulin-domains, a clear sequence "profile" for the RVC-fold has not yet been recognized.

Subunit Packing
Because of uniform geometry in RVC-domain frameworks, the packing of the subunits is similar in all structures studied. The variable parts of the loops and N- and C-terminal extensions "get out of the way" of the domain interactions. N-terminal extensions project internally. The C terminus of the RVC-domain itself is also internal, but in picornaviruses and plant viruses, the polypeptide chain turns immediately outwards, passing through one of the subunit interfaces. In the T = 3 insect virus, black beetle virus (BBV), the C-terminal extension remains internal.

VP1 of the picornaviruses corresponds to the A subunit of the T = 3 structures, VP2 to the C-subunit, and VP3 to the B-subunit; VP4 is really part of the N-terminal extension of VP2; it is cleaved from VP0 (=VP4 + VP2) during or after assembly [16]. If the structures of HRV14 and TBSV are superimposed so that centers and icosahedral axes coincide, then α-carbons of framework residues in VP1 of HRV14 and the A-subunit of TBSV fall within 2 to 3 Å of each other. Agreements for the other corresponding pairs are similar.

The outer shape of the virion depends not only on the packing of the RVC domains, but also on the way loops project from the framework. The domains themselves pack in such a way that the tips of the wedge-shaped structures protrude more at the fivefold than at the threefold. In HRV-14 and in poliovirus, the BC loops of VP1 (which lie at the outer corner of the wedge) are especially prominent, adding to the mass of the bulge around the fivefold axis. The depression encircling this bulge, bounded on the other side by the VP1 GH loops as they pack against the outer surface of VP2, has been called the "canyon" in HRV-14. Rossmann and colleagues [2] have suggested that it is the receptor binding site. In TBSV and TCV, the external aspect is dominated by C-terminal extensions, the P-domains, that form pairwise clustered projections.

Arms Coordinate Assembly

The most dramatic structural features of the icosahedral viruses are the elaborate networks and overlaps formed by the N- and C-terminal extensions.

In the T = 3 plant viruses, the N-terminal extensions have two parts: a positively charged, disordered "R-domain," for neutralizing RNA, and a connecting arm. The arm is ordered only on the C-subunits, where it folds along the base of the RVC-domain (known as the S-domain in these particles) and interdigitates with two others around the icosahedral threefold axis. The 60 ordered arms form an interconnected, internal framework. The viruses assemble from subunit dimers. The complete T = 3 particle may be described as an assembly of 60 A/B dimers (Blue/green in Figure 1.1) and 30 C/C dimers (red/red in Figure 1.1). Coordination of virion assembly involves sequential formation of the internal framework, which in turn determines whether a dimer adopts an A/B conformation, with disordered arms, or a C/C conformation, with ordered arms propagating the framework [17]. The C-terminal extensions of T = 3 plant virus subunits are highly variable. In TBSV, TCV, and their relatives, they form an entire folded domain (the P-domain); in SBMV, there is essentially no extension. The pairwise clustering of P-domains in TBSV and TCV clearly identifies the dimeric assembly unit.

In the picornaviruses, there is no R-domain (polyamines neutralize RNA), and there are ordered N-terminal extensions on all three subunits (see Figure 1.3). These extensions form a complex internal network. The arm of VP1 runs under the RVC-domain of a neighboring VP3. The arm of VP2 can be thought to include VP4, since the cleavage that produces them from VP0 occurs during or after assembly. VP4 is entirely internal. VP4 extends from the body of VP2 toward the icosahedral fivefold axis, where five, symmetrically related copies interact with five VP3 arms, forming an elaborate, inward-projecting knob. The N terminus of VP4 is myristylated: In poliovirus, the five myristyl groups form a little hydrophobic micelle [18]. The C-terminal extension of the subunits also overlaps neighbors. The pattern of overlap defines the VP0/1/3 "protomer," shown in Figure 1.3, and the interaction of VP3 arms and VP4 links five such protomers. The assembly of virions appears to involve these protomers and their pentamers as intermediates [16].

In T = 3 nodaviruses such as BBV, it appears to be the C-terminal extension that provides the regulatory switch. Both N- and C-terminal extensions are internal. The C-terminal arm lies in a cleft between RVC-

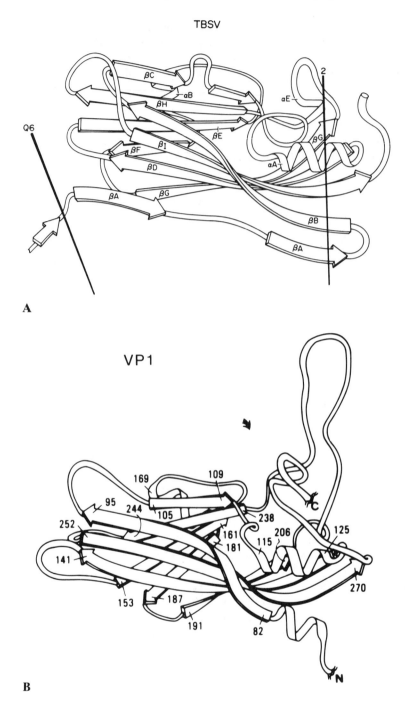

Figure 1.4. **A** Details of the S domain of TBSV. **B–D** Details of the RVC-domains from poliovirus VP1, VP2, and VP3. The arrow on the diagram of VP1 shows how

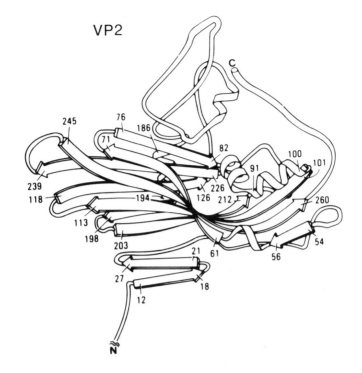

C

D

the prominence of the GH loop effectively creates a depression in the subunit surface. This is the "canyon" in HRV14 [2] and the opening of its drug-binding site [25].

domains of C/C dimers (like the N-terminal arm of the plant viruses). It is cleaved from the body of the protein after assembly, presumably by an autolytic process [19].

Altered Virions

When polio and other picornaviruses bind to cells, they undergo a structural transformation that leads to a loss of VP4 [20]. Particles eluted from the membrane before internalization are similar in diameter to normal virions, but they are no longer infectious. Since VP4 is entirely internal, the transformation must involve an expansion, presumably more or less reversed when the altered particles dissociate. There is an obvious parallel in the expansion that many T = 3 plant viruses undergo when divalent cations are removed and the pH is raised above neutrality [21–23]. Unlike the picornavirus transformation, the process is reversible, since no protein is lost. The structure of expanded TBSV shows that Ca^{2+} regulates expansion by stabilizing certain interfaces between RVC-domains (Figure 1.5) [23]. Expansion when Ca^{2+} is removed involves dissociation of these interfaces, which move apart under the repulsive force of negative charges previously neutralized by divalent cations. The arms of A- and B-subunits can loop out through the gaps thus created, becoming accessible to proteolytic degradation [24]. No model has yet been proposed for the way picornaviruses expand to release VP4. One piece of evidence may come from the location of the binding site on HRV14 for a series of antiviral drugs [25,26]. These drugs slip into the hydrophobic core of VP1, with one end against the GH loop. This GH loop overlaps VP2 and VP3, and it is possible that insertion of the drug reinforces the loop and stabilizes the VP1/VP2/VP3 interfaces. These interfaces are analogs of the Ca^{2+}-binding interfaces in TBSV.

Proposed Examples of Related Structures

The demonstrated examples of subunits with an RVC-fold are listed in Table 1.1. A search for sequence similarities has led to suggestions that at least two other classes of viruses contain this structure: the togaviruses [27] and the hepadnaviruses [28]. Viruses of both these classes have enveloped particles, and it is the core subunit that appears to show homology. In the togaviruses, the core subunit packages message-sense RNA. In mature hepadnavirus particles, it packages DNA, but replication of these viruses proceeds through an RNA intermediate [29]. Since sequence relationships among observed structures are modest, this search may well have missed other cases. Moreover, the alignments are of interest primarily because other features of the proposed folded subunits have plausible functional significance.

The core or nucleocapsid of Sindbis virus (and, by inference, of the other

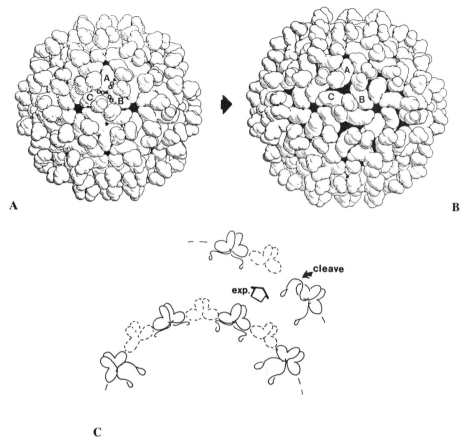

A

B

C

Figure 1.5. Expansion of TBSV [23]. **A** Compact structure. The positions of Ca^{++}
sites (two ions per interface) are shown for certain interfaces—similar sites are at
symmetry-equivalent locations. The Ca^{++} ligands are aspartic acid residues on the
EF loop of one subunit and on the CD and GH loops of the other. **B** Expanded
structure. When Ca^{++} is withdrawn and the pH raised to just above neutrality, the
net charge on the liganding side chains causes the Ca^{++}-binding interfaces to
separate from each other, forcing the particle to expand. **C** Arms of A- and
B-subunits loop out in the expanded structure, becoming accessible to protease [24].

togaviruses) is an icosahedral structure. Three-dimensional reconstruction
from electron micrographs of virions and cores embedded in vitreous ice
suggest that the core has a T = 3 arrangement of 180 subunits [27,30]. The
amino-acid sequence of the core protein reveals an obviously modular
organization, with a long (103-residue), positively charged N-terminal arm
appended to a 170-residue domain [31]. The sequence of this shell domain
can be aligned with the sequences of VP3s from various picornaviruses in

such a way that, in the predicted folded structure, residues believed to participate in a processing cleavage come together. That is, serine, histidine, and aspartic acid have been shown to be essential for cleavage of the core polypeptide from nascent polyprotein precursor [32], and in the proposed RVC-fold, these residues (not proximate in the sequence) are found clustered together in the GH and CD loops. If the principal domain of the core subunit is indeed an RVC-fold, and if the number of subunits is indeed 180, then the diameter of the core suggests a packing related to the expanded T = 3 plant virions. Togavirus cores have been recognized as fenestrated structures, and the packaged RNA is susceptible to RNase [33]. Because the envelope of the intact virion provides a protective barrier, there appears to be no requirement for an impenetrable shell.

The sequence of the core antigen of hepatitis B virus can also be aligned in a plausible way with picornavirus VP3 sequences [28]. The putative RVC-fold then places known antigenic sites at reasonable locations.

Common Elements of Function

Why are all these structures similar? It is logical to propose an evolutionary relationship. A common precursor could be either a primitive, positive-strand RNA virus or a cellular structure, perhaps one involved in packaging RNA. Other indications of a relationship among plant viruses, picornaviruses, togaviruses, hepadnaviruses, and retroviruses include similarities among RNA-dependent RNA polymerases and reverse transcriptases [34,35].

Whatever the evolutionary history, the structures share common functions: assembly and RNA packaging, RNA protection, attachment, entry, and uncoating. Not all types of particles must be able to perform all these functions. For example, togavirus cores must assemble and uncoat, but the envelope can take care of RNA protection and initial events in entry. The cores do not interact with a cell-surface receptor, but they must recognize the cytoplasmic tail of envelope glycoproteins. The T = 3 plant viruses enter cells directly, by injury or through cytoplasmic bridges, so that attachment and penetration may not be part of the functional repertoire of the virion. The picornavirus particles appear to carry out the full range of activities listed above. Table 1.2 lists these functions, together with their known, presumed, or proposed structural correlates.

We will discuss more extensively (and more speculatively) one set of functions—penetration and uncoating. Relatively little is known in general about these functions, but it is possible to propose some mechanisms based on a comparison of various RNA virus structures and their specific properties. The first step in penetration of a picornavirus appears to be endocytosis of the receptor-bound particle into coated vesicles and thence into endosomes [36]. A subsequent step must be escape of the virion (or

Table 1.2. Possible relationships between structural changes in positive-strand RNA viruses and steps in assembly or entry.

Virus	Assembly	Binding	Penetration	Uncoating
Picornaviruses	5S→14S→shell: Importance of N-terminal arms; VP0→VP2 + VP4 cleavage [16]	Receptor site; binding triggers transition to altered virion; some antiviral drugs may inhibit this transition [16,20,25]	Endocytosis: Release from endosome possibly mediated by VP4 or by exposed parts of VP1 [39]	?
T = 3 plant viruses	T = 3 pathway: Regulated by arms [17]	—	Direct entry into cytoplasm: Expansion ensues	Expansion can lead to uncoating by ribosome-associated activity [38]
Togavirus core (proposed)	?	—	(Released into cytoplasm after membrane fusion)	Already expanded [27]

some RNA-containing subviral particle) into the cytoplasm. It is not known how this occurs, but there is some evidence that another class of nonenveloped viruses (adenoviruses) can emerge from endosomes by destabilizing or lysing the endosomal membrane, carrying other molecules (e.g., toxins) with them from the endosomal compartment into the cytoplasm [37]. Receptor binding is known to trigger a conformational change in the picornavirus particle, leading to loss of VP4 [16,20]. Could the productive pathway involve loss of VP4 during or after engulfment, such that the myristylated VP4 might insert locally and at high concentration into the endosomal membrane? Could this insertion be part of the destabilization of the endosomal membrane that allows the escape of (altered) virions into the cytoplasm? Plant viruses, lacking a VP4 analog, seem to circumvent these steps altogether, since initial infection is thought to occur by direct cytoplasmic injection (e.g., by an insect vector) and cell-to-cell transfer by passage through cytoplasmic bridges (plasmodesmata). It might be significant that the $T = 3$ insect viruses, which presumably enter by receptor-mediated endocytosis, have a proteolytic cleavage step is their assembly pathway [19]. It is not yet known whether these viruses can expand or whether the cleaved C-terminal peptide can exist from an expanded particle. Nonetheless, the correlations suggest the following (testable) model: Receptor-triggered expansion of animal and insect viruses is followed by the exit of internal peptides, produced by post-assembly cleavage, one function of which is to facilitate escape of the altered virion into the cytoplasm. The high particle/PFU ratio generally observed implies that these steps need not be very efficient and that transport to the lysosomal "graveyard" may be the more frequent fate of a receptor-bound particle.

Uncoating of viral RNA may be coordinated with the events of penetration, but the properties of plant virions and of their infections suggest that it can also be an active cytoplasmic process. Several plant viruses can be uncoated in vitro by ribosome-associated factors, provided that they have been pretreated in ways that would lead to expansion [38]. The following mechanism immediately suggests itself: Virus particles, expanded under intracellular conditions, become the substrate for an appropriate, energy-dependent uncoating activity, involving either the ribosome itself or some other cytoplasmic factor that would be present in the usual *in vitro* protein synthesis reaction. There is no indication at present of what molecular signals the uncoating activity would recognize. If we extend this notion to mammalian and insect viruses, then the substrate for an uncoating activity would be the altered virion that escapes from the endosome, as suggested above. This mechanism, purely speculative at present, would uncouple penetration and uncoating. There is little available evidence bearing on this point.

In summary, the small, positive-strand RNA virions are not merely passive packages of the RNA genome. Common features in the design of plant, insect, and mammalian viruses suggest that they may have similar mechanisms of penetration and uncoating. The preceding discussion outlines a possible set of common mechanisms.

Acknowledgments

I am grateful to J.M. Hogle for comments on the manuscript. Work on virus structure in this laboratory has been supported by NIH Grant CA-13202 and by HHMI.

References

1. Hogle JM, Chow M, Filman DJ (1985) Three-dimensional structure of poliovirus at 2.9 Å resolution. Science 229:1358–1365
2. Rossmann MG, Arnold E, Erickson JW, Frankenberger EA, Griffith JP, Hecht H-J, Johnson JE, Kamer G, Luo M, Mosser AG, Rueckert RR, Sherry B, Vriend G (1985) Structure of a human common cold virus and functional relationship to other picornaviruses. Nature 317:145–153
3. Luo M, Vriend G, Kamer G, Minor I, Arnold E, Rossmann MG, Boege U, Scraba DG, Duke GM, Palmenberg AC (1987) The atomic structure of mengo virus at 3.0 Å resolution. Science 235:182–191
4. Hosur MV, Schmidt T, Tucker RC, Johnson FE, Gallagher TM, Selling BH, Rueckert RR (1987) Structure of an insert virus at 3.0 Å resolution. Proteins 2:167–176
5. Harrison SC, Olson A, Schutt CE, Winkler FK, Bricogne G (1978) Tomato bushy stunt virus at 2.9 Å resolution. Nature 276:368–373
6. Abad-Zapatero C, Abdel-Meguid SS, Johnson JE, Leslie AGW, Rayment I, Rossmann MG, Suck D, Tsukihara T (1980) Structure of southern bean mosaic virus at 2.8 Å resolution. Nature 286:33–39
7. Hogle JM, Maeda A, Harrison SC (1986) Structure and assembly of turnip crinkle virus, I. X-ray crystallographic structure analysis at 3.2 Å resolution. J Mol Biol 191:625–638
8. Stauffacher CV, Usha R, Harrington M, Schmidt T, Hosur MV, Johnson JE (1987) Structure of cowpea mosaic virus at 3.5 Å resolution. In Moras D, Drenth J, Strandberg B, Suck D, Wilson D (eds) Crystallography and Molecular Biology. Plenum Publishing, New York, pp 293–308
9. Liljas L, Unge T, Fridberg K, Jones TA, Lovgren S, Skoglund O, Strandberg B (1982) Structure of satellite tobacco necrosis virus at 3.0 Å resolution. J Mol Biol 159:93–108
10. Fukuyama K, Abdel-Meguid SS, Johnson JE, Rossmann MG (1983) Structure of a T=1 aggregate of alfalfa mosaic virus coat protein seen at 4.5 Å resolution. J Mol Biol 167:873–894
11. Erickson JW, Silva AM, Murthy MRN, Fita I, Rossmann MG (1985) The structure of a T=1 icosahedral empty particle from southern bean mosaic virus. Science 229:625–629
12. Caspar DDL, Klug A (1962) Physical principles in the construction of regular viruses. Cold Spring Harbor Symp Quant Biol 27:1–22
13. Cusack S, Miller A, Krijgsman PCJ, Mellema JE (1982) An investigation of the

structure of alfalfa mosaic virus by small-angle neutron scattering. J Mol Biol
125:525–543

14. Cremers AFM, Oostergetel GT, Schilstra MJ, Mellema JE (1981) An electron
    microscopic investigation of the structure of alfalfa mosaic virus. J Mol Biol
    125:545–561

15. Carrington JD, Morris TJ, Stockley PG, Harrison SC (1987) Structure and
    assembly of turnip crinkle virus, IV. Analysis of the coat protein gene and
    implications of the subunit primary structure. J Mol Biol 194:265–276

16. Rueckert RR (1985) Picornaviruses and their replication. *In* Fields BN, Knipe
    DM, Chanock RM, Melnick JL, Roizman B, Shope RE (eds) Virology. Raven
    Press, New York, pp 705–738

17. Sorger PK, Stockley PG, Harrison SC (1986) Structure and assembly of turnip
    crinkle virus, II: Mechanism of reassembly *in vitro*. J Mol Biol 191:639–658

18. Chow M, Newman JFE, Filman D, Hogle JM, Rowlands BJ, Brown F (1987)
    Myristylation of picornavirus capsid protein VP4 and its structural significance.
    Nature 327:482–486

19. Gallagher TM, Rueckert RR (1988) Assembly-dependent maturation cleavage in
    provirions of a small icosahedral insect ribovirus. J Virol (in press)

20. Guttman N, Baltimore D (1977) A plasma membrane component able to bind and
    alter virions of poliovirus type 1: Studies on cell-free alteration using a simplified
    assay. Virology 82:25–36

21. Incardona NL, Kaesberg P (1974) A pH-induced structural change in bromegrass
    mosaic virus. Biophys J 4:11

22. Kruse J, Kruse KM, Witz J, Chauvin C, Jacrot B, Tardieu A(1982) Divalent
    ion-dependent reversible swelling of tomato bushy stunt virus and organization
    of the expanded version. J Mol Biol 162:393–417

23. Robinson IK, Harrison SC (1982) Structure of the expanded state of tomato
    bushy stunt virus. Nature 279:563–568

24. Harrison SC, Sorger PK, Stockley PG, Hogle J, Altman R, Strong RK (1987)
    Mechanism of RNA virus assembly and disassembly. *In* Brinton MS, Rueckert
    RR (eds) Positive Strand RNA Viruses. Alan R. Liss, New York, pp 379–395

25. Smith TJ, Kremer MJ, Lou M, Vriend G, Arnold E, Kamer G, Rossmann MG,
    McKinlay MA, Diana G, Otto MJ (1986) The site of attachment in human
    rhinovirus 14 for antiviral agents that inhibit uncoating. Science 233:1286–1293

26. Badger J, Minor I, Kremer MJ, Oliveira MA, Smith TJ, Griffith JP, Guerin
    DMA, Krishnaswamy S, Luo M, Rossmann MG, McKinlay MA, Diana GD,
    Dutko FJ, Fancher M, Rueckert RR, Heinz G (1988) Structural analysis of a
    series of antiviral agents complexed with human rhinovirus 14. Proc Natl Acad
    Sci USA 85:3304–3308

27. Fuller SD, Argos P (1987) Is Sindbis a simple picornavirus with an envelope?
    EMBO J 6:1099–1105

28. Argos P, Fuller SD (1988) A model for the hepatitis B virus core protein:
    Prediction of antigenic sites and relationship to RNA virus capsid proteins.
    EMBO J 7:819–824

29. Ganem D, Varmus HE (1987) The molecular biology of the hepatitis B viruses.
    Ann Rev Biochem 56:651–693

30. Fuller SD (1987) The $T=4$ envelope of sindbis virus is organized by interactions
    with a complementary $T=3$ capsid. Cell 48:923–934

31. Strauss EG, Rice CM, Strauss JH (1984) Complete nucleotide sequence of the genomic RNA of sindbis virus. Virology 133:92–110

32. Hahn CS, Strauss EG, Strauss JH (1985) Sequence analysis of three sindbis virus mutants temperature-sensitive in the capsid protein autoprotease. Proc Natl Acad Sci USA 82:4648–4652

33. Söderlund H, Kääriäinen L, Von Bonsdorff CH, Weckstrom P (1978) Properties of Semliki forest virus nucleocapsid II. An irreversible contraction by acid pH. Virology 47:753–760

34. Kamer G, Argos P (1984) Primary structural comparison of RNA-dependent polymerases from plant, animal and bacterial viruses. Nucleic Acids Res 12:7269–7282

35. Ahlquist P, Strauss EG, Rice CM, Strauss JH, Haseloff J, Zimmern D (1985) Sindbis virus proteins nsP1 and nsP2 contain homology to nonstructural proteins from several RNA plant viruses. J Virol 53:536–542

36. Marsh M, Helenius A (1988) Virus entry into animal cells. Adv Virus Research (in press)

37. FitzGerald DLP, Padmanabhan R, Pastan I, Willingham M (1983) Adenovirus-induced release of epidermal growth factor and pseudomonas toxin into the cytosol of KB cells during receptor-mediated endocytosis. Cell 32:607–617

38. Brisco M, Hull R, Wilson TMA (1986) Swelling of isometric and of bacilliform plant virus nucleocapsids is required for virus-specific protein synthesis in vitro. Virology 148:210–217

39. Fricks, CE (1988). Studies of the conformational changes of poliovirus during virus neutralization, cell entry, and infection. PhD Thesis, Univ of California, San Diego

CHAPTER 2
Structural Determinants of Serotype Specificity and Host Range in Poliovirus

JAMES M. HOGLE, RASHID SYED, TODD O. YEATES, DAVID JACOBSON, TOD CRITCHLOW, AND DAVID J. FILMAN

In the second volume of this series, the three-dimensional structure of the Mahoney strain of type 1 poliovirus (P1/Mahoney) and the implications of the structure for the architecture, evolution, assembly, and immune recognition of polio and closely related picornaviruses were presented [1]. Since that time, we have solved the structure of the Sabin (attenuated vaccine) strain of type 3 poliovirus (P3/Sabin) [2] and, more recently, of an intertypic chimera (V510 constructed by Martin, Wychowski, and Girard at the Pasteur Institute) [3], in which a loop representing the major antigenic site of P1/Mahoney was replaced with the corresponding loop from the mouse-adapted Lansing strain of type 2 poliovirus [4]. Replacement of the loop confers mouse adaptation on the normally primate-specific Mahoney strain. Comparison of the three structures has provided insights into the structural determinants for serotype specificity and provided the first clues regarding the structural basis for host restriction and pathogenesis in poliovirus.

Poliovirus Structure

Poliovirus is a member of the picornavirus family, which also includes the closely related coxsackieviruses and rhinoviruses, the cardioviruses (encephalomyocarditis [EMC], Mengo, and Theiler's virus), the aphthoviruses (foot-and-mouth disease virus), and hepatitis A virus. Although there are a large number of poliovirus strains, all known isolates of poliovirus can be grouped into three serotypes based on their reactivity to panels of neutraliz-ing antisera. The poliovirion is approximately 310 Å in diameter. The external protein shell of the particle is comprised of 60 copies of each of four coat proteins (VP1 $Mr = 33,000$, VP2 $Mr = 30,000$, VP3 $Mr = 26,000$, and VP4 $Mr = 7,400$), arranged on a $T = 1$ icosahedral surface. The protein shell encapsidates a unique single-stranded molecule of positive-sense RNA, 7,500 nucleotides in length. The polio genome is translated from a single open reading frame to yield a large polyprotein. All viral proteins (including the capsid proteins and their immediate precursors) are derived from the polyprotein by posttranslation processing catalyzed by virally encoded proteases. Capsid protein processing is apparently coupled to viral assem-bly. Thus, cleavage of the capsid precursor P1 ($Mr = 100,000$) to VP0, VP1, and VP3 is coupled to the formation of a pentameric assembly interme-diate, and the final cleavage of VP0 to yield VP4 and VP2 occurs late in assembly and is associated with the encapsidation of the viral RNA (see ref. 5).

We have described the three major capsid proteins (VP1, VP2, and VP3) of P1/Mahoney as having similar conserved cores (eight-stranded, antiparal-lel β barrels shaped roughly like triangular wedges), dissimilar extensions at their amino and carboxyl termini, and dissimilar loops connecting the regular secondary structural elements of the cores (Figure 2.1) [2,6]. The wedge-shaped, eight-stranded β barrel is apparently a ubiquitous structural motif in simple RNA viruses. A very similar structural motif also has been observed in related picornaviruses (including rhinovirus 14 [7], mengo virus [8], and foot-and-mouth disease virus) [8a] and also in several unrelated viruses of plants (including tomato bushy stunt virus [9], southern bean mosaic virus [10], satellite tobacco mosaic virus [11], turnip crinkle virus [12], and cowpea mosaic virus [13]) and insects (black beetle virus [14]). In Chapter 1 of this volume, Harrison introduces a common nomenclature for this β barrel core, dubbed the RVC-fold (RNA virus capsid fold).

In virions the RVC-domains form the closed shell of the particle, packing together with the narrow ends of the wedges of VP1 pointing toward the particle fivefold axes and the narrow ends of the wedge-shaped cores of VP2 and VP3 alternating around the particle threefold axes. The tilts of the cores produce prominent radial protrusions at the fivefold and threefold axes, giving the particle the shape of the geometric solid shown in Figure 2.2. The amino-terminal extensions form a network on the inner surface of the protein shell, which may function to direct the assembly of the virus. The formation of this network must be preceded by proteolytic processing of the capsid protein precursors to free the chain termini, which could explain the apparent link between capsid protein processing and assembly. The connecting loops and carboxyl termini contribute to the major features exposed on the outer surface of the virion and are the major contributors to the antigenic sites of the virus [6,15].

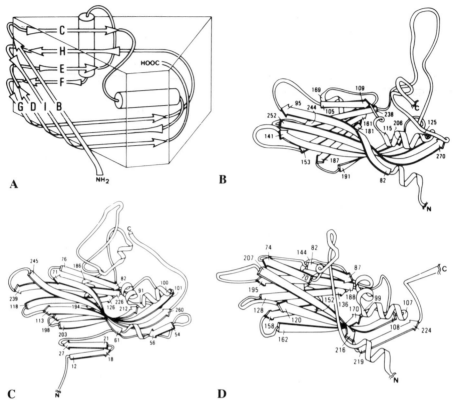

Figure 2.1. The structure and organization of the major capsid proteins of poliovirus. **A.** Schematic representation of the conserved, wedge-shaped, eight-stranded antiparallel β barrel core motif shared by VP1, VP2, and VP3. Individual β strands are shown as arrows and are labeled alphabetically. Flanking helices are indicated by cylinders. Ribbon diagrams of **B** VP1, **C** VP2, and **D** VP3. Residue numbers have been included as landmarks. Extensions at the amino and carboxyl termini of VP1 and VP3 have been truncated for clarity.

Comparison of the Structures of P1/Mahoney and P3/Sabin

P3/Sabin and P1/Mahoney share 83% sequence identity at the amino-acid level in the capsid protein region. Consistent with the high degree of sequence homology between the two strains, the structures of P3/Sabin and P1/Mahoney are strikingly similar. This structural similarity is especially pronounced in the cores of the capsid protein subunits, but is also seen in the amino-terminal extensions and in many of the connecting loops. In several places in the structures, similarity of the main chain conformation is maintained despite significant local sequence differences, with the sequence changes being accommodated through localized adjustments of side chain conformations or by compensating changes in neighboring side chains.

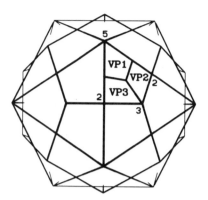

Figure 2.2. A geometric representation of the outer surface of the poliovirion, generated by superimposing an icosahedron and a dodecahedron. The symmetry axes of the particle and the positions of VP1, VP2, and VP3 in one protomer are indicated. Like the virion, the geometric figure has large radial projections at the fivefold axes and somewhat smaller projections at the threefold axes.

Significant conformational differences in the main chain are confined to the exposed loops and to the chain termini of the virus. These structural differences fall into three general categories: (1) Differences in loop conformation due to insertions in one strain relative to the other. The observable insertions/deletions cause limited, highly localized structural perturbations that have little effect other than one or two residues on either side of the insertion. (2) Loops with several sequence differences *including the replacement of a proline residue.* (3) Differences observed at points of transition between ordered and disordered structure. It may be relevant that significant structural differences occur in all three of the major antigenic sites of the virion (as defined in Page et al. [15]). In particular, structural changes are seen in the BC loop of VP1 (which constitutes a major portion of antigenic site 1), the insertion in the GH loop of VP1 (site 2), the insertion at position 289 of VP1 (site 3A), and the large structural difference in the BC loop of VP3 (site 3B). This suggests that three-dimensional structural differences, as well as simple sequence changes, might play an important role in determining serotypes.

The most significant conformational difference between the P3/Sabin and P1/Mahoney structures occurs in the loop connecting the B and C strands of the RVC-domain of VP1 (residues 95–104), where the distance between equivalent α carbon positions is as large as 8 Å (Figure 2.3). Of the six sequence differences in this loop, residues that appear to contribute most significantly to the conformational difference are the substitution of glutamic acid (Glu) (P3/Sabin) for proline (Pro) (P1/Mahoney) at position 95 and the substitution of Pro (P3/Sabin) for serine (Ser) (P1/Mahoney) at position 97. Large structural differences associated with the substitution of Pro in exposed loops with considerable local sequence divergence are also seen in the BC loop of VP3 and in the HI loop of VP2. In contrast, the EF loop of VP2 (residues 160–170, part of antigenic site 2), an area of high sequence variability that does not involve the substitution of Pro, is nearly identical in structure in the two poliovirus strains.

An examination of the known sequences shows that Pro at position 97 (Pro 97) and an acidic amino acid at position 95 is a common feature of all known, naturally occurring strains of type 2 and type 3 poliovirus [16], suggesting that the structure observed for the BC loop in P3/Sabin may be general to type 3 polioviruses [17] and that the structure of this loop in type 2 strains will be closer to the structure of the BC loop of P3/Sabin than it is to the loop structure in P1/Mahoney. In contrast, Pro 95 in P1/Mahoney has been seen in only two of the type 1 strains for which sequences are known, suggesting that the conformation of the BC loop observed in the P1/ Mahoney structure may be atypical for type 1 polioviruses [17].

Intertypic Hybrids of Poliovirus

Recently, three research groups have reported the construction of viable intertypic chimeras in which the BC loop from one strain of poliovirus has been replaced with the corresponding sequence from a different serotype. These chimeras include two examples in which a type 3 loop [18,19], and two examples in which a type 2 loop [3,20], has been constructed into a type 1 background. In every case, the hybrid displays the appropriate mosaic antigenicity and is able to induce neutralizing antibodies against both parental serotypes. In the type 2/type 1 chimeras, replacement of the BC loop in the primate-specific P1/Mahoney strain with the corresponding loop from the mouse-adapted P2/Lansing strain produces a hybrid that is able to cause fatal paralysis in mice [3,20]. This demonstrates that the replacement of this 10 amino-acid loop (which includes only six sequence changes) is sufficient to confer mouse adaptation on P1/Mahoney. One of the type 2/type 1 chimeras [3] grows to very high titers in cultured cells. We have recently determined the structure of this chimera by molecular replacement methods, and a preliminary atomic model has been built to fit the electron-density map. Details of the structure will be reported elsewhere [4].

Consistent with the prediction based on sequence comparisons, the conformation of the BC loop in the chimera (Particularly at the termini of the loop) is more similar to that observed in the P3/Sabin model than it is to the

$\longrightarrow$

Figure 2.3. Structural differences in three strains of poliovirus in the vicinity of the particle fivefold axis. This stereo representation shows main chain atoms from five symmetry-related copies of the narrow end of the β barrel of capsid protein VP1 viewed from outside the virion. Note that only a small portion of VP1 is included. Structurally conserved atoms (principally in the eight strands of the antiparallel β barrel) are shown in white. Conformational differences (which are observed in three of the loops that connect the beta strands) are indicated in cyan (P1/Mahoney), yellow (P3/Sabin), and magenta (type 2/type 1 chimera). Structural differences as large as 8 Å are seen in the BC loops (the colored loops furthest from the center of view). Smaller, but significant, structural differences are also seen in the DE loops (the colored loops closest to the center of view).

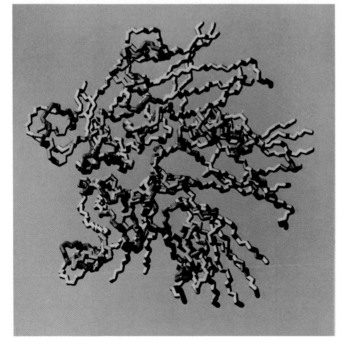

Figure 2.3

conformation observed in the parental P1/Mahoney strain. The overall conformation of this loop in the chimera (especially for residues 98–103 at the center of the loop), however, is significantly different than that observed in either of the previously determined structures (see Figure 2.3).

In previously reported studies, Vincent Racaniello and his colleagues have demonstrated that the ability of the Lansing strain of type 2 poliovirus to infect mice maps to the capsid region of the viral genome [21] and that mutations in the BC loop of VP1 attenuate mouse virulence [22]. In particular they have shown that mutations at lysine (Lys) 99 and arginine (Arg) 100 of VP1 significantly attenuate mouse virulence in the Lansing strain [22]. It is interesting to note that the altered conformation of the BC loop in the chimera results in a significant increase in the exposure of Lys 99 and Arg 100, and that Lys 99, along with asparine (Asp) 95, Pro 97, and Ser 101, undergo specific interactions that may stabilize the loop structure in the chimera. Because the interactions of residues in the BC loop tend to be primarily with other residues in the loop, these same interactions may specify a very similar loop structure within the parental Lansing 2 strain.

One particularly striking feature of the chimera structure is that in addition to the conformational changes in the heterologous BC loop, there are significant conformational changes in the DE loop of VP1 (the third loop down at the narrow end of the wedge-shaped core) where the sequences of the chimera and the parental P1/Mahoney are identical (Figure 2.3). Interestingly, the structure observed in the DE loop of the chimera is very similar to that observed in P3/Sabin (Figure 2.3), suggesting that there may be a coupling between conformational changes in the BC and the DE loops of VP1.

What Are the Determinants of Mouse Virulence?

The observation of significant structural differences outside the heterologous BC loop raises the interesting possibility that the BC loop itself may not be the sole or dominant determinant of mouse adaptation. Other factors, including increased exposure of neighboring residues (e.g., in the HI or DE loops of VP1) or the observed conformational alteration in the DE loop, may play critical roles as well.

The mechanisms controlling the ability of the chimera and the parental Lansing strain to infect mice are as yet poorly understood. Indeed, the problem has proved particularly difficult to analyze, since neither the chimera nor the Lansing 2 strain are able to infect mice by routes other than intracerebral or intraspinal inoculation, nor is either strain able to infect common mouse cell lines in culture. Given the demonstrated importance of the highly exposed BC loop (regardless of whether the contribution is direct or indirect), the simplest (but as yet unsubstantiated) model is that mouse virulence is determined at the level of the ability of the virion to interact with the viral receptor (or associated proteins) in the mouse central nervous system (CNS). There is a widely publicized hypothesis, originally suggested by Rossmann and his colleagues [7], that the receptor-binding sites of polio and rhinoviruses are located in the deep depressions (or "canyons") that surround the fivefold axes of the virions. The residues implicated as important in controlling mouse virulence, and particularly Lys 99, Arg 100, and the DE loop of VP1, however, are located very near the apex of the large protrusion at the fivefold axis, well up the canyon wall. This suggests that residues near the fivefold axes may play an important role in receptor binding in the CNS, at least in the mouse model.

The "mutagenesis cartridges" that have been developed for the construction of the intertypic chimeras provide a general mechanism for preparing substitutions within the BC loop of VP1. We anticipate that in the future these constructs and similar constructs in other loops in the capsid proteins will be used to probe the genetic determinants of host range in poliovirus. Furthermore, by providing the opportunity to screen for neurovirulence in mice rather than in monkeys, the mouse-adapted chimeras significantly increase the feasibility of determining the genetic determinants of neurotropism within the capsid coding region of the poliovirus genome. It is our hope that interpretation of these genetic studies in the context of the three-dimensional structures of both wild type and genetically altered virions will eventually allow us to define the structural features that contribute to the determination of host range and neurotropism in poliovirus.

Acknowledgments

This work was supported by NIH Grant AI-20566. The authors thank Marc Girard, Annette Martin, and Czeslaw Wychowski (Pasteur Institute) for supplying seed stocks of the type 2/type 1 chimera prior to its published description. The authors also acknowledge the contributions of a number of collaborators, particularly Dr. Marie Chow (Massachusetts Institute of Technology) and Dr. Philip D. Minor (National Institute of Biological Standards and Control, London). We thank Margaret Graber for assistance in the preparation of the manuscript. This is publication no. 5564-MB of the Research Institute of Scripps Clinic.

References

1. Hogle JM, Chow M, Filman DJ (1986) The three-dimensional structure of poliovirus: Implications for virus evolution, assembly and immune recognition. *In* Notkins AL, Oldstone MBA (eds) Concepts in Viral Pathogenesis. Springer-Verlag, New York, pp 3–14
2. Filman DJ, Syed R, Chow M, Macadam AJ, Minor PD, Hogle JM (1988) Structural factors that control conformational transitions and serotype specificity in type 3 poliovirus. EMBO J (in press)

3. Martin A, Wychowski C, Couderc T, Crainic R, Hogle JM, Girard M (1988) Engineering a poliovirus type 2 antigenic site on a type 1 capsid results in a chimaeric virus which is neurovirulent for mice. EMBO J 7:2839–2847

4. Yeates TO, Syed R, Filman DJ, Martin A, Wychowski C, Girard M, Hogle JM (1988) Structure of a mouse virulent intertypic chimera of type 2 and type 1 poliovirus. (In preparation)

5. Rueckert RR (1985) Picornaviruses and their replication. In Fields B (ed) Virology. Raven Press, New York, pp 705–738

6. Hogle JM, Chow M, Filman DJ (1985) The three-dimensional structure of poliovirus at 2.9 Å resolution. Science 229:1358–1365

7. Rossmann MG, Arnold E, Erickson JW, Frankenberger EA, Griffith JP, Hecht H-J, Johnson JE, Kamer G, Luo M, Mosser AG, Rueckert RR, Sherry B, Vriend G (1985) Structure of a human common cold virus and functional relationship to other picornaviruses. Nature 317:145–153

8. Luo M, Vriend G, Kamer G, Minor I, Arnold E, Rossmann MG, Boege U, Scraba DG, Duke GM, Palmenberg AC (1987) The atomic structure of Mengo virus at 3.0 Å resolution. Science 235:182–191

8a. Acharya R, Fry E, Stuart D, Fox G, Rowlands D, Brown F (1989) The three-dimensional structure of foot-and-mouth disease virus at 2.9 Å resolution. Nature 337:709–716

9. Harrison SC, Olson AJ, Schutt CE, Winkler FK, Bricogne G (1978) Tomato bushy stunt virus at 2.9 Å resolution. Nature 317:368–373

10. Abad-Zapatero A, Abdel-Meguid SS, Johnson JE, Leslie AGW, Rayment I, Rossmann MG, Suck D, Tsukihara T (1980) Structure of southern bean mosaic virus at 2.8 Å resolution. Nature 286:33–39

11. Liljas L, Unge T, Jones TA, Fridborg K, Lovgren S, Skoglund U, Strandberg B (1982) Structure of satellite tobacco necrosis virus at 3.0 Å resolution. J Mol Biol 159:93–108

12. Hogle JM, Maeda A, Harrison SC (1986) The structure of turnip crinkle virus at 3.2 Å resolution. J Mol Biol 191:625–638

13. Stauffacher C, Usha R, Harrington M, Schmidt T, Hosur MV, Johnson JE (1987) The structure of cowpea mosaic virus at 3.5 Å resolution. In Moras D, Drenth D, Strandberg B, Suck D, Wilson K (eds) Crystallography in Molecular Biology. Plenum Publishing, New York, pp 293–308

14. Hosur MV, Schmidt T, Tucker RC, Johnson JE, Gallagher TM, Selling BH, Rueckert RR (1988) Structure of an insect virus at 3.0 Å resolution. Proteins 2:167–176

15. Page GS, Mosser AG, Hogle JM, Filman DJ, Rueckert RR, Chow M (1988) Three-dimensional structure of the poliovirus serotype 1 neutralizing determinants. J Virol 62:1781–1794

16. Minor PD, Ferguson M, Phillips A, McGrath DI, Huovilainen A, Hovi T (1987) Conservation in vivo of protease cleavage sites in antigenic sites of poliovirus. J Gen Virol 68:1857–1865

17. Hogle JM, Filman DJ, Syed R, Chow M, Minor PD (1988) Structural basis for serotypic differences and thermostability in poliovirus. In Semler B, Ehrenfeld E (eds) Molecular Aspects of Picornavirus Infection and Detection. ASM Publications, New York, pp 125–137

18. Burke KL, Dunn G, Ferguson M, Minor PD, Almond JW (1988) Antigen chimaeras of poliovirus: Potential novel vaccines against picornavirus infections. Nature 332:81–82

19. Murray MG, Kuhn RJ, Arita M, Kawamura N, Nomoto A, Wimmer E (1988) Poliovirus type 1/type 3 antigenic hybrid virus constructed *in vitro* elicits type 1 and type 3 neutralizing antibodies in rabbits and monkeys. Proc Natl Acad Sci USA 85:3203–3207
20. Murray MG, Bradley J, Yang X-F, Wimmer E, Moss EG, Racaniello VR (1988) Poliovirus host range is determined by a short amino acid sequence in neutralization antigenic site I. Science 241:213–215
21. La Monica N, Meriam C, Racaniello VR (1986) Mapping of sequences required for mouse neurovirulence of poliovirus type 2 Lansing. J Virol 57:515–525
22. La Monica N, Kupsky WJ, Racaniello VR (1987) Reduced mouse neurovirulence of poliovirus type 2 Lansing antigenic variants selected with monoclonal antibodies. Virology 161:429–437

CHAPTER 3

The Molecular and Functional Basis of Poliovirus Attenuation and Host Range

VINCENT R. RACANIELLO

Poliovirus is the causative agent of poliomyelitis, an acute disease of the central nervous system (CNS). For 50 years after its isolation in 1909, research on the virus was aimed at providing the necessary information on antigenic types, pathogenesis, and immunity so that vaccines could be developed. After two effective vaccines were produced, research on poliovirus turned toward understanding events that occur in infected cells, such as viral RNA and protein synthesis and inhibition of host macromolecular synthesis. Three developments of the early 1980s once again altered the direction of poliovirus research: the determination of the nucleotide sequence of the poliovirus genome and mapping of the viral polypeptides [1], the demonstration that cloned poliovirus cDNA is infectious in cultured cells [2], and the resolution of the three-dimensional structure of the poliovirion [3]. Although these findings have clearly facilitated the study of poliovirus replication, they have also made it possible to address previously unanswered questions about poliovirus neurovirulence. This chapter will summarize and evaluate recent progress made in understanding the molecular and functional basis for the attenuation of the poliovirus vaccine strains and for the restricted host range of the virus.

Poliovirus is a member of the Picornaviridae family, which contains an array of human pathogens that share similar capsid structure, genomic organization, and strategy of gene expression. The poliovirion is an icosahedral particle composed of four capsid proteins, VP1, VP2, VP3, and VP4, which contains a single-stranded, positive-sense RNA genome approximately 7.5 kb in length. In the infected cell, the genomic RNA is translated into a polyprotein of 250,000 kd, which is processed by two viral proteases to form functional viral proteins. Features of the viral genome include a 5'-linked protein, VPg, a 5'-noncoding region of approximately 743 nucleotides, a 3′-noncoding region of approximately 73 nucleotides, and a poly(A) tail.

Attenuation of Neurovirulence: The Molecular Basis

One currently available vaccine against poliomyelitis is a live, attenuated preparation containing all three viral serotypes. The vaccine is taken orally and replicates in the gut, inducing local and humoral antibodies but failing to invade the CNS. The attenuated poliovirus strains were originally isolated in the 1950s by Dr. A. B. Sabin by multiple passage of neurovirulent viruses in various primate cells in vivo and in vitro [4]. Virus strains possessing the lowest neurovirulence when inoculated intraspinally into cynomolgous monkeys (*Macaca irus*) and inducing an immune response when taken orally, but not producing a CNS disease, were selected for vaccine use.

Since their isolation, the Sabin vaccine strains have been studied to determine the basis for their attenuated phenotype, but only in the past 5 years has much progress been made. Sequence analysis of the cloned cDNAs derived from the genomes of the vaccine strains and their neuro-virulent parents revealed that the process of attenuation is accompanied by a relatively small number of base changes. For example, the P1/Sabin viral genome differs from its neurovirulent parent, P1/Mahoney, by 55 nucleotide substitutions out of a total genome length of 7441 [5], whereas the genome of P3/Sabin differs from that of P3/Leon by only 10 point mutations [6].

The general approach to identifying mutations responsible for the attenuated phenotype has been to construct viral recombinants between neuro-virulent and attenuated strains and to determine the neurovirulence of these recombinants in monkeys. The first such recombinants examined were isolated by coinfecting cells with two viruses and taking advantage of the natural ability of poliovirus to undergo recombination [7]. In subsequent work viral recombinants have been isolated by exchanging DNA restriction fragments between infectious cDNA clones and recovering virus by trans-fection of cultured cells with recombinant plasmids.

Analysis of the neurovirulence of recombinants between P1/Sabin and P1/Mahoney by intrathalamic inoculation of cynomolgous monkeys showed that attenuating determinants are scattered throughout the genome [8]. However, a strong attenuating mutation is located in the 5′-noncoding region, at nucleotide 480, which is an A in P1/Mahoney RNA and a G in P1/Sabin RNA (A. Nomoto, personal communication), [9]. Analysis of recombinants between P3/Sabin and P3/Leon by intraspinal inoculation of monkeys indicates that two mutations account for the attenuated phenotype of P3/Sabin: a base change from C to U at nucleotide 472 in the 5′-noncoding region, and a change from serine (Ser) to phenylalanine (Phe) in VP3, which confers a temperature-sensitive phenotype on the virus [10].

A different approach has been used to identify attenuating mutations in the type 2 vaccine strain, which was derived from a naturally attenuated isolate. This approach is based on the observation that the 472 mutation is able to attenuate the P2/Lansing strain of poliovirus in mice. P2/Lansing, originally isolated from a case of poliomyelitis, was adapted to grow in mice by repeated passage [11], and it induces a fatal paralytic disease in mice after intracerebral inoculation. A recombinant of P2/Lansing, in which the 5'-noncoding region was replaced with that of P3/119 (a virus isolated from a vaccine-associated case of poliomyelitis) containing a C at base 472, was neurovirulent in mice [12]. A similar recombinant, in which the 5'-noncoding region was derived from P3/Sabin and differed only by a U at base 472, was dramatically attenuated in mice [12]. Therefore, the mouse model identifies mutations known to attenuate monkey intraspinal neurovirulence and can be used to identify regions of the P2/Sabin genome that attentuate P2/Lansing in mice. Recombinants were constructed in which segments of the P2/Lansing genome were replaced by homologous sequences from P2/P712, the attenuated parent of P2/Sabin. The results of neurovirulence tests indicate that the 5'-noncoding region of P2/P712, between nucleotides 457 and 630, as well as a central portion encoding part of VP3, VP1, 2A, 2B, and part of 2C, can attenuate intracerebral neurovirulence of P2/Lansing in mice (E. Moss, R. O'Neill, and V.R. Racaniello, J Virol, in press).

Attenuation of Neurovirulence: Functional Basis

It therefore appears that all three poliovirus vaccine strains probably have in common at least one mutation in the 5'-noncoding region, around nucleotide 480, that can attenuate the neurovirulence of polioviruses inoculated into the CNS of experimental animals. A partial answer to the question of how this mutation reduces neurovirulence comes from experiments described above that showed that the 472 mutation attenuates poliovirus neurovirulence in mice [12]. The attenuated P2/Lansing–P3/Sabin viral recombinant, containing a U at 472, replicates poorly if at all in the mouse CNS, whereas the neurovirulent virus with a C at this position replicates to high titers as expected [12]. In contrast, viruses with either a C or a U at this position replicate equally well in cultured HeLa cells. The change from C to U at base 472 is therefore a host range mutation that in some way limits viral replication in neurons but not in HeLa cells. Thus the 5'-noncoding region attenuates intracerebral or intraspinal neurovirulence in experimental animals by reducing or abolishing replication in the CNS.

Recent experiments provide information on the specific mechanism by which the 472 mutation limits replication in the CNS. A human neuroblastoma cell line, SKNSH, has been identified in which the attenuated P3/Sabin–P2/Lansing recombinant (see above) with a U at base 472 replicates more slowly and to lower titer than neurovirulent recombinants containing a C at this position (N. La Monica and V.R. Racaniello, J Virol, in press). Analysis of viral replication in SKNSH cells indicates that viral

RNA with a U at 472 enters cells normally but is translated at a slower rate than RNA containing a C at this position. As a result, fewer viral capsids are produced, and hence, virus yield is decreased. As expected, there is no difference in the rate at which RNAs with C or U at base 472 are translated in HeLa cells. These results are consistent with the previous observation that in vitro translation of P3/Sabin viral RNA is less efficient than that of P3/Leon or P3/119 RNA [13]. It appears that, after inoculation into the CNS, the presence of a U at 472 results in reduced viral translation, failure to replicate efficiently, and absence of disease. However the manner in which this mutation reduces translation in a cell-type–specific manner is not yet known.

Less is known about the mechanism by which the mutation in VP3 of P3/Sabin attenuates intraspinal neurovirulence. The Ser to Phe change appears to destabilize protomer interactions in the virion, leading to a temperature-sensitive phenotype (J. Almond, personal communication). This property might attenuate neurovirulence by reducing the efficiency of viral replication in the CNS. The mutation might also affect a step in virion binding, entry, or uncoating in a cell-type–specific manner that is not related to the *ts* phenotype. An example of such a mechanism is provided by mutations in VP1 of P2/Lansing that reduce its replication and neuro-virulence in the mouse CNS, but do not alter replication in cultured cells [14] (see below).

The only specific base changes that have been associated with the attenuated phenotype are those around 480 of all three vaccine serotypes and the mutation in VP3 of P3/Sabin. Further work is required to identify mutations responsible for attenuation of the type 1 and type 2 strains and to clarify their functional basis.

Unanswered Questions

Are mutations known to reduce poliovirus neurovirulence after inoculation directly into the CNS also responsible for failure of the vaccine strains to cause paralysis when taken orally? Although it is clear that presence of a U at base 472 in the type 3 strain prevents replication in the CNS, it is not clear how attenuation of oral neurovirulence can be explained by a mechanism in which this mutation limits virus replication to the gut. This conclusion is based on the isolation from the gut of infants, two to three days after vaccination, of type 3 viruses containing a C at 472 [15], indicating that a C at 472 is required for replication in the gut as well as in the CNS. These fecal isolates are of significantly higher monkey spinal neurovirulence than the ingested vaccine, a fact that was appreciated long before the vaccines were put into use [16]. Having lost a major determinant of its spinal neuro-virulence, why does the virus found in the infants' gut cause only the rare case of paralytic disease in these infants or in their contacts? One possibility is that the presence of a U in the ingested vaccine sufficiently delays replication so that rising levels of antibodies in the gut may more effectively limit virus yields, preventing spread to the CNS. Therefore, an important question is whether the 472 mutation reverts as rapidly in the gut of infants other than those reported [15]. At least for the type 3 vaccine, virus replication in the gut may also be reduced by the attenuating mutation in VP3, which is present in intestinal virus for several weeks after vaccination [15]. Although the VP3-attenuating mutation subsequently reverts, resulting in excreted viruses that are nearly as neurovirulent as the parent of the vaccine (P. Minor, personal communication), spread to the CNS may be precluded at that time by a vigorous antibody response.

Alternatively, the vaccine strains might contain mutations that specifically attenuate oral neurovirulence, which have not been detected in the studies summarized above that assay for monkey spinal neurovirulence. It is known that in order to invade the CNS the poliovirus must replicate first in the gut and then at a second extraneural site, the identity of which has not been established [17]. P3/Sabin differs from P3/Leon by 10 base changes, only 2 of which are considered important for attenuation of monkey intraspinal neurovirulence. Some of these differences might be responsible for reducing replication in the gut or at secondary sites in the human, thus preventing the persisting viremia that leads to invasion of the CNS. Existence of mutations that specifically attenuate oral neurovirulence might also explain why the intraspinally neurovirulent viruses excreted several weeks after vaccination are not responsible for more contact cases of poliomyelitis.

Sabin concluded many years ago that the viral determinants of oral and intraspinal neurovirulence are not linked, based on studies of the P2/YSK strain, which is highly paralytogenic in monkeys when administered orally or intracerebrally. A variant of this strain obtained by passage in mice was still intracerebrally neurovirulent but had lost most of its neurovirulence by the oral route [18].

The study of type 3 polioviruses, isolated from cases of vaccine-associated poliomyelitis, provides information relevant to this discussion. One such isolate, P3/119, which was obtained from the CNS of a fatal case, is neurovirulent in monkeys after intraspinal inoculation [10]. Determination of the complete sequence of the P3/119 viral genome indicates that it is vaccine-like at 8 of 10 base positions where P3/Sabin differs from P3/Leon [19]. P3/119 contains a C at base 472 and also a second site mutation that suppresses the *ts* phenotype, changes expected in a virus that has replicated in the human gut. More importantly, P3/119 differs from P3/Sabin at 7 positions. Is it possible that some of these base changes are suppressors of mutations in P3/Sabin that specifically attenuate oral neurovirulence? To provide a final answer to this question, it will be necessary to employ viral recombinants to determine whether any of the base differences between P3/Sabin and P3/Leon specifically attentuate oral neurovirulence in monkeys.

Host Range

Although the host range of poliovirus is extremely narrow—the virus infects mainly primates and only primate cell cultures—some strains that grow in nonprimate hosts have been identified. An example is the mouse-adapted P2/Lansing strain of poliovirus described above. Other strains have been adapted to such hosts as the chick embryo and suckling hamsters. Although these unusual strains have existed for many years, the basis for their expanded host range is not known.

Recently it was determined that the ability of P2/Lansing to induce paralysis in mice maps to a small region of capsid polypeptide VP1. By constructing defined viral recombinants, it was found that the phenotype of mouse intracerebral neurovirulence could be conferred to the mouse-avirulent P1/Mahoney by substitution of its capsid coding sequences with those of P2/Lansing [20]. Subsequently it was found that certain antigenic variants of P2/Lansing, selected for their resistance to neutralization with monoclonal antibodies, possessed reduced neurovirulence in mice [14]. The mutations that lead to neutralization resistance and reduced neurovirulence were located between amino acids 93 to 101 of capsid polypeptide VP1, previously identified as antigenic site I, a polypeptide loop on the surface of the virion [3]. To establish the role of antigenic site I in mouse neuro-virulence, a viral recombinant was constructed in which the eight amino-acid sequence PASTTNKD in antigenic site I of P1/Mahoney was replaced with the sequence DAPTKRAS from P2/Lansing. The resulting antigenic chimera was neurovirulent in mice [21]. Similar results have been reported by others (M. Girard, personal communication) [22]. Therefore, the host range of poliovirus may be determined by an eight amino-acid sequence of VP1 on the virion surface.

Any of several viral functions, such as receptor binding, entry into the cell, or uncoating of the viral genome, might be provided by antigenic site I of P2/Lansing to allow replication to occur in mice. Antigenic site I is located at the fivefold axis of symmetry of the poliovirion, just above a groove known as the "canyon" that circles each vertex [3,23]. It has been proposed that the canyon is involved in binding to a host cell receptor [23]; if this hypothesis is true, it is conceivable that the nearby antigenic site I may be involved in receptor binding. Antigenic site I of P2/Lansing may possess, for example, a particular amino-acid sequence and shape that enables the virus to bind to a receptor on mouse neurons. It seems likely, however, that the function served by antigenic site I of P2/Lansing in infection of mice has no correlate during replication in cultured primate cells. For example, the ability of P2/Lansing to replicate in cultured cells is no different from that of mouse-avirulent viruses. Even more compelling is the recent observation that antigenic site I of P1/Sabin may be replaced by a completely different amino-acid sequence (J.W. Almond, personal communication). It will therefore be of great interest to identify the function, provided by antigenic site I, that enables P2/Lansing to infect mice.

Conclusions

Although many attenuated viral vaccines are in use throughout the world, the basis for their reduced virulence is unknown. Studies on poliovirus have provided the most information on this subject, due to our extensive physical and genetic knowledge of this virus, the ability to manipulate its genome, and the existence of three excellent attenuated vaccines. Most progress has been made toward understanding the molecular basis for the attenuation of monkey spinal neurovirulence of the Sabin vaccine strains and the reasons why such viruses fail to replicate in the CNS. However, we still do not know why orally inoculated poliovaccine produces a good immune response but fails to invade the CNS. Whether attenuation results from cell-type–specific restriction of replication or a general reduced efficiency of multiplication in tissues or both remains to be determined. The mouse-adapted phenotype of P2/Lansing has been mapped to an eight amino-acid sequence in capsid polypeptide VP1, thus providing information on the basis for the ability of certain poliovirus strains to infect hosts other than primates. The functional basis for this expanded host range, however, awaits elucidation. The flexibility possible in the study of poliovirus will permit a greater understanding of these and many other aspects of viral neurovirulence, limited only by the imagination of the investigator.

References

1. Kitamura N, Semler BL, Rothberg PG, Larsen GR, Adler CJ, Dorner AJ, Emini EA, Hanecak R, Lee JJ, van der Werf S, Anderson CW, Wimmer E (1981) Primary structure, gene organization and polypeptide expression of poliovirus RNA. Nature 291:547–553
2. Racaniello VR, Baltimore D (1981) Cloned poliovirus complementary DNA is infectious in mammalian cells. Science 214:914–919
3. Hogle JM, Chow M, Filman DJ (1985) Three-dimensional structure of poliovirus at 2.9 Å resolution. Science 229:1358–1365
4. Sabin AB (1985) Oral poliovirus vaccine: History of its development and use and current challenge to eliminate poliomyelitis from the world. J Inf Dis 151:420–436
5. Nomoto A, Omata T, Toyoda H, Kuge S, Horie H, Kataoka Y, Genba Y, Imura N (1982) Complete nucleotide sequence of the attenuated poliovirus Sabin 1 strain genome. Proc Natl Acad Sci USA 79:5793–5797
6. Stanway G, Hughes PJ, Mountford RC, Reeve P, Minor PD, Schild GC, Almond JW (1984) Comparison of the complete nucleotide sequences of the genomes of the neurovirulent poliovirus P3/Leon/37 and its attenuated vaccine derivative P3/Leon 2a₁b. Proc Natl Acad Sci USA 81:1539–1543
7. Agol VI, Drozdov SG, Frolova MP, Grachev VP, Kolesnikova MS, Kozlov VG, Ralph NM, Romanova LI, Tolskaya EA, Viktorova EG (1985) Neurovirulence of the intertypic poliovirus recombinant v3/a1-25: Characterization of strains isolated from the spinal cord of diseased monkeys and evaluation of the contribution of the 3′-half of the genome. J Gen Virol 65:309–316

8. Omata T, Kohara M, Kuge S, Komatsu T, Abe S, Semler BL, Kameda A, Arita M, Wimmer E, Nomoto A (1986) Genetic analysis of the attenuation phenotype of poliovirus type 1. J Virol 58:348–358

9. Nomoto A, Kohara M, Kuge S, Kawamura N, Arita M, Komatsu T, Abe S, Semler BL, Wimmer E, Itoh H (1987) Study on virulence of poliovirus type 1 using in vitro modified viruses. *In* Brinton MA, Rueckert RR (eds) Positive Strand RNA Viruses, UCLA Symposia on Molecular and Cellular Biology, Vol 54, Alan R Liss, New York, pp 437–452

10. Westrop GD, Evans DMA, Minor PD, Magrath D, Schild GC, Almond JW (1987) Investigation of the molecular basis of attenuation in the Sabin type 3 vaccine using novel recombinant polioviruses constructed from infectious cDNA. *In* Rowlands DJ, Mayo MA, Mahy BWJ (eds) The Molecular Biology of the Positive Strand RNA Viruses, Fed Europ Microbiol Soc Symp 32:53–60, Alan R. Liss, New York

11. Armstrong C (1939) Successful transfer of the Lansing strain of poliomyelitis virus from the cotton rat to the white mouse. Pub Health Rep 54:2302–2305

12. La Monica N, Almond JW, Racaniello VR (1987) A mouse model for poliovirus neurovirulence identifies mutations that attentuate the virus for humans. J Virol 61:2917–2920

13. Svitkin YV, Maslova SV, Agol V (1985) The genomes of attenuated and virulent poliovirus strains differ in their in vitro translation efficiencies. Virology 147:243–252

14. La Monica N, Kupsky W, Racaniello VR (1987) Reduced mouse neurovirulence of poliovirus type 2 Lansing antigenic variants selected with monoclonal antibodies. Virology 161:429–437

15. Evans DMA, Dunn G, Minor PD, Schild GC, Cann AJ, Stanway G, Almond JW, Currey K, Maizel JV (1985) Increased neurovirulence associated with a single nucleotide change in a noncoding region of the Sabin type 3 poliovaccine genome. Nature 314:548–550

16. Sabin AB (1956) Present status of attenuated live-virus poliomyelitis vaccine. J Amer Med Assoc 162:1589–1596

17. Bodian D, Horstmann DH (1965) Polioviruses. *In* Horsfall, FL, Tamm, I (eds) Viral and Rickettsial Infections of Man, Lippincott, Philadelphia, pp 430–473

18. Sabin AB (1955) Characteristics and genetic potentialities of experimentally produced and naturally occurring variants of poliomyelitis virus. Ann NY Acad Sci 61:924–938

19. Cann AJ, Stanway G, Hughes PJ, Minor PD, Evans DMA, Schild GC, Almond JW (1984) Reversion to neurovirulence of the live-attenuated Sabin type 3 oral poliovirus vaccine. Nucleic Acids Res 12:7787–7792

20. La Monica N, Meriam C, Racaniello VR (1986) Mapping of sequences required for mouse neurovirulence of poliovirus type 2 Lansing. J Virol 57:515–525

21. Murray MG, Bradley J, Yang X-F, Wimmer E, Moss EG, Racaniello VR (1988) Poliovirus host range is determined by the eight amino acid sequence in neutralization antigenic site I. Science 241:213–215

22. Martin A, Wychowski C, Benichou D, Crainic R, Girard M (1988) Construction of a chimaeric type 1/type 2 poliovirus by genetic recombination. Ann Inst Pasteur/Virol 139:79–88

23. Rossmann MG, Arnold E, Erickson JW, Frankenberger EA, Griffith JP, Hecht H-J, Johnson JE, Kamer G (1985) Structure of a human common cold virus and functional relationship to other picornaviruses. Nature 317:145–153

Gene Regulation

CHAPTER 4
Role of Enhancer Regions in Leukemia Induction by Nondefective Murine C Type Retroviruses

NANCY HOPKINS, ERICA GOLEMIS, AND NANCY SPECK

Nondefective, C type, murine retroviruses induce leukemias or lymphomas with long latencies following injection into newborn mice. As a group, they are often referred to as murine leukemia viruses (MuLVs). Transformation by MuLVs is mediated by the integration of proviruses adjacent to one or more protooncogenes, thus altering the transcriptional regulation of these cellular genes. Tumors result from the clonal or oligoclonal outgrowth of such cells, probably after the accumulation of additional genetic alterations.

An intriguing aspect of MuLV pathogenesis is the distinct disease specificity of different isolates. Although the majority of oncogenic MuLVs induce T cell tumors, others induce erythroleukemias, myeloid tumors, B cell tumors, or mixtures of these types. In some cases the type of tumor induced depends on the particular inbred strain of mouse injected [1].

Genetic studies to localize the viral genes responsible for disease specificity identified the U3 region of the LTR (long terminal repeat), and, more specifically, the viral transcriptional enhancer region, as a potent determinant of this phenotype [2–4]. For example, the Moloney MuLV, which induces T cell lymphomas following injection in newborn NFS mice, can be converted to a virus that almost exclusively induces erythroleukemias by replacing approximately 200 bases, including the transcriptional enhancer, with the corresponding sequence from the erythroleukemogenic Friend virus or a Friend MCF (mink cell focus-forming) virus [5–7]. Reciprocally, the Moloney enhancer region converts Friend MuLV or Friend MCF to a virus that induces T cell lymphomas in NFS mice. The finding that the LTRs of T cell lymphoma inducing retroviruses also encode thymotropism [8], and earlier studies that had shown the U3 region to be an important determinant of leukemogenic potential [9,10], combined to focus attention on the mechanism by which transcriptional regulation of MuLVs shapes their disease-inducing phenotypes.

This chapter focuses on the enhancer region as a determinant of disease specificity. We will review recent genetic studies, consider enhancer elements as determinants of tissue-specific replication of MuLVs, and examine attempts to understand how the interaction of nuclear factors with viral enhancers determines their transcriptional behavior.

Genetic Analysis of Enhancer Regions as Determinants of Disease Specificity

Two groups have tried to identify the smallest segment within the enhancer region that can determine disease specificity. These studies need to be viewed in terms of the structure of typical MuLV enhancers (Figure 4.1). The best-defined enhancers, those of Moloney MuLV and the nearly identical Moloney sarcoma virus (MSV), contain a 75-bp direct repeat approximately 150 nucleotides 5' to the viral cap site [11]. Transcriptional regulatory sequences probably extend both 3' and 5' of the direct repeat itself, and include at least a GC-rich sequence immediately 3' to the direct repeat [12].

Genetic studies to dissect the enhancer region involved Moloney virus as one parent and either Friend (E. Golemis, Y. Li, J. Hartley, and H. Hopkins (1989) J Virol 63:328–337) or a Friend MCF virus [7] as the other. The Friend and Friend MCF enhancer regions are about 85% homologous to that of Moloney. In both studies, it was found that determinants of disease specificity reside in more than one segment of the enhancer regions. Viruses with enhancers that are derived partly from Moloney virus and partly from Friend or Friend MCF induce a mixture of T cell lymphomas and erythroleukemias. In studies by Golemis and colleagues, determinants of disease specificity mapped to three distinct segments of the Friend and Moloney enhancer regions: both the 5' and the 3' halves of each copy of the direct repeat sequence and the 3'-adjacent GC-rich region. Single segments (for example both 3' halves of the direct repeat) were weak or negligible determinants of disease specificity when placed in the genome of the other parental virus; pairs of segments (e.g., the 5' halves of each direct repeat plus the GC-rich segment) were much stronger determinants. The most effective pair was the two halves of the direct repeat; when derived from the same parental virus, they were almost as potent as the direct repeat plus GC-rich segment at determining disease specificity.

Other recent genetic studies have provided additional examples in which exchanging MuLV LTRs or enhancer regions exchanges disease specificity [13]. In a somewhat different type of study, recombinants with heterologous enhancer regions have also been constructed. Such viruses can have unexpected disease specificities. For example, replacing the Moloney virus enhancer with the 72 base-pair direct repeat and adjacent GC-rich promoter segment of SV40 yields a recombinant that induces B cell lymphomas and some myeloid leukemias, tumors rarely if ever seen with either parental virus [14].

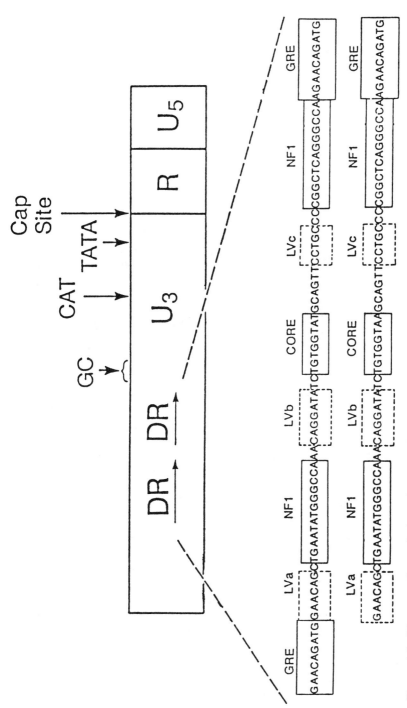

Figure 4.1. Diagram of a typical murine retrovirus LTR showing positions of some transcriptional elements. The enhancer is usually present as a direct repeat sequence. The nucleotide sequence of the Moloney virus enhancer is shown. Binding sites for nuclear factors that have been identified within this enhancer are boxed (from ref. 27).

LTRs as Determinants of Organ Tropism and Tissue-Specific Transcription

One might imagine that disease specificity is correlated with efficient viral replication in appropriate target organs, and there is evidence to support this in the case of lymphomagenic viruses. Several weeks after infection of newborn mice with MuLVs, virus can be detected in many tissues at a low level. However, the most extensive replication occurs in the lymphatic organs, spleen, and thymus [15]. Much evidence has confirmed the correlation, originally noted by Kaplan and his colleagues, between the ability of MuLVs to replicate in the thymus (thymotropism) and their ability to induce T cell lymphomas [16].

Genetic studies have shown that a potent determinant of thymotropism maps in the U3 region of several thymotropic MuLVs: the B-tropic virus of Balb/c mice, Gross passage A, SL3-3, and Moloney virus [13,17,18]. Sequence comparisons with the nonthymotropic parent used in these genetic studies, or analysis of appropriate recombinants, argues that it is the direct repeat region that confers the phenotype in these cases. It should be noted that additional determinants of thymotropism map outside U3 in some viruses, specifically in gp70 and p15E in the cases studied so far.

In some thymotropic MuLVs, such as SL3-3, certain MCF viruses, and the radiation leukemia virus, thymotropism reflects a strong organ preference and viral replication in the thymus is several logs higher than in the spleen, where it is barely detectable. In contrast, Moloney virus replicates to high titers in both organs, so although it is very thymotropic, there is no striking organ preference.

No one has reported an organ tropism that reliably predicts erythroleukemogenicity. Friend virus replicates in spleen with an efficiency comparable to that of Moloney virus. Friend virus is unusual in that it replicates quite well in thymus, too, even in strains of mice where it only causes erythroleukemias. Analysis of the organ tropism of LTR recombinants between Friend and Moloney reveals that the Moloney LTR confers about a 10-fold increase in thymotropism relative to that of Friend [18].

A better correlate of T cell lymphomagenesis in the case of most Friend–Moloney LTR recombinants is the appearance of significant amounts of MCF virus in the thymus (E. Golemis, N. Hopkins, and J. Hartley, unpublished results). Many lines of evidence have implicated thymic MCF viruses as the proximal agents in rapid T cell leukemogenesis [19]. With some Friend–Moloney enhancer recombinants that differ in disease specificity, the number of thymocytes infected four to five weeks after injection differed by less than a log, while differences in the number of thymocytes producing MCF viruses differed by several logs. Thus, thymotropism, but even more dramatically the ability to generate thymic MCFs, correlated with T cell lymphomagenesis. Further complicating this situation are studies suggesting that the time at which a large number of thymocytes are infected, either by ecotropic viruses or by the MCF viruses that arise after infection, is important for leukemia induction. It seems that high virus titers in the thymus are not enough; they must occur within the first few weeks after infection if T cell tumors are to be induced [18,20].

Attempts to find in vitro correlates for the contribution of LTR sequences to organ tropism have relied on either transient or stable transcription assays using enhancer sequences derived from various viral isolates to drive the expression of such an indicator gene as CAT or *neo* [21–24]. In general, results from these assays reflect the genetic studies of LTR-mediated organ tropism. The enhancers of the thymotropic viruses SL3-3 and MCF 247 showed a clear transcriptional preference in T cells over fibroblasts or erythroid cells compared to the enhancer from the nonthymotropic, non-leukemogenic Akv virus. Comparisons of Moloney and Friend MuLVs have also demonstrated a transcriptional preference for Moloney in T cells and Friend in erythroid cells. Here the differences have been less dramatic, and significant variability in the relative differences was seen in different cell lines and by independent investigators, perhaps reflecting the smaller differences seen between theses viruses in their organ tropism in vivo.

A particularly interesting assay for cell type-specific transcription by MuLV U3 regions involves the use of long-term bone marrow cultures infected with vectors carrying the neomycin resistance gene but differing in the U3/R portions of their LTRs [25]. Vectors with enhancer–promoter derived from Friend or Moloney virus were both able to confer G418 resistance on multipotential progenitor cells, but only the viruses with Friend sequences yielded G418-resistant erythroid or myeloid colonies following selection for these cell types in appropriate conditioned media. This result was interpreted to mean that although both viruses can direct gene expression in more primitive multilineage hematopoietic progenitor cells, the Moloney LTR is unable to promote sufficient expression of the *neo*R gene to confer G418 resistance during differentiation to committed myeloid or erythroid cells. This result could help to explain why Moloney virus fails to induce erythroleukemia. However, one is left to explain experiments indicating that Moloney transcriptional signals are active in erythroid progenitors [26].

Nuclear Factors That Interact With MuLV Enhancers

A goal of research in this area now is to identify nuclear factors that interact with MuLV enhancers in such a way as to explain their cell type prefer-ences, organ tropism, and disease specificity. Biochemical analyses of nuclear factors that interact with MuLV enhancers have revealed a complex organization that rivals and is, in fact, similar to that of SV40. The Moloney MuLV enhancer, the best studied so far, is assembled from multiple adjacent, often overlapping sequence motifs that are binding sites for nuclear factors (Figure 4.1) [27]. There are three glucocorticoid regulatory elements (GREs) [28,29], two copies of a conserved "core" motif, found also in the SV40 and polyoma enhancers and shown to bind a 42 K protein (EBP) [30], and four binding sites for nuclear factor 1 (NF1) [31]. In addition, there are binding sites for three factors designated LVa, LVb, and LVc, which have been thus far identified only in MuLV enhancers. Many of the binding sites for nuclear factors found in the Moloney enhancer, particularly (5')-LVb/core/NF1/GRE-(3'), are highly conserved in a large number of murine, feline, and primate C type retroviral enhancers. A comparison of the binding sites for nuclear factors on the Friend enhancer shows that there are sites shared with Moloney, whereas areas of sequence divergence contain sites specific to Friend (N. Manley, M. O'Connell, P.A. Sharp, and N. Hopkins, unpublished results).

A simple explanation for the tropism and disease specificity conferred by the Moloney and Friend enhancer regions is not readily apparent from the biochemical analyses to date. None of the factors that bound to these viral enhancers was found exclusively in T cells, erythroid cells, or any other hematopoietic cell type examined [27]. However, meaningful differences in amounts of factors between cell types could be difficult to detect in these studies, which employ mostly tumor cell lines. Furthermore, as discussed above, the tropisms of Friend and Moloney enhancers for the cell types studied so far are relative, not absolute, and one might not expect to find dramatic differences in the tissue distribution of factors that mediate their activity. The possibility must also be considered that what appear to be ubiquitously expressed factors are in fact tissue-specific forms of nuclear factors. There are precedents for such complexity in the case of both NF1 and EBP [30,31]. Another possibility is that the important tissue-specific factors have not yet been detected using the gel shift assay. Finally, tropism could be mediated by tissue-specific induction of transcription above the basal level in the appropriate cell. Steroid hormones are the best candidates, thus far, for this mechanism. The binding site for the hormone receptor GRE is highly conserved among retroviral enhancers, and transcription from several retroviral enhancers, including those of MMTV, Moloney-MSV, SL3-3, Akv, and Moloney MuLV, is moderately to significantly inducible by dexamethasone. Such induction could well occur preferentially in certain cell types.

In the case of the SL3-3 enhancer, a difference in the relative amounts of a core-binding factor was detected in nuclear extracts from T versus B cells [32]. This difference correlated with a preferential decrease in transcription in T cells when this site was mutated and the resulting enhancer placed in front of a reporter gene in transcription assays. This factor is thus a candidate for a tissue-specific factor that might mediate the tissue tropism of an MuLV enhancer. It will be important to see whether mutations in this site alter the thymotropism or T cell lymphomagenesis of SL3-3.

It is apparent that a combination of genetic, biochemical, and functional analyses, perhaps for each enhancer in question, will be needed to elucidate the mechanisms by which preferential tissue-specific transcription, organ tropism, and disease specificity are achieved.

Summary

The majority of oncogenic MuLVs induce T cell lymphomas/leukemias. In a number of viruses, the LTR is an important determinant of this specificity. In some cases, the viral enhancer has a significant transcriptional preference for T cells. In contrast, the Moloney enhancer, although thymotropic, is more subtle in its transcriptional preference, being quite effective in many hematopoietic cells. It seems clear that a very restricted cell type specificity is not necessary to produce the sharp disease specificity seen for T lymphotropic nondefective MuLVs in NFS mice.

How do the Friend or Friend MCF enhancer regions encode erythro-leukemogenicity? Are these enhancers able to function preferentially in a critical population of erythroid precursors? If so, is it because they bind factors in these cells that Moloney virus cannot bind or because they fail to bind negative factors that suppress expression of Moloney and other T cell leukemia virus enhancers? These are all questions for future study.

A more detailed genetic analysis of MuLV enhancer regions will be needed to define the protein-binding sites that play a key role in determining disease specificity. Identification and ultimately biochemical purification of the cellular factors that bind to these sites will probably be necessary to explain transcriptional preferences in molecular detail. Even such an analysis, however, may not be sufficient to explain the role of enhancers in disease specificity fully.

As indicated in this review, transformation is an indirect result of the initial infection by MuLVs, and many host and viral genetic factors influence the outcome of an infection. For example, if the ability to generate MCF viruses is important in leukemogenesis, then the efficiency of viral transcriptional signals in cell types in which MCFs arise may be important, as well as the ability to achieve high titers in the vicinity of appropriate target cells and to activate protooncogenes in target cells. It is questionable whether any single in vitro assay could be a sufficient correlate for disease, adequately reflecting the progression and complexity inherent in the system. Progress in this field has relied, and will most likely continue to rely, on genetic analyses in intact animals.

References

1. Weiss R, Teich N, Varmus H, Coffin J (eds) (1982) RNA Tumor Viruses, Cold Spring Harbor Laboratory, Cold Spring Harbor, NY
2. Laimins L, Khoury G, Gorman C, Howard B, Gruss P (1982) Host-specific activation of transcription by tandem repeats from simian virus 40 and Moloney sarcoma virus. Proc Natl Acad Sci USA 79:6453–6457

3. Chatis PA, Holland CA, Hartley JW, Rowe WP, Hopkins N (1983) Role for the 3' end of the genome in determining disease specificity of Friend and Moloney murine leukemia viruses. Proc Natl Acad Sci USA 80:4408–4411

4. Ishimoto A, Adachi A, Sakai K, Matsuyama M (1985) Long terminal repeat of Friend-MCF virus contains the sequence responsible for erythroid leukemia. Virology 141:30–42

5. Chatis PA, Holland CA, Silver JE, Frederickson TN, Hopkins N, Hartley JW (1984) A 3' end fragment encompassing the transcriptional enhancer of nondefective Friend virus confers erythroleukemogenicity on Moloney leukemia virus. J Virol 52:248–254

6. Li Y, Golemis E, Hartley JW, Hopkins N (1987) Disease specificity of nondefective Friend and Moloney leukemia viruses is controlled by a small number of nucleotides. J Virol 61:696–700

7. Ishimoto A, Takimoto M, Adachi A, Kakuyama M, Kato S, Kakimi K, Fukuoka K, Ogiu T, Matsuyama M (1987) Sequences responsible for erythroid and lymphoid leukemia in the long terminal repeats of Friend mink cell focus-forming and Moloney murine leukemia viruses. J Virol 61:1861–1866

8. DesGroseillers L, Rassart E, Jolicoeur P (1983) Thymotropism of murine leukemia virus is conferred by its long terminal repeat. Proc Natl Acad Sci USA 80:4203–4207

9. Lung ML, Hartley JW, Rowe WP, Hopkins NH (1983) Large RNase T1-resistant oligonucleotides encoding p15E and the U3 region of the long terminal repeat distinguish two biological classes of mink cell focus-forming type C viruses of inbred mice. J Virol 45:275–290

10. Lenz J, Celander D, Crowther RL, Patarca R, Perkins DW, Haseltine WA (1984) Determination of the leukemogenicity of a murine retrovirus by sequences within the long terminal repeat. Nature 308:467–470

11. Levinson B, Khoury G, Vande Woude G, Gruss P (1982) Activation of SV40 genome by 72-base–pair tandem repeats of Moloney sarcoma virus. Nature 295:568–572

12. Laimins LA, Gruss P, Pozzatti R, Khoury G (1983) Characterization of enhancer elements in the long terminal repeat of Moloney murine sarcoma virus. J Virol 49:183–189

13. DesGroseillers L, Jolicoeur P (1984) Mapping the viral sequences conferring leukemogenicity and disease specificity in Moloney and amphotropic murine leukemia virus. J Virol 52:448–456

14. Hanecak R, Pattengale P, Fan H (1988) Addition or substitution of simian virus 40 enhancer sequences into Moloney murine leukemia virus (M-MuLV) long terminal repeat yields infectious M-MuLV with altered biological properties. J Virol 62:2427–2436

15. Jaenisch R (1980) Retroviruses and embryogenesis: Microinjection of Moloney leukemia virus into midgestation mouse embryos. Cell 19:181–188

16. DeCleve A, Sato C, Liebermann J, Kaplan H (1974) Selective thymic localization of murine leukemia virus-related antigens in C57BL/Ka mice after inoculation with radiation leukemia virus. Proc Natl Acad Sci USA 71:3124–3128

17. Rosen C, Haseltine W, Lenz J, Ruprecht R, Cloyd MW (1985) Tissue selectivity of murine leukemia virus infection is determined by long terminal repeat sequences. J Virol 55:862–866

18. Evans L, Morrey J (1987) Tissue-specific replication of Friend and Moloney murine leukemia viruses in infected mice. J Virol 61:1350–1357
19. Hartley JW Wolford NK, Old LJ, Rowe WP (1977) A new class of murine leukemia virus associated with development of spontaneous leukemias. Proc Natl Acad Sci USA 78:6023–6027
20. Davis BR, Brightman BK, Chandy KG, Fan H (1987) Characterization of a preleukemic state induced by Moloney murine leukemia virus: Evidence for two infection events during leukemogenesis. Proc Natl Acad Sci USA 84:4875–4879
21. Celander D, Haseltine WA (1984) Tissue-specific transcription preference as a determinant of cell tropism and leukemogenic potential of murine retroviruses. Nature 312:159–162
22. Short MK, Okenquist SA, Lenz J (1987) Correlation of leukemogenic potential of murine retroviruses with transcriptional tissue preference of the viral long terminal repeats. J Virol 61:1067–1072
23. Yoshimura F, Davison B, Chaffin K (1985) Murine leukemia virus long terminal repeat sequences can enhance gene activity in a cell-type-specific manner. Mol Cell Biol 5:2832–2835
24. Bosze Z, Thiesen HJ, Charnay P (1986) A transcriptional enhancer with specificity for erythroid cells is located in the long terminal repeat of the Friend murine leukemia virus. EMBO J 5:1615–1623
25. Holland CA, Anklesaria P, Sakakeeny MA, Greenberger JS (1987) Enhancer sequences of a retroviral vector determine expression of gene in multipotent hematopoietic progenitors and committed erythroid cells. Proc Natl Acad Sci USA 84:8662–8666
26. Wolfe L, Ruscetti S (1986) Tissue tropism of a leukemogenic murine retrovirus is determined by sequences outside of the long terminal repeats. Proc Natl Acad Sci USA 83:3376–3380
27. Speck NA, Baltimore D (1987) Six nuclear factors interact with the 75-base–pair repeat of the Moloney murine leukemia virus enahancer. Mol Cell Biol 7:1101–1110
28. Miksicek R, Heber A, Schmid W, Danesch U, Posseckert G, Beato M, Schulze G (1986) Glucocorticoid responsiveness of the transcriptional enhancer of Moloney murine sarcoma virus. Cell 46:283–290
29. DeFranco D, Yamamoto K (1986) Two different factors act separately or together to specify functionally distinct activities at a single transcriptional enhancer. Mol Cell Biol 6:993–1001
30. Landschulz WH, Johnson PF, Adshi EY, Graves BJ, McKnight SL (1988) Isolation of a recombinant copy of the gene encoding C/EBP. Genes Develop 2:786–800
31. Santoro C, Mermod N, Andrews PC, Tjian R (1988) A family of human CCAAT-box-binding proteins active in transcription and DNA replication: Cloning and expression of multiple cDNAs. Nature 334:218–224
32. Thornell A, Hallberg B, Grundstrom T (1988) Differential protein binding in lymphocytes to a sequence in the enhancer of the mouse retrovirus SL3-3. Mol Cell Biol 8:1625–1637

CHAPTER 5

Molecular Aspects of Pathogenesis in the Polyoma Virus–Mouse System

ROBERT FREUND, THOMAS W. DUBENSKY, DAVID A. TALMAGE, CLYDE J. DAWE, AND THOMAS L. BENJAMIN

Polyoma virus normally establishes a silent persistent infection in its natural host, the mouse, although, in the laboratory, the virus can become a powerful pathogen. Given appropriate selections of virus and host strains, inoculation of newborn mice results in the rapid development of multiple tumors [1,2]. Tumors may arise from any one of a dozen different cell types, and some may appear grossly as early as 6 weeks. In this setting, polyoma virus is probably the most potent, broadly acting experimental oncogen known. Virus replication is also widespread under these experimental conditions. The kidney is the major site of virus amplification, although significant replication can occur in the lung and skin, as well as in the tumors themselves. Histological examination reveals cytopathic effects indicative of replication in as many as 30 different cell types [2].

Molecular biological and genetic studies of polyoma have flourished because of the use of cell culture systems for propagating the virus and for studying cell transformation [3]. The middle T protein is the major, if not the only viral gene product necessary to induce phenotypic changes associated with neoplastic transformation. Although apparently devoid of intrinsic enzymatic activity, this protein intervenes as a regulatory factor in protein and lipid kinase pathways in the cell. Two key cellular enzymes that directly interact with the middle T protein are the tyrosine-specific protein kinase $pp60^{c-src}$ [4] and a phosphatidylinositol kinase [5,6]. These molecular interactions and their relationships to cell transformation have been studied almost exclusively using established rat or mouse fibroblast cell lines, leaving unanswered a variety of questions concerning interactions of the virus with different cell types in the animal.

Questions arise when an attempt is made to relate the molecular biology of the virus to natural or experimental infections of the host. Certainly, the humoral and cellular immune systems of the host play major roles in determining the progression and outcome of infection [7,8]. Here, however, we address two areas in which recent evidence has provided some insights into the molecular aspects of infection in the host. The first area concerns tumor induction and attempts to define viral genetic determinants that affect the frequency and spectrum of tumors and to compare tumor induction and cell transformation in vitro. The second area concerns virus replication in the mouse. The bulk of evidence suggests that polyoma replication occurs by lytic or cytocidal infection. At the level of the individual infected cell, replication would therefore be incompatible with tumor induction. The relationship between these two kinds of infection in a permissive host is complex and poorly understood. Here we discuss some recent experiments relevant to this issue and also summarize findings on the pathology related directly or indirectly to lytic viral infections.

Effects of Virus Genotype on the Tumor Profile

Subtle differences in virus genotype can have profound effects on the tumor profile. This has been established through experiments using different cloned, wild type virus strains and a single, highly susceptible mouse strain as host. Two of four virus strains tested induce high tumor profiles, characterized by the appearance of multiple tumors in single animals, which led to a moribund condition within 2 to 4 months. The tumors are of both epithelial and mesenchymal origin. The other two strains induce low tumor profiles in which only a small fraction of the animals come down with predominantly single tumors. These tumors are consistently only of mesenchymal origin, and the affected animals typically survive to more than one year of age.

These results, which have been verified using molecularly cloned viruses of high and low tumor strains, justify the following conclusions [2]:

1. Viruses of different genotype induce remarkably different tumor profiles.

2. The ability to induce multiple types of tumor is inherent in a single virus genotype and is not a consequence of variations of enhancer sequences in the virus population [9,10].

3. Viral genetic determinants for tumor induction are more complex than for cell transformation, since high and low tumor strains are equally efficient at inducing transformation in vitro.

Coding and Noncoding Determinants Affect the Tumor Profile

Both regulatory and structural determinants in the high tumor strain contribute to a high tumor profile. Construction of a series of recombinant viruses derived from prototype high and low tumor strains, which were tested in animals, together with sequencing of the parental viral DNA, established the following points:

1. The ability to induce any and all tumors of epithelial origin depends absolutely on a coding determinant(s) present in the high tumor strain [11].

2. The middle T of the low tumor strain, when transferred to the background of the high tumor strain, induces a high tumor profile [12]. Structural determinants outside of middle T in the high tumor strain must therefore be essential for the induction of epithelial tumors and a high tumor profile.

3. Noncoding sequences from the high tumor strain act to increase frequencies of epithelial tumors. This effect is seen only in conjunction with coding sequences of the high tumor strains, which, as stated above, are absolutely required. The noncoding sequences that act in this manner are located on the early side of the replication origin and are distinct from the known enhancer region [11]. Enhancer regions, differing in sequence, appear thus far to have little or no differential effects on tumor induction [11,13].

4. A specific effect of a noncoding sequence is seen with respect to induction of thymic epitheliomas. This determinant consists of a duplication of 40 base pairs present in the high tumor strain just upstream of the early promoter. Thus, a strain of virus having the requisite structural determinants for inducing epithelial tumors but carrying only one copy of this element will induce only rare small thymic tumors, along with an otherwise high tumor profile. Introduction of a second copy of the 40 base-pair sequence results in 100% of the animals developing large overt thymic epitheliomas [14].

5. Virus–cell interactions leading to development of epithelial tumors show points of contrast as well as aspects of similarity with transformation of fibroblasts in vitro. Unlike transformed fibroblasts that typically contain integrated viral DNA at one or a few copies per cell, epithelial tumors appear to harbor free viral DNA in high copy numbers [13], in the range of 5,000 to 10,000 per cell (R. Freund and colleagues, unpublished results). The presence of integrated viral DNA in tumors cannot be ruled out. Middle T is recovered from extracts of epithelial tumors in complexes with pp60[c-src] [2] and phosphatidylinositol kinase [15], in line with the findings in infected or transformed fibroblasts in culture.

Induction of Tumors in Different Tissues May Require Different Molecular Interactions of Middle T

A mutation introduced into the middle T gene of a high tumor strain lowers the frequency and limits the spectrum of tumors induced by the virus. This mutant middle T containing a substitution of phenylalanine (Phe) for tyrosine (Tyr) results in a normal interaction with and activation of pp60[c-src], and at the same time in a failure to bind phosphatidylinositol kinase activity [15]. Biochemical evidence suggests that phosphorylation by $pp60^{c-src}$ of the Tyr normally present in middle T constitutes a molecular switch for binding of the phospholipid kinase.

Initial experiments evaluating this mutant in animals have revealed several interesting points. The mutant virus fails to induce tumors at several sites, such as the salivary gland and the kidney, whereas the wild type virus routinely induces these tumors. This suggests that binding and activation of the phospholipid kinase by middle T are required to induce neoplastic transformation in these tissues. In other sites, such as bone and hair follicles, the mutant induces tumors, suggesting either no requirement for this binding or a reduced threshhold in the degree of binding and activation of phosphatidylinositol kinase.

Changes in the thymus and mammary glands are seen in a fraction of mice inoculated with the mutant. In contrast to the wild type virus that induces overt thymic epitheliomas in a very high percentage of animals, the mutant induces only small nests of transformed epithelium accompanied by hyperplasia of cortical thymic lymphocytes. The wild type virus induces mammary adenocarcinomas in male as well as female mice at a ratio of 1 : 4, whereas the mutant induces this tumor type at lower frequencies and only in females. The tumors that develop in females are similar histologically to those induced by the wild type but arise in animals roughly twice as old, which are usually multipara. These results and others [13] suggest a possible role of hormonal stimulation in bypassing or complementing the molecular defect of the mutant middle T [15]. It seems possible that thymic and mammary epithelium that are neoplastically transformed by the mutant virus have slower growth rates than when they are transformed by the wild type virus. Further investigations with this mutant will be necessary to confirm and extend these findings.

High Tumor Strain Versus Low Tumor Strain Replication

High tumor strains but not low tumor strains of virus are able to replicate well in the animal. In particular, they are able to establish disseminated infections involving the kidney as a major site of amplification. Replication patterns of parental and recombinant strains have been studied following either subcutaneous or intranasal inoculation using "whole mouse blots" along with direct quantitation of infectious virus and viral DNA in various organs [16,17]. Following intranasal inoculation, both high and low tumor strains replicate well initially in the lung, but only the high tumor strain is able to infect and achieve high titers in the kidney.

A role for enhancer sequences in tissue specificity of polyoma replication has been shown using a "mosaic" enhancer containing sequences from the Mo-MLV LTR [18]. Interestingly, results with recombinants between the high and low tumor strains point to an important determinant for replication in the kidney residing in coding sequences, but not in the enhancer region [16]. Thus, it is possible that one and the same structural determinant— either large T or VP1—is essential for both efficient replication and induction of a high tumor profile. Such a single determinant could act first to allow boosting of virus titers by replication in the kidney, with the amplified virus then being responsible for efficient induction of tumors.

Although virus amplification in the kidney appears to be necessary, it may or may not be sufficient for induction of a broad tumor profile. Several factors distal to amplification in the kidney could be involved. The ability to establish productive infection directly in the target tissue for tumor induction might be important. This would be consistent with the fact that lytic lesions are found in every tissue belonging to the polyoma tumor constellation and with the occurrence of lytic lesions in a small but variable fraction of cells within the tumor mass itself [2]. It is also possible that the virus has transformation-specific determinants that act intracellularly in the target tissue.

Pathology Related to Virus Replication

Virulence of polyoma virus in newborn mice, assessed by various criteria of morbidity as well as mortality, has been correlated with the ability to establish lytic infection in the kidney [19]. Two strains of virus that differ greatly in their abilities to cause acute morbidity, runting, and early death were studied with respect to their effects on the kidney. Virulence was clearly associated with the ability to establish lytic infection in the kidney, leading to widespread destruction of tubules, renal failure, and death within a few weeks. Hemorrhaging in the brain as the result of lytic infection and destruction of vascular endothelium of small blood vessels were also deemed to cause death in a fraction of the animals inoculated with the virulent strain. The viral genetic determinants of virulence remain largely unknown.

In the course of establishing tumor profiles of various strains of polyoma virus, vascular lesions in tumor-bearing mice were noted [20]. These necrotizing arterial lesions were of two distinct types—one designated BLAND for bland lytic arterial necrotizing disease, the other PANoid for polyarteritis nodosa-like. The former disease was more common, being found in roughly one-half of the animals examined. BLAND lesions were found predominantly in the aorta and pulmonary artery and their secondary branches. They consisted of focal necrosis of the muscular layer of these vessels. These lesions showed clear evidence of lytic infection of smooth muscle cells between layers of the elastic lamina, based on immunoperoxidase staining for viral antigen. PANoid lesions were found with a similar anatomical distribution, but in only 11% of the mice examined. These lesions differ from BLAND lesions in having an extensive inflammatory component, with infiltration of leukocytes into some or all layers of the arterial wall and also in their negative staining for viral capsid antigen. Lytic infection cannot be ruled out as a triggering event that, however, is accompanied by masking of viral antigen in immune complexes. Arterial lesions can be correlated with viral genotype only to the extent that these lesions were found in tumor-bearing animals inoculated with various wild type or recombinant strains that induce a high tumor profile.

Directions for the Future

Attempts undertaken thus far to define the molecular aspects of tumor induction in the polyoma–mouse system have given rise to new and interesting questions that could not have been anticipated from studies of cell transformation in vitro. Molecular biological tools should continue to be useful in approaching specific questions within two broad areas of virus–host interaction.

Areas Concerning Viral Factors

1. Molecular interactions of the middle T protein that lead to tumor induction in different tissues
2. The natural role of middle T in the regulation of replication and establishment of persistent infections
3. Other viral determinants—both structural and regulatory—that affect tumorigenesis and replication

Areas Concerning Host Factors

1. The role of host immune systems in changing polyoma virus from a powerful pathogen into a silent symbiont. What viral antigens are being recognized and through what cellular mechanisms?
2. Host genetic factors, nonimmunological as well as immunological in nature, that confer resistance to tumor induction by polyoma virus
3. Developmental and behavioral aspects of various common and rare polyoma-induced tumors, for example, evaluation of invasive properties and metastasis
4. Identification of host cell mechanisms that lead to nonproductive viral infection in tumor cells, and the absence of such mechanisms in tissues where only lytic infection occurs
5. Pathogenesis related to lytic viral infections—viz., kidney and arterial lesions.

Acknowledgment

Research by the authors has been supported by a Grant 1 R35 CA44343 from the National Cancer Institute.

References

1. Eddy BE (1969) Polyoma virus. *In* Gard S, Hallauer C, Meyer KF (eds) Virol Monogr 7, Springer-Verlag, New York, pp 1–114
2. Dawe CJ, Freund R, Mandel G, Ballmer-Hofer K, Talmage DA, Benjamin TL (1987) Variations in polyoma virus genotype in relation to tumor induction in mice: Characterization of wild type strains with widely differing tumor profiles. Amer J Pathol 127:243–261
3. Tooze J (ed) (1981) DNA Tumor Viruses. Cold Spring Harbor Laboratory, Cold Spring Harbor, New York
4. Courtneidge SA, Smith AE (1983) Polyoma virus transforming protein associates with the product of the c-src cellular gene. Nature 303:435–439
5. Kaplan DR, Whitman M, Schaffhausen B, Pallas DC, White M, Cantley L, Roberts TM (1987) Common elements in growth factor stimulation and oncogenic transformation: 85 kd phosphoprotein and phosphatidylinositol kinase activity. Cell 50:1021–1029
6. Courtneidge SA, Heber A (1987) An 81 kd protein complexed with middle T antigen and $pp60^{c-src}$: A possible phosphatidylinositol kinase. Cell 50:1031–1037
7. Sjogren HO, Hellstrom I, Klein G (1961) Resistance of polyoma immunized mice against transplantation of established polyoma tumors. Exp Cell Res 23:204–208
8. Habel K (1962) Immunological determinants of polyoma virus oncogenesis. J Exp Med 115:181–193
9. Ruley HE, Fried M (1983) Sequence repeats in a polyoma virus DNA region important for gene expression. J Virol 47:233–237
10. Amati P (1985) Polyoma regulatory region: A potential probe for mouse cell differentiation. Cell 43:561–562
11. Freund R, Mandel G, Carmichael GG, Barncastle JP, Dawe CJ, Benjamin TL (1987) Polyomavirus tumor induction in mice: Influences of viral coding and noncoding sequences on tumor profiles. J Virol 61:2232–2239
12. Freund R, Dawe CJ, Benjamin TL (1988) The middle T proteins of high and low tumor strains of polyoma virus function equivalently in tumor induction. Virology 167:657–659
13. Berebbi M, Dandolo L, Hassoun J, Bernard AM, Blangy D (1988) Specific tissue targeting of polyoma virus oncogenicity in a thymic nude mice. Oncogene 2:149–156
14. Freund R, Dawe CJ, Benjamin TL (1988) A duplication of noncoding sequences in polyoma virus specifically augments the development of thymic tumors in mice. J Virol 62:3896–3899
15. Talmage DA, Freund R, Young AT, Dawe CJ, Benjamin TL Phosphorylation of middle T by $pp60^{c-src}$ is a molecular switch for binding of phosphatidylinositol kinase and tumorigenesis (manuscript submitted)
16. Dubensky TW, Freund R, Barncastle JP, Dawe, CJ, Benjamin TL Polyoma virus tumor induction in mice: Influences of viral replication and route of inoculation on tumor profiles (in preparation)
17. Dubensky TW, Villarreal LP (1984) The primary site of replication alters the eventual site of persistent infection by polyomavirus in mice. J Virol 50:541–546
18. Rochford R, Campbell BA, Villarreal LP (1987) A pancreas specificity results from the combination of polyomavirus and Moloney murine leukemia virus enhancer. Proc Natl Acad Sci USA 84:449–453

19. Bolen JB, Fisher SE, Chowdhury E, Shan T-C, Willison JE, Dawe CJ, Israel MA (1985) A determinant of polyoma virus virulence enhances virus growth in cells of renal origin. J Virol 53:335–339
20. Dawe CJ, Freund R, Barncastle JP, Dubensky TW, Mandel G, Benjamin TL (1987) Necrotizing arterial lesions in mice-bearing tumors induced by polyoma virus. J Exp Pathol 3:177–201

CHAPTER 6
Activation of Adenovirus Gene Expression

THOMAS SHENK

Adenovirus (Ad) is a DNA tumor virus whose chromosome is a linear, double-stranded DNA molecule about 36,000 base pairs in size. Six early transcription units (E1A, E1B, E2, E3, E4, and major late L1) become active very soon after the viral chromosome reaches the nucleus of the infected cell. Several hours later, concomitant with the onset of DNA replication, three late units (IX, IVa2, and major late) are expressed at high levels. Expression of these viral units is highly regulated at the level of transcription and involves a complex interplay between viral gene products and host transcription factors. This chapter examines the cascade of events that first activate the adenovirus E1A gene, then turn on the remainder of the early viral genes, and finally lead to full expression the late viral genes.

Activation of the Adenovirus E1A Trans-Activator Gene

The E1A gene is the first viral gene to become active subsequent to Ad infection. Its products then function to activate transcription of the other early viral genes [1,2], and E1A proteins also further stimulate transcription of the E1A gene itself. Activation of the E1A gene requires that the infecting chromosome associate with the nuclear matrix. Then a *cis*-acting enhancer element functions to stimulate transcription of the E1A gene.

Nuclear Matrix Association of the Viral Chromosome
After the virion reaches the nucleus, its chromosome is unpackaged and becomes associated with the nuclear matrix. This association is mediated by the virus-coded terminal protein, a product of the E2B gene. Earlier work

has shown that this protein functions to prime DNA replication and that it remains covalently attached to the 5' ends of the viral chromosome. A series of mutant Ad viruses with lesions in their terminal protein gene were employed to demonstrate that the protein plays a key role in matrix association (J. Schaack and T. Shenk, unpublished observation). Whereas Ad DNA from wild-type, virus-infected cells fractionates with the nuclear matrix, viral DNA from cells infected with terminal protein mutants does not associate with the matrix. This association is important for efficient and timely transcriptional activation. The E1A and other early viral genes are transcribed very poorly in cells infected with terminal protein mutants (J. Schaack and T. Shenk, unpublished observation). Thus, the viral gene product known to function first in the activation of Ad transcription is the terminal protein, which enters the infected cell in the virion.

Activation of the E1A Gene

Since no E1A protein can be detected in Ad virions, the E1A gene itself is very likely activated by cellular transcription machinery. Transcription is mediated, at least in part, through an enhancer element. Several regions located in the 5' -flanking sequences of the E1A gene exhibit enhancer activity in transfection assays. However, only one of these domains is required for optimal E1A expression when assays are performed in infected cells using mutated viruses [3]. This region stimulates transcription in the absence of viral gene products. However, all the viral genes can activate slowly and inefficiently in the absence of E1A proteins, so it is not clear just how unique the E1A control region is in this respect. It would make sense for the E1A enhancer element to function especially well in the absence of E1A proteins, since its gene product must be made before any viral proteins are synthesized in the newly infected cells.

A variety of cellular factors bind within the 5'-flanking region of the E1A gene, including E2F [4] and the cyclic AMP (cAMP) response element-binding protein (CREB) [5]. The binding sites for both of these factors can stimulate transcription of reporter genes under appropriate conditions. These cellular activities could contribute to the initial, basal level of E1A transcription, which is achieved independently of the expression of additional viral gene products, and as discussed below, they may also mediate enhanced transcription of the E1A gene in response to virus-coded products.

Activation of Early Genes

E1A-Mediated Activation

Mutational analysis has localized the E1A activation function to a region of the E1A protein that is conserved among different adenovirus serotypes. In fact, a 49 amino-acid peptide corresponding to this conserved region can activate an E1A-inducible transcriptional control region [6].

How do the E1A proteins activate transcription? Some activating proteins bind to specific DNA sequences within control regions. This does not appear to be the case for E1A proteins. Although an E1A polypeptide can be photo-crosslinked to DNA fragments in vitro [7] and although the protein contains a Zn^{2+}-finger motif characteristic of DNA-binding proteins within its conserved activating domain, the protein exhibits no detectable sequence specificity in its binding. Given the lack of sequence-specific binding, the favored hypothesis is that E1A products act to increase the concentration or activity of a set of transcription factors. It appears that, at least under certain conditions, E1A proteins can accomplish this directly. Both the E1A protein produced in *Escherichia coli* [8] and the 49 amino-acid peptide with activating activity [6] can stimulate activity of E1A-dependent transcriptional control regions. Perhaps E1A proteins modify the activities of cellular factors by binding to them and forming a functionally altered complex. E1A protein-containing complexes that have been indentified by immunoprecipitation [9] are excellent candidates for such complexes. Alternatively, E1A products could physically modify cellular factors. Although no enzymatic activities have been reported for purified E1A proteins, this is a reasonable hypothesis.

A variety of cellular transcription factors appear to be influenced by E1A proteins. One is CREB, or activating transcription factor (ATF). Binding sites for this activity reside in the 5'-flanking region of all early Ad genes except the E1B gene [5,10]. This factor very likely plays a role in E1A activation, since cAMP acts in synergy with E1B proteins to stimulate transcription of early viral genes [11]. Mutational analysis indicates that another factor, binding to the TATA motif, endows E1A responsiveness on the viral E1B gene [12]. E1A proteins can also enhance RNA polymerase III transcription (the Ad chromosome contains two VA RNA genes, which are transcribed by RNA polymerase III) by increasing the activity of transcription factor IIIC [13,14].

Thus, E1A gene products can activate transcription by both RNA polymerase II and III through several different factors, and at least in some cases, the E1A proteins influence the activity of cellular factors directly.

Additional Viral Activating Functions

E1A proteins are not the only Ad transcriptional-activating gene products. An E4 gene product that can stimulate transcription of the E2 gene [15] is of special interest. A well-studied cellular factor, E2F, that binds upstream of the E2 start site [16] is functionally modified by an E4 product so that its DNA-binding behavior is altered (S. Hardy and T. Shenk, unpublished observations). This suggests that the E4 gene product probably stimulates E2 transcription by altering E2F activity. This link between transcriptional activation and modification of a DNA-binding factor provides an opportunity to probe a mechanism of transcriptional activation. In addition to the E4 product, the E1B–21-kd polypeptide [17,18] and the E2A protein (L.-S. Chang and T. Shenk, unpublished observations) can also stimulate transcription by viral control regions.

A complex network of viral functions interact with cellular factors to fully activate transcription of the early viral genes.

Activation of Late Genes

Of the three Ad late transcription units, the major late unit is by far the best studied. This unit generates a primary transcript of about 29,000 nucleotides that is subsequently processed by differential polyadenylation and splicing to produce five families of late mRNAs [L1 through L5 families, grouped on the basis of common poly(A) site utilization]. Early after infection, the major late transcriptional control region is active, but only the 5′ proximal mRNAs (primarily L1 mRNAs) are expressed because the RNA polymerase terminates at multiple sites before reaching the end of the unit [19]. Late after infection, the premature termination is somehow relieved, and all the late mRNA families are expressed. So, part of the apparent activation of the major late transcription unit results from a posttranscriptional event.

The apparent frequency of initiation at the major late promoter is greatly increased with the onset of viral DNA replication. This increase probably results in part from a substantial increase in template copy member. However, it appears that the DNA replication process itself also contributes to the activation by somehow inducing a cis-acting alternation in the viral chromosome [20].

A factor, the upstream stimulatory factor (USF) [21] or the major late transcription factor (MLTF) [22], binds to a recognition site in the 5′-flanking region of the major late unit. This factor is essential for activity of the major late control region both early and late after infection. It does not appear to be altered during infection. A second activity apparently interacts with sequences 3′ to the major late initiation site [23]. This factor is either encoded by Ad or induced by Ad infection and probably contributes to the increase in activity of the major late transcriptional control region late after infection.

Conclusion

Transcriptional regulation plays a major role in the activation of Ad genes. Many of the players that mediate this regulation, both virus-coded proteins and cellular factors, have now been identified. The next goal is to understand in mechanistic terms how these viral and cellular gene products interact to orchestrate the highly ordered, regulated expression of the Ad chromosome.

Acknowledgments

Unpublished work quoted from the author's laboratory was supported by a Grant (CA 38965) from the National Cancer Institute. T.S. is an American Cancer Society Research Professor.

References

1. Berk AJ, Lee F, Harrison T, Williams J, Sharp PA (1979) Pre-early adenovirus 5 gene product regulates synthesis of early viral gene mRNAs. Cell 17:935–944
2. Jones N, Shenk T (1979) An adenovirus type 5 early gene function regulates expression of other early viral genes. Proc Natl Acad Sci USA 76:3665–3669
3. Hearing P, Shenk T (1983) The adenovirus type 5 E1A transcriptional control region contains a duplicated enhancer element. Cell 33:695–703
4. Kovesdi I, Reichel R, Nevins JR (1987) Role of an adenovirus E2 promoter binding factor in E1A-mediated coordinate gene control. Proc Natl Acad Sci USA 84:2180–2184
5. Hardy S, Shenk T (1988) Adenovirus E1A gene products and adenosine 3′, 5′-cyclic monophosphate activate transcription through a common factor. Proc Natl Acad Sci USA 85:4171–4175
6. Green M, Loewenstein PM, Pusztai R, Symington JS (1988) An adenovirus E1A protein domain activates transcription *in vivo* and *in vitro* in the absence of protein synthesis. Cell 53:921–926
7. Chatterjee PK, Bruner M, Flint SJ, Harter ML (1988) DNA-binding properties of an adenovirus 289R E1A protein. EMBO J 7:835–841
8. Spangler R, Bruner M, Dalie B, Harter ML (1987) Activation of adenovirus promoters by the adenovirus E1A protein in cell-free extracts. Science 237:1044–1046
9. Harlow E, Whyte P, Franza BR, Schley C (1986) Association of adenovirus early region 1A proteins with cellular polypeptides. Mol Cell Biol 6:1579–1589
10. Lee KAW, Hai T-Y, SivaRaman L, Thimmappaya B, Hurst HC, Jones NC, Green MR (1987) A cellular protein, activating transcription factor, activates transcription of multiple E1A-inducible adenovirus early promoters. Proc Natl Acad Sci USA 84:8355–8359
11. Engel DA, Hardy S, Shenk T (1988) Cyclic AMP acts in synergy with E1A protein to activate transcription of the adenovirus early genes E4 and E1A. Genes Dev 2:1517–1528
12. Wu L, Rosser DS, Schmidt MC, Berk A (1987) A TATA box implicated in E1A transcriptional activation of a simple adenovirus 2 promoter. Nature 326:512–515
13. Hoeffler WK, Roeder RG (1985) Enhancement of RNA polymerase III transcription by the E1A gene product of adenovirus. Cell 41:955–963
14. Yoshinaga S, Dean N, Han M, Berk AJ (1986) Adenovirus stimulation of transcription by RNA polymerase III: Evidence for an E1A-dependent increase in transcription factor III C concentration. EMBO J 5:343–354
15. Goding C, Jalinot P, Zajchowski D, Boeuf H, Kedinger C (1985) Sequence-specific transactivation of the adenovirus IIa early promoter by the viral E1V transcription unit. EMBO J 4:1523–1528

16. Kovesdi I, Reichel R, Nevins JR (1986) Identification of a cellular transcription factor involved in E1A *trans*-activation. Cell 45:219–228

17. Herrmann CH, Dery CV, Mathews MB (1987) Transactivation of host and viral genes by the adenovirus E1B 19K tumor antigen. Oncogene 2:25–35

18. Yoshida K, Venkatesh L, Kuppuswamy M, Chinnadurai G (1987) Adenovirus transforming 19-kd T antigen has an enhancer-dependent trans-activation function and relieves enhancer repression mediated by viral and cellular genes. Genes Dev 1:645–658

19. Iwamoto S, Eggerding F, Falck-Pederson E, Darnell JE (1986) Transcription unit mapping in adenovirus: Regions of termination. J Virol 59:112–199

20. Thomas GP, Mathews MB (1980) DNA replication and the early to late transition in adenovirus infection. Cell 22:523–533

21. Sawadogo M, Roeder RG (1985) Interaction of a gene-specific transcription factor with the adenovirus major late promoter upstream of the TATA box region. Cell 43:165–175

22. Chodosh LA, Carthew RW, Sharp PA (1986) A single polypeptide possesses the binding and transcription activities of the adenovirus major late transcription factor. Mol Cell 6:4723–4733

23. Mansour S, Grodzicker T, Tjian R (1986) Downstream sequences affect transcription initiation from the adenovirus major late promoter. Mol Cell Biol 6:2684–2694

CHAPTER 7
Cis- and *Trans*-Activation of HIV

FLOSSIE WONG-STAAL AND JAY RAPPAPORT

Human immunodeficiency virus (HIV) is the causative agent of acquired immunodeficiency syndrome (AIDS). Infection via cell-free virus particles is restricted to CD4-positive cells, including macrophages, helper T - cells, and glial cells. In contrast to the specificity of binding to the cell surface receptor, HIV LTR-directed gene expression is carried out in a variety of cell types upon DNA transfection. Regulation of HIV gene expression, as discussed below, involves transcriptional and posttranscriptional events, including RNA processing.

The HIV LTR consists of a 453 base-pair U3 region, a 98 base-pair R region, and an 83 base-pair U5 region, similar to those of other known retroviruses [1] (Figure 7.1**B**). Within the U3 and R regions are multiple *cis*-acting elements involved in HIV gene expression (Figure 7.1**A**). Like most eukaryotic cellular and viral protein coding genes, HIV contains a TATA box 27 base pairs upstream from the transcription initiation site. As demonstrated in other eukaryotic transcription systems, this element plays a role in positioning the transcription initiation site as well as contributing to the level of promoter efficiency [2]. Cellular factors interacting with this sequence may be general components of RNA polymerase II preinitiation complexes. These factors apparently enable the RNA polymerase II enzyme to recognize and bind to the promoter and may stabilize the binding of upstream activator proteins [3].

Upstream of the HIV TATA box lie three tandem GC-rich sequences (-46 to -77). These sequences have been shown to bind transcription factor Sp1 [4]. Base-pair substitution within Sp1 binding region III (-68 to -77) has little effect on transcription in vitro whereas residues within sites I (-46 to -55) and II (-57 to -66) appear to be more critical [4]. Oddly enough, purified Sp1 binds to site III more strongly than to sites I and II. In contrast

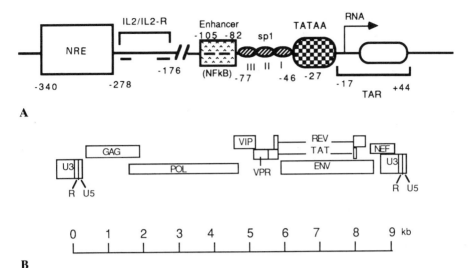

Figure 7.1. **A** Diagram of the HIV-1 LTR/Promoter, *cis-*regulatory, elements, and *trans-*acting cellular factors. **B** Genetic organization of HIV. We have adopted here a newly proposed nomenclature for the HIV accessory genes (R.C. Gallo, W. Haseltine, L. Montagnier, F. Wong-Staal, and M. Yoshida, *Science,* in press; *Nature,* in press; *AIDS Res. Hum. Retroviruses,* in press). Correspondence of new names with previous names are as follows:

tat	=	*tat-3, TA*
rev	=	*art, trs*
vip	=	*sor,* A, P', Q
vpr	=	(R)
nef	=	3' *orf,* B, E', F

to the findings from experiments using purified Sp1, DNase footprinting experiments using extracts from HeLa cells failed to protect sites I and II [5], suggesting that proteins present in extracts, either in solution or bound to other LTR elements, interfere with binding to this region. Since *cis-*acting sequences corresponding to Sp1 binding sites I and II are important in determining the level of transcription in vivo [5] and in vitro [4], *trans-*acting factors may interact weakly or transiently with this region. The role of Sp1 in HIV transcription is supported by competition-footprinting experiments. Oligonucleotides to a high affinity, Sp1-binding site, as demonstrated by Garcia, et al. [5], reduce the binding of factors to the TATA region, suggesting that Sp1 stabilizes the binding of the TATA factors through a protein–protein interaction. Sp1 oligonucleotide competition also results in the loss of protection of region −78 to −83, which lies immediately upstream of site III and overlaps with the enhancer region. Sp1 might then also interact with the enhancer factor(s) through protein–protein interaction.

The control of HIV gene expression may directly relate to the long latency period and disease progression. Long-term cultures of HIV-infected T cells produce little virus replication, but stimulation of these cultures with the mitogen phytohemagglutinin A (PHA) [6] or phorbol ester [7] resulted in dramatic increases in virus production and cytopathology. The HIV enhancer, first localized within the region −137 to −17 [8] has been further localized to the region −82 to −105 [9]. Within this element are two 11 base-pair imperfect direct repeats. The sequence GGGGACTTTCC, found in the downstream repeat, corresponds exactly to the inducible transcription factor NFκB-binding site found in the κ light-chain immunoglobulin enhancer, and this sequence indeed binds to NFκB in vitro [10]. Since NFκB is stimulated in activated T cells, this observation may provide a molecular basis of HIV induction by T cell activation.

Activation of T cells by mitogens is thought to occur via the pathway utilized during antigen activation [11]. In both cases, stimulation results in elevated protein kinase C activity. Phosphorylation of preexisting nuclear factors could provide a mechanism for a rapid response to events occurring at the cell surface. This model is supported by studies demonstrating that activation of NFκB occurs via a posttranslational mechanism [12]. Presumably, multiple transcriptional factors could be activated at various levels, reprogramming a resting T cell to secrete lymphokines that, in turn, could provide additional stimulus through autocrine mechanisms. HIV has apparently tapped into this circuitry. Interestingly, sequences homologous to the IL-2 and IL-2 receptor control regions are found at −274 and −221 within the HIV LTR [1]. DNA sequences between −176 and −278, which include these homologies, have been demonstrated by Siekevitz et al. [13] to be important in contributing to the magnitude of the mitogenic response. Presumably inducible protein factors interact with these domains as well. These authors have also shown that the negative regulatory element (NRE) 8 upstream of this region functions to mollify the mitogen response.

In addition to the ensemble of cellular factors that orchestrate the control of viral gene expression, HIV encodes *trans*-acting protein factors that control both the quantity and quality of viral proteins produced. The genetic organization of HIV is shown in Figure 7.1**B**. The HIV *tat* protein activates gene expression directed from the viral LTR and is required for virus replication [14]. The 1.8-kb *tat* mRNA is produced by a double splicing event and contains one 5′ noncoding exon and two exons encoding an 86 amino-acid protein [15]. *Tat* minus mutant proviruses fail to produce significant levels of mRNA and proteins [16,17]. Deletion analysis within the HIV-1 LTR has revealed that sequences −17 to +80 are required, in *cis* for *tat* responsiveness [8]. The 3′ border of this element, TAR, has recently been more closely mapped to +44 [18]. The TAR region itself can confer *tat* responsiveness when placed downstream of a heterologous promoter, although with reduced efficiency (5% to 10% of the level of activation seen with the homologous promoter) [19]. Upstream sequences, therefore, may also participate in the *tat* response.

The mechanism of *tat* activation is not well understood. Evidence has been presented implicating *tat* in transcriptional and/or posttranscriptional events. *Tat* increases steady state levels of viral mRNA [19–21] primarily at the transcriptional level as demonstrated by nuclear run-on experiments [17,18]. Koa et al. [22] have demonstrated that prematurely terminated transcripts accumulate in the absence of *tat*, with 3′ ends mapping to nucleotide position +59. *Tat*, therefore, appears to relive a block in elongation. This model, however, does not completely explain the mechanism of *tat* function, since deletions within the TAR region have not led to increases in basal level activity [23].

Since *tat* exerts its effects via the TAR region, it is possible that *tat* binds to this region by itself or in combination with cellular factors. In support of this mode, *tat* contains seven cysteine (Cys) residues that could form a metal-binding domain similar to that of other DNA-binding proteins. Frankel et al. [24] have recently demonstrated that *tat* binds two molecules of zinc or cadmium per *tat* monomer and forms a metal-linked dimer. The seven Cys may therefore coordinate intra- and intermolecular complexes with metals. This region is essential for *tat* function, since point mutations in all but one of these Cys residues abolish *tat* activity [25]. Although there is as yet no direct evidence of *tat* binding to TAR DNA, a cellular factor has been shown to bind to the TAR region by DNAse footprinting [26]. It remains to be determined if this protein plays a role in the *tat* response.

The location of the TAR region downstream from the transcription initiation site supports the notion that *tat* acts posttranscriptionally via sequences within the RNA. The sequence +1 to +59 within the 5′ end of HIV-1 RNA contain inverted repeats as well as imperfect direct repeats. This sequence has been proposed to form an unusual RNA stem loop structure [27]. Free energy estimates predict that this structure is extremely stable (G - −37 Kcal/mole). These predictions have been confirmed in vitro by RNAse protection studies [28]. The 5′ stem loop, present in all viral *m*RNA species, may explain the poor translational efficiency of these messages [16,18]. *Tat* could conceivably melt this structure, direct the formation of alternative structures, or adapt these unusual structures to the translational machinery. Although *tat* is primarily nuclear [29], a small percentage of the total *tat* protein in the cytoplasm could conceivably carry out the latter possible function. However, the role of *tat* at the translational level remains controversial.

Although *tat* influences virus gene expression quantitatively, another viral protein, *rev* (regulator of expression of virion proteins), previously designated *trs* or *art*, has a qualitative effect. In the absence of *rev*, viral structural proteins are not synthesized [16,30]. Analysis of mRNA species produced by transfected mutant proviruses has revealed an altered mRNA profile.

Rev− mutants fail to produce stable messages for *gag–pol* (9.2 kb) and *env* (4.3 kb) and only multiply spliced 1.8 to 2-kb mRNA species accumulate [16,31]. These data suggest that *rev* enables viral RNA to evade the cellular splicing mechanisms or, alternatively, stabilizes the 9.2- and 4.3-kb messages. This latter possibility seems unlikely, given the concomitant increase in the 1.8- to 2-kb MRNA in *rev−* transfectants. RNA profiles similar to *rev−* are seen in HTLV-I in the absence of a viral gene product, pp27 [32]. It is likely that this protein plays an analogous role. As for *tat*, a controversy exists as to the mechanism of action of the *rev* proteins. Sodroski et al. [30] suggested that this protein relieves a block in translation of the structural genes. Using cDNA subclones expressing *tat*, *rev*, and *env*, Knight, et al. [33] have reported that the *rev* protein dramatically increased the level of the *env* protein (gp120) with only slight increases in the level of *env* mRNA. They have not, however, delineated the splicing pattersn of viral RNA in the presence or absence of *rev* or ruled out the use of cryptic splice sites in the absence of *rev*. Such an analysis will be important in view of the numerous splice donor and acceptor sites in the HIV genome. Recently, *cis*-acting sequences within the *env* gene have been reported to be involved in both repression and antirepression of translation [34]. In addition to the splicing and translational effects that have been attributed to *rev*, this protein has also been demonstrated to have a suppressive role in HIV-1 transcription [31]. Further studies will be required to understand the function of this interesting viral *trans*-regulatory protein that may be acting at multiple levels.

The events that result in virus reactivation from a resting infected T cell presumably begin with binding of cellular factors that recognize the HIV-1 LTR. The level of virus gene expression is further elevated by mitogen (antigen) stimulation, which results in the activation of inducible protein factors such as NF B, which recognizes the viral enhancer. This level of virus gene expression is further augmented by the synthesis of *tat* protein. At the same time, the *rev* protein reaches a critical level and allows the accumulation of structural mRNA species and proteins.

In this way, *rev* may act as a trigger for late gene expression. A third regulatory gene of HIV is *nef* (retrovirus inhibitory factor), previously designated 3′ *orf* or F. It has been shown that mutants deficient in *nef* replicate to higher titers than wild-type mutants [35], suggesting that the product of this gene is a negative regulator for virus expression. Recent studies demonstrated that *nef* is a membrane-associated, GTP-binding protein [36], possibly regulating gene expression by signal transduction.

Given the numerous *cis*-acting elements and *trans*-acting factors that function to regulate HIV-1 gene expression positively or negatively, multiple control points exist that could be useful targets for therapeutic intervention. In particular, the *tat* and *rev* proteins, both essential for virus replication, appear to be attractive candidates for antiviral therapy.

References

1. Starich B, Ratner L, Okamoto T, Gallo RC, Wong-Staal F (1985) Characterization of the longer terminal repeat sequences of the HTLV-III LTR. Science 227:538–540
2. Breathnach R, Chambon P (1981) Organization and expression of eukaryotic splitgenes coding for proteins. Annu Rev Biochem 50:349–393
3. Reinberg D, Horikoshi M, Roeder RG (1987) Factors involved in specific transcription in mammalian RNA polymerase II: Functional analysis of initiation factors IIA and IID and identification of a new factor operating at sequences downstream of the initiation site. J Biol Chem 262:3322–3330
4. Jones KA, Kadonaga JT, Luciw PA, Tijian R (1986) Activation of the AIDS retrovirus promoter by the cellular transcription factor Sp1. Science 232:755–759
5. Garcia et al
6. Zagury D, Bernard J, Leonard R, Cheynier R, Feldman M, Sarin PS, Gallo RC (1986) Long-term cultures of HTLV-III infected T-cells: A model of cytopathology of T-cell depletion in AIDS. Science 231:850–853
7. Harada S, Koyanagi Y, Nakashima H, Kobayashi N, Yamomoto N (1986) Tumor promoter, TPA, enhances replication of HTLV-III/LAV. Virology 154:249–258
8. Rosen CA, Sodroski JG, Haseltine WA (1985) Location of *cis*-acting regulator sequences in the T-cell lymphotropic virus type III (HTLV-III/LAV) long terminal repeat. Cell 41:813–823
9. Kaufman JD, Valendra G, Roderiquez G, Bushar G, Giri C, Norcross MA (1987) Phorbol ester enhances human immunodeficiency virus-promoted gene expression and acts on a repeated 10-base-pair functional enhancer element. Mol Cell Biol 7:3759–3766
10. Nabel G, Baltimore D (1987) An inducible transcription factor activates expression of human immunodeficiency virus in T-cells. Nature 326:711–713
11. Nishizuka Y (1985) Studies and perspectives of protein kinase C. Science 233:305–311
12. Sen R, Baltimore D (1986) Inducibility of kappa immunoglobulin enhancer-binding protein NF B by a post-translational mechanism. Cell 47:921–928
13. Siekevitz M, Josephs SF, Dukovich M, Peffer N, Wong-Staal F, Green WC (1987) Activation of the HIV-1 LTR by T-cell mitogens and the *trans*-activator protein of HTLV-I. Science 238:1575–1578
14. Fisher AG, Feinberg MB, Josephs SF, Harper MD, Marselle LM, Reyes G, Gonda MA, Aldovini A, Debouk C, Gallo RC, Wong-Staal F (1986a) The *trans*-activator gene of HTLV-III is essential for virus replication. Nature 320:367–373
15. Arya SK, Guo C, Josephs SF, Wong-Staal F (1985) Trans-activator gene of human T-Lymphotropic Virus type III (HTLV-III). Science 229:69–73
16. Feinberg MB, Jarrett RF, Aldovini A, Gallo RC, Wong-Staal F (1986) HTLV-III expression and production involve complex regulation at the levels of splicing and translation of viral mRNA. Cell 46:807–817
17. Sadaie MR et al (1987)
18. Hauber J, Cullen BR (1988) Mutational analysis of the *trans*-activation–

responsive region of the human immunodeficiency virus type 1 long terminal repeat. J Virol 62:673–679

19. Cullen BR (1986) Trans-activation of human immunodeficiency virus occurs via a bimodal mechanism. Cell 46:973–982

20. Wright C, Felber B, Paskalis H, Pavlakis G (1986) Expression and characterization of the *trans*-activator of HTLV-III/LAV virus. Science 234:988–992

21. Peterlin M, Luciw P, Barr P, Walker M (1986) Elevated levels of mRNA can account for the *trans*-activation of human immunodeficiency virus. Proc Natl Acad Sci USA 83:9734–9738

22. Kao S-Y, Calman AF, Luciw PA, Peterlin BM (1987) Antitermination of transcription within the long terminal repeat of HIV-1 by *tat* gene product. Nature 330:489–493

23. Cullen et al (1988)

24. Frankel AD, Bredt DS, Pabo CO (1988) *Tat* protein from immunodeficiency virus forms a metal-linked dimer. Science 240:70–73

25. Sadaie MR, Rappaport J, Benter T, Josephs SF, Willis R, Wong-Staal F Missense mutations in an infectious HIV genome: Functional mapping of *tat* and demonstration of a novel *trs* splice acceptor (Submitted)

26. Garcia JA, Wu FK, Mitsuyasu R, Gaynor RB (1987) Interactions of cellular proteins involved in the transcription regulation of human immunodeficiency virus. EMBO J 6:3761–3770

27. Okamoto T, Wong-Staal F (1986) Demonstration of virus specific transcription activator(s) in cells infected with HTLV-III by an *in vitro* cell-free system. Cell 47:29–35

28. Muesing M, Smith D, Capon D (1987) Regulation of mRNA accumulation by a human immundeficiency virus thrans-activator protein. Cell 48:691–701

29. Hauber J, Perkins A, Hermir EP, Cullen BR (1987) *Trans*-activation of human immunodeficiency virus gene expression is mediated by nuclear events. *Proc Natl Acad Sci USA* 84:6384–6368

30. Sodroski J, Goh WC, Rosen C, Dayton A, Terwilliger E, Haseltine W (1986) A second post-transcriptional *trans*-activator gene required for HTLV-III replication. Nature 321:412–417

31. Sadaie MR, Benter T, Wong-Staal F (1988) Site directed mutagenesis of two *trans*-reglatory genes (tat-iii, trs) of HIV-1. Science 239:910–913

32. Inoue J, Yoshida M, Seiki M (1987) Transcriptional (p40^x) and post-transcriptional (p27X^{III}) regulators are required for the expression and replication of human T-cell leukemia virus type I. Proc Natl Acad Sci 84:3653–3657

33. Knight DM, Flomerfelt FA, Ghrayeb J (1987) Expression of the *art*/*trs* protein of HIV and study of its role in viral envelope synthesis. Science 236:837–840

34. Rosen CA, Tenwillinger E, Dayton A, Sodroski JG, Haseltine WA (1988) Intragenic *cis*-acting *art* gene responsive sequence of HIV. Proc Natl Acad Sci USA 85:2071–2075

35. Fisher AG, Ratner L, Mitsuya H, Marselle LM, Harper ME, Broder S, Gallo RC, Wong-Staal F (1986b) Infectious mutants of HTLV-III with changes in the 3′ region and markedly reduced cytopathic effects. Science 233:655–659

36. Guy B, Kieny MP, Reviere Y, Le Peuch C, Dott K, Girard M, Montagnier L, Lecocq J-P (1987) HIV F/3′ orf encodes a phosphorylate GTP-binding protein resembling an oncogene product. Nature 330:266–269

CHAPTER 8
Herpes Virus Vectors

Minas Arsenakis and Bernard Roizman

The size, structure, and wide host range of herpes simplex viruses 1 and 2 (HSV-1 and HSV-2) make them suitable for use as vectors of genetic material both their own and that of foreign genomes. This chapter will describe in brief the unique structure of the HSV genome and the procedures used for selecting recombinant genomes, and illustrate the utility of this technology with specific examples of application.

Structure of the HSV-1 Genome

The HSV-1 genome is a linear, double-stranded DNA molecule approximately 150 kbp in size [1]. It consists of two covalently linked components designated L (long) and S (short), each consisting of unique sequences (U_L and U_S, respectively) flanked by inverted repeats [2]. The inverted repeats flanking the L component, designated **ab** and **b′a′**, are each 9 Kbp in size, and those flanking the S component, termed **ca** and **a′c′**, are each 6.5 Kbp [2,3]. One consequence of this arrangement is that the L and S components invert relative to each other and give rise to four equimolar isomers. The inverted repeats of the L component each contain in their entirety two genes; those of the S component each contain one copy of the $\alpha4$ gene, the major viral regulatory protein, and an origin of replication [4–6].

Features of the HSV-1 Genome
That Make It Useful as a Vector

The HSV-1 genome can be used as a vector for the expression of non-HSV genetic material because it contains "dispensable" genes and DNA sequences that allow insertions to be made at several locations. For example, because the internal repeats **b'a'** and **ac** are equivalent to those present at the termini of the DNA, they can be deleted without grossly affecting the viability of the virus [7]. Dispensable domains are also present in the unique sequences of the L and S components. Second, it is estimated that at least 25 Kbp can be inserted into the viral genome. Thus, whereas the unmodified HSV-1 genome can accept as up to 9.6 Kbp of additional genetic material [8], viral mutants with the internal repeats deleted should be able to accept at least 25 Kbp. Recent evidence that all but the gene specifying the glycoprotein D can be deleted from the unique sequences of the S component without affecting the viability of the virus suggests that additional DNA sequences could be inserted into the genome [9–11]. Third, the HSV genome specifies both spliced and unspliced mRNA. This may be important in certain cases in which the cDNA copy is not expressed as efficiently as the genomic copy and in cases in which the cDNA is not available. Last, non-HSV genes fused to HSV promoters and appropriate other viral regulatory signals and inserted into the HSV genome are regulated as bona-fide HSV genes. This is a general requirement for the expression of non-HSV gene expressions, but cellular genes driven by their own promoters have been expressed off the viral genome in a few instances (R.D. Everett, personal communication).

Principles of Engineering Insertions and Deletions in the HSV-1 Genome

The viral recombinants produced to date took advantage of site-directed recombination between a cloned fragment of DNA and the intact HSV DNA vector cotransfected into appropriate cells. For the recombination to occur, the DNA fragment must be constructed so as to contain a target sequence in which an appropriate deletion has been made, or one that contains an insert. The target sequence is flanked by sequences homologous to the corresponding sequences surrounding the unmodified target site in the viral genome.

Because the frequency of recombination between the cloned DNA fragment and the intact viral DNA may be low and because screening for recombinants is a slow and tedious job that can be further complicated when the desired recombinant is at a growth disadvantage relative to the parental virus, the employment of a selectable marker was found to be extremely useful. The selection adopted [12,13] centers on the HSV-1 thymidine kinase **(TK)** and is based on two considerations:

1. The enzyme has a broad substrate specificity and phosphorylates a variety of thymidine and other nucleoside analogs. Some of these analogs are preferentially phosphorylated by the viral **TK** and are not toxic to uninfected cells, but are lethal when phosphorylated by the viral **TK** in infected cells (e.g., Ara T and Acyclovir). In the case of analogs phosphory-

lated by cellular **TK** (e.g., BUdR), uninfected cells and cells infected with **TK⁻** virus survive. In both cases phosphorylation of the analogs by the viral enzyme results in the destruction of the infected cell and in preferential survival of the cell infected with **TK⁻** virus. Thus, to select recombinants in which the **tk** gene is inactivated by insertion or deletion, the progeny of the transfection are plated either on wild-type cells (**TK⁺**) in the presence of Ara T or Acyclovir or, alternatively, on **TK⁻** cells in the presence of any of the analogs (including BUdR). The net result in both cases is that only cells infected with **TK⁻** viruses survive to produce infectious virus progeny.

2. The second important consideration is that the viral **TK** is not essential for virus multiplication in cells that maintain adequate pools of deoxynucleotides. In **TK⁻** cells the major and only primary pathway for the generation of TdRMP is through the conversion of UdRMP by thymidilate synthetase (**TS**). This pathway can be blocked by methotrexate, and therefore, its addition to **TK⁻** cells is lethal. Thus, to select recombinants in which a functional **tk** gene was inserted, the progeny of the cotransfection is plated on **TK⁻** cells in the presence of methotrexate. Under these conditions only cells infected by the **TK⁺** viruses survive to produce infectious virus progeny.

The construction of recombinant viruses can follow two different routes depending on the desired final product. The simplest case is the insertion of additional DNA into the domain of the viral **tk** gene at its natural location. To make an insertion at a different location, start with a **TK⁻** virus into which a functional **tk** gene has been inserted at the desired location. Next, by recombination, the second copy of the **tk** gene with the desired DNA insert is exchanged. The product is the desired recombinant, but it is **TK⁻**. If necessary or desirable, the **tk** gene can be restored at its natural location. The second route is used to construct recombinants carrying deletions in specific genes. In this instance the parental virus is also a **TK⁻** virus into which is introduced a functional copy of the **tk** gene within the sequences of the target gene. At this point the target gene has been interrupted, but not necessarily inactivated. Next, the **tk** gene is replaced with a deleted version of the target gene. In the final step, we replace the **tk** gene at its natural location. The two schemes are presented in Figure 8.1.

The HSV-1 Genome as Vectors for the Solution of Specific Problems

The use of HSV vectors is of interest for two reasons. First, it has been used to probe the function of specific genes or DNA sequences. For example, insertion of the terminal **a** sequence and components thereof at a new location in the HSV genome led to the identification of the *cis*-acting site for the inversion of the L and S components. Substitution or insertion of

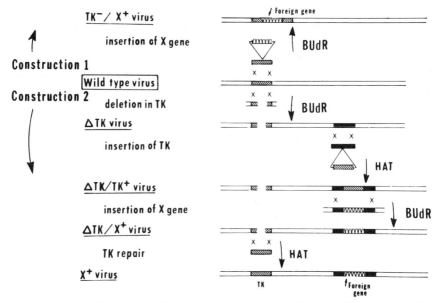

Figure 8.1. Flow diagram for the construction of recombinant HSV genomes. The HSV-1 DNA is represented by double lines. The viral **tk** gene is represented by hatching, the target gene to be deleted by the solid bars, and the foreign gene X by the triangled area. The symbol delta (Δ) denotes deletion. HAT and BUdR refer to growth media containing hypoxanthine, aminopterin, and thymidine and bromo-deoxyuridine, respectively.

new promoters led to the identification of the domains of the promoters of α and γ2 genes [14–16]. The technology for producing recombinants has been used to identify genes that are not required for growth in cell culture.

The most significant, long-term potential of HSV as a vector is in connection with the development of live attenuated vaccines against HSV infections. Such recombinants are under development [17]. A major component of these recombinants is the design of specific sites for insertion of genes specifying antigenic determinants to induce protective immunity against other viruses infecting humans. HSV as a vector for immunization against infectious diseases has three potential uses. First, vaccines against HSV infections are long overdue and are likely to emerge in the near future. Current data indicate that subunit vaccines are not effective in protecting against HSV infections [18]. The possibility that a live vaccine will confer protection against HSV infections rests on the observation that an earlier HSV-1 infection tends to attenuate the course of HSV-2 infections [19]. Second, as noted earlier in the text, non-HSV genes inserted into the HSV genome are readily expressed, especially if driven by competent HSV promoters. Among the non-HSV genes expressed to date are those specifying the HIV-1 glycoprotein (M. Arsenakis, B. Roizman, and W. Haseltine, unpublished observations), the hepatitis B virus S antigen [20], the EBNA 1 protein of the Epstein–Barr virus [21], and chicken ovalbumin. The actual use of the HSV genome as a vector hinges, however, on the demonstration of the safety and efficacy of the HSV recombinant vector and on the development of powerful promoters capable of expressing the non-HSV genes at a high level. Work in these areas is in progress.

References

1. Kieff ED, Bachenheimer SL, Roizman B (1971) Size, composition, and structure of the deoxyribonucleic acid of herpes simplex virus subtypes 1 and 2. J Virol 8:125–132
2. Wadsworth S, Jacob RJ, Roizman B (1975) Anatomy of herpes simplex virus DNA. II. Size, composition, and arrangement of inverted terminal repetitions. J. Virol 15:1487–1497
3. Sheldrick B, Berthelot N (1975) Inverted repetitions in the chromosome of herpes simplex virus. Cold Spring Harbor Symp Quant Biol 39:667–668
4. Mackem S, Roizman B (1980) Regulation of herpesvirus macromolecular synthesis: Transcription initiation sites and domains of α genes. Proc Natl Acad Sci USA 77:7122–7126
5. Mocarski ES, Roizman B (1982) Herpes virus dependent amplification and inversion of cell associated viral thymidine kinase gene flanked by a sequences and linked to an origin of DNA replication. Proc Natl Acad Sci USA 79:5626–5630
6. Watson RJ, Preston CM, Clements JB (1979) Separation and characterization of herpes virus type 1 immediate early mRNAs. J Virol 31:42–52
7. Poffenberger KL, Tabares E, Roizman B (1983) Characterization of a viable, noninverting herpes simplex virus 1 genome derived by deletion of sequences at the junction of components L and S. Proc Natl Acad Sci USA 80:2690–2694
8. Roizman B, Jenkins FJ (1985) Genetic engineering of novel genomes of large DNA viruses. Science 229:1208–1214
9. Longnecker R, Roizman B (1986) Generation of an inverting herpes simplex virus 1 mutant lacking the L–S junction a sequences, an origin of DNA synthesis, and several genes including those specifying glycoprotein E and the α47 gene. J Virol 58:583–591
10. Longnecker R, Roizman B (1987) Clustering of genes dispensable for growth in culture in the S component of the HSV-1 genome. Science 236:573–576
11. Post LE, Roizman B (1981) A generalized technique for deletion of specific genes in large genomes: α Gene 22 of herpes simplex virus 1 is not essential for growth. Cell 25:227–232
12. Mocarski ES, Post LE, Roizman B (1980) Molecular engineering of the herpes simplex virus genome: Insertion of a second L–S junction into the genome causes additional inversions. Cell 22:243–255
13. Post LE, Mackem S, Roizman B (1981) Regulation of α genes of herpes simplex virus: Expression of chimeric genes produced by fusion of thymidine kinase with a gene promoters. Cell 24:555–565

14. Mackem S, Roizman B (1982) Regulation of α genes of herpes simplex virus: The α27 gene promoter–thymidine kinase chimera is positively regulated in converted L cells. J Virol 43:1015–1023

15. Mackem S, Roizman B (1982) Structural features of the α gene 4, 0, and 27 promoter regulatory sequences which confer α regulation on chimeric thymidine kinase genes. J Virol 44:939–949

16. Silver S, Roizman B (1985) γ_2-thymidine kinase chimeras are identically transcribed but regulated as γ_2 genes in herpes simplex virus genomes and as β genes in cell genomes. Mol Cell Biol 5:518–528

17. Meignier B, Longnecker R, and Roizman B (1988) In vivo behavior of genetically engineered herpes simplex viruses R7017 and R7020. I. Construction and evaluation in rodent animal models. J Infect 158:602–614

18. Meignier B, Jourdier TM, Norrild B, Pereira L, Roizman B (1987) Immunization of experimental animals with reconstituted glycoprotein mixtures of herpes simplex virus 1 and 2: Protection against challenge with virulent virus. J Infect Dis 155:921–930

19. Corey L (1985) The natural history of genital herpes simplex virus: Perspectives on an increasing problem. *In* Roizman B, Lopez C (eds) The Herpesviruses, vol 4. Plenum Publishing, New York, pp 1–35

20. Shih M-F, Arsenakis M, Tiollais P, Roizman B (1984) Expression of hepatitis B virus S gene by herpes simplex virus vectors carrying α- and β-regulated gene chimeras. Proc Natl Acad Sci USA 81:5867–5870

21. Hummel M, Arsenakis M, Marchini A, Lee L, Roizman B, Kieff E (1986) Herpes simplex virus expressing Epstein–Barr virus nuclear antigen 1. Virology 148:337–348

CHAPTER 9
Recombinant Proteins Expressed by Baculovirus Vectors

Max D. Summers

One of the most valuable tools in molecular biology is the ability to clone and to construct hybrid genes by recombinant DNA techniques and to express abundantly hybrid gene products with an appropriate heterologous expression system. The helper-independent baculovirus expression vector [1–4] is a new, highly versatile tool in the limited spectrum of eucaryotic expression systems. Compared to bacterial, vertebrate, and yeast systems, it is unique because (1) foreign genes are expressed under the transcription control of the very strong baculovirus polyhedrin gene promoter and (2) recombinant polypeptides are produced in lepidopteran (moth) insect cells during lytic infection with a DNA viral vector. The baculovirus vector has been used to express large amounts of viral, mammalian, plant, and procaryotic genes, and nearly all of the recombinant protein products are produced in a functionally authentic form [1]. This brief review emphasizes the current state of baculovirus vector development and describes several applications that have not been previously reported.

Baculovirus Biology: The Role of Polyhedrin

Baculoviruses constitute a very large group of more than 450 DNA viruses pathogenic to a comparable number of species in seven orders of insects and some noninsect arthropods. Viral replication occurs in the infected cell nucleus, and two forms of mature viral progeny are produced. Early in infection, progeny virus migrates from the nucleus and buds from the infected cell surface to produce the extracellular form of the virus. This form is responsible for cell-to-cell or secondary infection in the organism. Later in infection, nucleocapsids obtain a viral envelope in the cell nucleus and become embedded in a protein crystal. This second form of progeny is referred to as *occluded virus*. The viral occlusion is a highly ordered protein crystal assembly of the viral-encoded protein polyhedrin.

The genome of the *Autographa californica* nuclear polyhedrosis virus (AcMNPV), the prototype of the baculovirus family, is approximately 130,000 base pairs and is a double-stranded, covalently closed DNA molecule with superhelical configuration. The DNA is packaged as a nucleoprotein complex in a rod-shaped nucleocapsid, which must obtain an envelope before becoming incorporated into the viral occlusion. Although it is the most studied baculovirus, very little is known of the molecular biology of AcMNPV infection, and replication and the functional organization of the genome [5].

The ability to produce a viral occlusion, a highly ordered protein crystal composed of a major viral encoded protein, is unique to insect viruses. The regulation of polyhedrin gene expression is late to very late in the infection cycle, and the degree of expression can vary significantly from very little [3] to more than 50% of the total cell proteins. The crystal of polyhedrin protein functions to protect the occluded viral form in an extracellular state until the virus is ingested by a susceptible insect. When the insect feeds upon a contaminated food source, the viral occlusions enter the midgut where the crystal structure of the polyhedrin protein is dissociated in the high-pH environment to release the infectious occluded virus from which natural infection occurs [6].

The polyhedrin protein is not essential for viral infection or replication. This important property, coupled with the strength of the polyhedrin promoter, represents two major advantages of the baculovirus system. Other advantages include working with a nonpathogenic virus and non-transformed cells, in addition to the highly expressed baculovirus polyhedrin gene. Several factors can affect the level of polyhedrin or polyhedrin promoter-linked, foreign gene expression. The molecular basis for this is not known, but other factors include the viability of the insect cells, temperature, type of insect cell line used, or tissue infected in the insect. Also of importance is a change in cell culture media components or the quality of the fetal bovine serum. The insect *Spodoptera frugiperda* 1PLB-Sf21-AC clonal isolate Sf9 cells are very shear sensitive, and shear stress can directly affect cell viability. Changes or alterations in recommended procedures [3] can easily result in a two- to tenfold reduction of expression of polyhedrin or any foreign gene linked to its promoter.

Baculovirus Transfer Vectors and Selection for Recombinant Baculoviruses

Although baculoviruses are as genetically complex as pox and herpes viruses [7], the technology associated with the development of transfer vectors and the selection of a recombinant baculovirus containing the hybrid gene are now routine in many laboratories [1,3].

There are several vector constructs available for making recombinant proteins as fused or nonfused forms. After a review of the published literature [1], it was decided that any transfer vector with a polyhedrin nontranslated leader sequence in its natural form including all or part of the polyhedrin initiator methionine codon could be recommended. The vectors pAcYM1 [8] or pVL941 [9] (Luckow and Summers, unpublished observations) or their equivalents are recommended for the abundant expression of nonfused recombinant proteins. In most cases these vectors produce the most abundant level of steady-state recombinant mRNA for a foreign gene that approaches, or is equivalent, to that of polyhedrin mRNA.

Only genomic DNAs containing genes without introns or cDNA clones are currently recommended for abundant expression in existing baculovirus vectors, and it is suggested that as much of the 5' nontranslated DNA sequences of the foreign gene as possible be removed.

Once a transfer vector has been selected and the foreign DNA sequence properly inserted and varified for location and orientation, the insertion of the hybrid gene into the wild-type polyhedrin region (which is not essential for infection or replication) of the genome is carried out by homologous recombination. Screening for the recombinant baculovirus can be accomplished by DNA or RNA hybridization with a probe specific to the foreign gene insert, or recombinant proteins can be detected with an antibody, or enzymatic assay, or other biological/functional or structural property for which there is a sensitive detection procedure [1] (Table 9.1). The recombinant progeny forms of the virus resulting from the transfection assay can vary from 0.1 to 5% of the total output. The amount of foreign DNA that can be inserted into a nonessential region of AcMNPV is not known. However, sequences as long as 4.1 kb have been inserted successfully [40]. Because of the size of the genome and the rod-shaped nucleocapsid, it might not be unrealistic to expect this to be in excess of 25,000 base pairs of foreign DNA.

Authentic Properties of Recombinant Proteins

Luckow and Summers [1] reviewed approximately 34 foreign genomic or cDNA clones of virus, plant, and mammalian genes expressed with the baculovirus vector. Since the submission of that manuscript (November 1987), the expression of more than 50 genes has now been reported (Table 8.1).

All foreign genes expressed under the transcriptional regulation of the polyhedrin gene promoter are produced maximally by 48 to 72 hours postinfection. This is because the highly regulated transcription and translation of polyhedrin is initiated late and continues very late in the infection compared to other baculovirus genes.

Table 9.1. BEV recombinant protein expression and processing.

Recombinant protein	Cellular location			Other Processing	References
	Nucleus	Cytosol	Membrane associated		
Bovine eye-lens protein rhodopsin receptor	—	—	Membrane	—	10
CD4 (T4) (hydrophilic extracellular element)	—	—	Secreted	Glycosylated	11
Denge envelope glycoprotein (E)	—	—	Plasma membrane	Glycosylated	12
Dengue NS1	—	—	—	Glycosylated	12
Endogenous human gastrin-releasing peptide precursor	—	—	Secreted Pro GRP → GRP 1-125 1-27	No amidation	13
Epidermal growth factor (human)	—	—	Plasma membrane	Glycosylated	14
Hepatitis B virus surface antigen	—	—	Luman, RER	Glycosylated	15–17
Hepatitis B virus surface antigen	—	—	Secreted	Glycosylated	18–20
Hepatitis B-M protein	—	—	Secreted	—	21
HIV gp 160	—	—	Cell surface	Glycosylated	22–23
Human a-interferon	—	—	Secreted	—	24–26
Human acid b-glucosidase lysosomal hydrase	—	—	Not secreted	Glycosylated	27
Human b-interferon	—	—	Secreted; glycosylated	Signal peptide	3
Human colony-stimulating factor I	—	—	Secreted	Glycosylated/ dimeric assembly	28
Human erythopoeitin	—	—	Secreted	Glycosylated	29
Human glucocerebrosidase (lysosmal)	—	—	Secreted (40%)	Glycosylated/ signal cleaved	30
Human interleukin-2	—	—	Secreted	—	31
Human transferrin receptor	—	—	Plasma membrane	Glycosylated/ palmitylation	32
Influenza virus hemagglutinin	—	—	Integral membrane protein	Acylated/oligomeric assembly/ glycosylated/ HA → HA$_1$ + HA$_2$/ hemadsorption	33–36

					Ref.
Myelin-associated glycoprotein	—	—	Secreted	Glycosylated	37
Para influenza (type 3) hemagglutinin neuraminidase	—	—	Plasma membrane	Glycosylated	38
Patatin	—	—	Secreted	—	39
Phaseolin	—	—	Secreted	Glycosylated	40
Sindbis virus E1	—	—	Plasma membrane	Glycosylated	41
Sindbis virus E2	—	—	Plasma membrane	Glycosylated	41
Tissue plasminogen activator	—	—	Secreted	Glycosylated/ signal cleaved/ prosequence cleaved	42,43,58
FMDV capsid proteins	—	+		—	44
Hantaan virus capsid	—	+ (?)		—	45
Hepatitis A virus capsid	—	+	Insoluble aggregate	—	46
Hepatitis B-S protein	—	+	Partially secreted	—	21
Human insulin receptor (protein-tyrosine kinase domain)	—	+		Phosphorylated	47,48
Human insulin receptor (cytoplasmic domain)	—	+		glycosylated Phosphorylated	49
Hunan terminal transferase	—	+		Phosphorylated	50
Human tyrosine hydroxylase	—	+ (?)		(?)	51
PT phlebovirus NS_s protein	—	+		—	52
Punta Toro phlebovirus nucleocapsid (N) protein	—	+		—	52
Sindbis virus capsid	—	+		—	41
Drosophila Krüppel gene product	+	—		Phosphorylated	53
Hepatitis B virus core ag	+	—		—	16,17
HTLV-I p40ˣ	+	—		Phosphorylated	54
Human c-myc protooncogene	+	—		Phosphorylated	55
N. crassa activator protein	+	—		Phosphylated/ glycosylated	56
Polyoma virus large T	+	—		Phosphorylated	57
SV40 large T	+	—		Oligomerization/ palymitylation/ glycosylated/ phosphorylated	15

Regardless of their source, the majority of recombinant proteins produced in recombinant baculovirus-infected *Spodoptera frugiperda* Sf9 cells are functionally authentic. Except for complex N-glycosylation, most co- and posttranslational modifications are apparently quite similar, if not authentic. Recombinant proteins are proteolytically processed, N-glycosylated, secreted, and inserted into the plasma membrane, or targeted to other Sf9 cellular organelles. Homooligomer assembly apparently occurs by a process similar to the process in mammalian cells, as does the processing of polyproteins (Sindbis 26S transcription unit). Recombinant proteins are antigenic and immunogenic, and induce productive immunity.

Summary

The results obtained with a wide variety of recombinant proteins expressed in baculovirus-infected cells continue to support the broad generalization that the insect virus-based expression system can express large amounts of functionally authentic recombinant proteins. The rapidity by which the recombinant baculovirus containing the foreign gene of choice can be isolated is also an important reason for using baculovirus vector and has enhanced the ability to assess functional aspects of protein structure by selected mutations within genes quickly. It is also a vector system that has a demonstrated potential use in medical and agricultural basic research and for the production of commercial products for human and veterinary medicine. It is the only DNA viral vector for the cloning and expression of genes in insect cells or insects. In the latter respect, it is a major tool for study of the molecular biology of insect systems.

References

1. Luckow VA, Summers MD (1988a) Trends in the development of baculovirus expression vectors. Bio/Technology 6:47–55
2. Smith GE, Summers MD Fraser MJ (1983) Production of human beta interferon in insect cells infected with a baculovirus expression vector. Mol Cell Biol 3:2156–2165
3. Summers MD, Smith GE (1987) A manual of methods for baculovirus vectors and insect cell culture procedures. Texas Agricultural Experiment Station Bulletin No. 1555
4. Doerfler W (1986) Expression of the *Autographa californica* nuclear polyhedrosis virus genome in insect cells: Homologous viral and heterologous vertebrate genes—the baculovirus vector system *In* Doerfler W, Böhm P (eds) The Molecular Biology of Baculoviruses, Current Topics in Microbiology and Immunology, vol. 131. Springer-Verlag, New York, pp 51–68
5. Doerfler W, Böhm P (1986). *In* Doerfler W, Böhm P (eds) The Molecular Biology of Baculoviruses, Current Topics in Microbiology and Immunology vol 131. Springer-Verlag, New York, p 168

6. Granados RR, Williams KA (1986) *In vivo* infection and replication of baculo-viruses. *In* Granados RR, Federick BA (eds) The Biology of Baculoviruses, vol I, Biological Properties and Molecular Biology. CRC Press pp 89–108

7. Moss B Flexner C (1987) Vaccinia virus expression vectors. Ann Rev Immunol 5:305–324

8. Matsuura Y, Possee RD, Overton HA, and Bishop DHL (1987) Baculovirus expression vectors: The requirements for high level expression of proteins, including glycoproteins. J Gen Virol 68:1233–1250

9. Summers MD (1988) Baculovirus-directed foreign gene expression. Banbury Report: Current Communications in Molecular Biology. Cold Spring Harbor Laboratory, Cold Spring Harbor, New York (in press)

10. Janssen JJM, van de Ven WJM, van Groningen-Luyben WAHM, Roosien J, Vlak JM, de Grip WJ (1988) Synthesis of functional bovine opsin in insect cells under control of the baculovirus polyhedrin promoter. Mol Biol Rep (in press)

11. Hussey RE, Richardson NE, Kowalski M, Brown NR, Chang H-C, Siliciano RF, Dorfman T, Waker B, Sodroski J, Reinherz EL (1988) A soluble CD4 protein selectively inhibits HIV replication and syncytium formation. Nature 331:78–81

12. Zhang Y-M, Hayes EP, McCarty TC, Dubois DR, Summers PL, Eckels KH, Chanock RM, Lai C-J (1988) Immunization of mice with dengue structural proteins and non-structural protein NS1 expressed by baculovirus recombinant induces resistance to dengue virus encephalitis. J Virol (in press)

13. Lebacq-Verheyden A-M, Kasprzyk PG, Raum MG, Coelingh KVW, Battey JF (1988) Post-translational processing of endogenous and of baculovirus-expressed human gastrin-releasing peptide precursor. Mol Cell Biol (submitted)

14. Greenfield C, Patel G, Clark S, Jones N, Waterfield MD (1988) Expression of human EGF receptor with ligand-stimulatable kinase activity in insect cells using a baculovirus vector. EMBO J 7:139–146

15. Lanford RE (1988) Expression of simian virus 40 T antigen in insect cells using a baculovirus expression vector Virology (in press)

16. Lanford RE, Eichberg JW, Dreesman GR, Notvall LM, Luckow VA, Summers MD (1987a) Expression of hepatitis B virus surface and core antigens in bacuolovirus: Abst. International Symposium on Viral Hepatitis and Liver Disease. London, England

17. Lanford RE, Kennedy RC, Dreesman GR, Eichberg JW, Notvall L, Luckow VA, Summers MD (1987b) Expression of hepatitis B virus surface and core antigens using a baculovirus expression vector. *In* Zukerman AJ (ed) Viral Hepatitis and Liver Diseases. Alan R. Liss, New York

18. Kang CY, Bishop DHL, Seo J-S, Matsuura Y, Choe M (1987) Secretion of particles of hepatitis B surface antigen from insect cells using a baculovirus vector J Gen Virol 68:2607–2613

19. Kang CY (1988) Baculovirus vectors for expression of foreign genes. Adv Virus Res (in press)

20. Mohamad AA, Price PM (1987) A baculovirus expression system to study the assembly of HBV in insect cells: Co-expression of the three NBV envelope proteins results in their co-assembly and secretion as HBsAg particles. Cold Spring Harbor Conference on Hepatitis B Viruses, Cold Spring Harbor, New York

21. Price PM, Mohamad A, Zelent A, Neurath AR, Acs G (1988) Translational selection in the expression of the hepatitis B virus envelope proteins. *DNA* (in press)

22. Rusche JR, Lynn DL, Robert-Guroff M, Langlois AJ, Lyerly HK, Carson H, Krohn K, Ranki A, Gallo RC, Bolognesi DP, Putney SD, Matthews TJ (1987) Humoral immune response to the entire human immunodeficiency virus envelope glycoprotein made in insect cells. Proc Natl Acad Sci USA 84:6924–6928

23. Rusche JR, Javerherian K, McDanal C, Petro J, Lynn DL, Grimaila R, Langlois A, Gallo RC, Arthur LO, Fischinger PJ, Bolognesi DP, Putney SD, Matthews TJ (1988). Antibodies that inhibit fusion of human immunodeficiency virus-infected cells bind 24-amino acid sequence of the viral envelope, gp 120. Proc Natl Acad Sci USA 85:3198–3202

24. Maeda S, Kawai T, Obinata M, Chika T, Horiuchi T, Maekawa K, Nakasuji K, Saeki S, Sato Y, Yamada K, Furusawa M (1984) Characteristics of human interferon-a produced by a gene transferred by a baculovirus vector in the silkworm, *Bombyx mori*. Proc Japan Acad 60:423–426

25. Maeda S, Kawai T, Obinata M, Fujiwara H, Horiuchi T, Saeki Y, Sato Y, Furusawa M (1985) Production of human a-interferon in silkworm using a baculovirus vector. Nature 315:592–594

26. Horiuchi T, Marumoto Y, Saeki Y, Sato Y, Furusawa M, Kondo A, Maeda S (1987) High-level expression of the human-a-interferon gene through the use of an improved baculovirus vector in the silkworm, *Bombyx mori*. Agri Biol Chem 51:1573–1580

27. Grabowski GA, White WR, and Grace ME (1988) Expression of functional human acid β-glucosidase in COS-1 and *Spodoptera frugiperda* cells. Enzyme (in press)

28. Inlow D, Harano D, Maiorella B (1987) Large-scale insect culture for recombinant protein production. Symposium on Strategies in Cell-Culture Scale-Up. American Chemical Society National Meeting, New Orleans, Louisiana

29. Wojchowski DM, Orkin SH, Sytkowski AJ (1987). Active human erthropoietin expressed in insect cells using a baculovirus vector: A role for N-linked oligosaccharide. Biochim Biophys Acta 910:224–232

30. Martin BM, Tsuji S, LaMarca ME, Maysak K, Eliason W, Ginns EI (1988) Glycosylation and processing of high levels of active human glucocerebrosidase in invertebrate cells using a baculovirus expression vector. DNA7:99–106

31. Smith GE, Ju G, Ericson BL, Moschera J, Lahm H, Chizzonite R, Summers MD (1985) Modification and secretion of human interleukin 2 produced in insect cells by a baculovirus expression vector. Proc Natl Acad Sci USA 82:8404–8408

32. Domingo DL, Trowbridge IS (1988) Characterization of the human transferrin receptor produced in a baculovirus expression system. J Chem Biol (in press)

33. Kuroda K, Hauser C, Rudolf R, Klenk HD, Doerfler W (1986) Expression of the influenza virus haemagglutinin in insect cells by a baculovirus vector EMBO J 5:1359–1365

34. Kuroda K, Hauser C, Rott R, Doerfler W, Klenk H-D (1988a) Processing of the hemagglutinin of influenza virus expressed in insect cells by a baculovirus vector *In* Invertebrate Cell System Applications (in press)

35. Kuroda K, Gröner A, Frese K, Hauser C, Rott R, Doerfler W, Klenk H-D (1988b) Synthesis of biologically active influenza virus hemagglutinin in insect larvae (submitted)

36. Possee RD (1986) Cell-surface expression of influenza virus haemagglutinin in insect cells using a baculovirus vector. Virus Res 5:43–59

37. Johnson PW, Richardson CD, Roder JC, Dunn RJ (1988) Expression of myelin

associated glycoprotein as a soluble extracellular domain J Biol Chem (submitted)

38. Coelingh KLVW, Murphey BR, Collins PL, Lebacq-Verheyden AM, Battey JF (1987) Expression of biologically active and antigenically authentic parainfluenza type 3 hemagglutinin-neuraminidase glycoprotein by a recombinant baculovirus. Virology 160:465–472

39. Andrews DL, Beams B, Summers MD, Park WD (1987) Characterization of the lipid acyl hydrolase activity of the major tuber protein, patatin, by cloning and abundant expression in a baculovirus vector. Biochem J 252:199–206

40. Bustos M, Luckow VA, Griffing L, Summers MD, Hall TC (1987) Expression, glycosylation, and secretion of phaseolin in a baculovirus system. Plant Mol Bio 10:475–488

41. Oker-Blom C, Strauss JH, Summers MD (1989) Expression of Sindbis virus 26S cDNA in *Spodoptera frugiperda* (Sf9) cells using a baculovirus expression vector (submitted)

42. Furlong AM, Thomsen DR, Marotti KR, Adams LD, Post LE, Sharma SK (1988) Purification and characterization of recombinant tissue plasminogen activator secreted from baculovirus infected insect cells. 72nd Annual meeting of the Federation of American Societies for Experimental Biology, Las Vegas, Nevada, May 1–5, 1988 (Fed Amer Soc Exp Biol) J 2:275 (Abs)

43. Jarvis DL, Summers MD (1988) Glycosylation and secretion of human tissue plasminogen activator in recombinant baculovirus-infected insect cells. Mol Cell Biol (submitted)

44. Roosien J, Ryan M, Belsham G, Vlak JM (1988) Expression of FMDV capsids using a baculovirus expression vector (in preparation)

45. Schmaljohn CD, Sugiyama K, Schmaljohn AL, Biship DHL (1988) Baculovirus expression of the small genome segment of Hantaan virus and potential use of the expressed nucleocapsid protein as a diagnostic antigen J Gen Virol 69:8777–8786

46. Harmon S, Johnston JM, Ziegelhoffer T, Richards OC, Summers DF, Ehrenfeld E (1988) Expression of hepatitis A virus capsid sequences in insect cells. Virus Res (in press)

47. Ellis L, Levitan A, Cobb MH, Ramos P (1988) Efficient expression in insect cells of a soluble, active human insulin receptor protein-tyrosine kinase domain by the use of a baculovirus vector. J Virol 62:1634–1639

48. Sissom J, Ellis L (1988) Secretion from insect cells of a soluble high affinity human insulin receptor ligand binding domain by use of a baculovirus vector. J Virol (Submitted)

49. Herrera R, Lebwohl D, de Herreros AG, Kallen RG, Rosen OM (1988) Synthesis, purification and characterization of the cytoplasmic domain of the human insulin receptor using a baculovirus expression system J Biol Chem 263:5560–5568

50. Chang LMS, Rafter E, Rusquet-Valerius R, Peterson RC, White ST, Bollum FJ (1988) Expression and processing of recombinant human terminal transferase in the baculovirus system. J. Biol. Chem (submitted)

51. Ginns EI, Rehavi M, Martin BM, Weller M, O'Malley KL, LaMarca ME, McAllister CG, Paul SM (1988) Expression of human tyrosine hydroxylase cDNA in invertebrate cells using a baculovirus vector. J Biol Chem 263:7406–7410

52. Overton HA, Ihara T, Bishop DHL (1987) Identification of the N and NS$_s$

proteins coded by the ambisense S RNA of punta toro phlebovirus using monospecific antisera raised to baculovirus expressed N and NS_s proteins. Virology 157:338–350

53. Ollo R, Maniatis T (1987) *Drosophila* Krüppel gene product produced in a baculovirus expression system is a nuclear phosphoprotein that binds to DNA. Proc Natl Acad Sci USA 84:5700–5704

54. Jeang K-T, Giam C-Z, Nerenberg M, Khoury G (1987a) Abundant synthesis of functional human T-cell leukemia virus type I $p40^x$ protein in eucaryotic cells by using a baculovirus expression vector J. Virol 61:708–713

55. Miyamoto C, Smith GE, Farrell-Towt J, Chizzonite R, Summers MD, Ju G (1985) Production of human c- *myc* protein in insect cells infected with a baculovirus expression vector. Mol Cell Biol 5:2860–2865

56. Baum JA, Geever R, Giles NH (1987) Expression of qa-1F activator protein: Identification of upstream binding sites in the *qa* gene cluster and localization of the DNA-binding domain. Mol Cell Biol 7:1256–1266

57. Rice WC, Lorimar HE, Prives C, Miller LK, (187) Expression of polyomavirus large T antigen by using a baculovirus vector J. Virol 61:1712–1716

58. Luckow VA, Summers MD (1988b) Signals important for high level expression of foreign genes in *Autographa california* nuclear polyhedrosis virus epxression vectors Virology (in Press)

CHAPTER 10
Transforming Genes of Bovine and Human Papillomaviruses

DOUGLAS R. LOWY AND JOHN T. SCHILLER

Papillomaviruses (PVs) induce benign epithelial tumors in their natural hosts. The best-characterized members of this virus group are the human papillomaviruses (HPV), bovine papillomaviruses (BPV), and cottontail rabbit papillomavirus (CRPV). PVs have also been isolated from many other mammals, as well as from birds. More than 50 different HPV genotypes have now been identified [1–3].

HPVs have recently attracted considerable attention medically because, in addition to being the causative agent of warts, some HPV types have been closely associated with certain human epithelial malignancies. In particular, HPV DNA has been found in about 90% of cervical carcinomas and in virtually all malignant cutaneous tumors that occur in epidermodysplasia verruciformis (EV), a rare condition of widespread chronic cutaneous HPV lesions associated with a high incidence of malignant conversion [4,5]. A causative role for HPVs in the development of these and other malignant tumors that contain HPV DNA has not been definitively established, but some animal PVs have been shown to induce malignant tumors under natural as well as experimental conditions.

The growing awareness of the potential medical importance of PVs has fostered efforts to understand their functional organization at the genetic level. Although PVs cannot be propagated in vitro, assays have been developed to study their role in the induction of cellular transformation. This brief review focuses on the identification of the PV genes involved in viral transformation and the characterization of their protein products.

Genome Structure

All PVs share certain genetic and biologic features [6]. Their genomes are closed, circular, doubled-stranded DNA about 8 kb in length, and the viral RNAs are all transcribed from the same strand. The viral genome (Figure 10.1) can be divided into three parts: (1) a segment of about 1 kb devoid of any long open reading frames (ORF), which has been called the upstream regulatory region (URR) or long control region (LCR); (2) a 4-kb segment with several ORFs (E1–E8) that encodes nonstructural (early) viral proteins—some function has been localized to each of these ORFs, except for E3; and (3) a 3-kb segment that encodes the structural (late) viral proteins (L1 and L2).

PVs undergo vegetative replication only in stratified squamous epithelia, in which progeny virions can be identified in differentiating, suprabasal epidermal cells. Most PVs, including CRPV and all HPVs, induce papillomas because they only infect epidermal cells. Bovine papillomavirus type 1 (BPV-1) is the prototype of a group of animal PVs that induce fibropapillomas [1,3]; in addition to productively infecting the epidermis, BPV-1 also nonproductively infects and transforms the underlying dermal fibroblasts. This expanded infectivity of BPV-1 is correlated with its ability to induce fibroblastic tumors in heterologous hosts, an activity that strictly epitheliotropic viruses lack.

Genetics of BPV-1 Transformation

BPV-1 virions, or BPV-1 DNA, are quite efficient in inducing focal transformation of certain established rodent cell lines, such as C127 and NIH 3T3, whereas the strictly epitheliotropic viruses induce morphologic alterations less efficiently than BPV-1 [7–9]. Hence, the BPV-1 transformation system has been used most extensively to study PV genetics. In vitro transformation by BPV-1 of established rodent cells appears to be similar to the

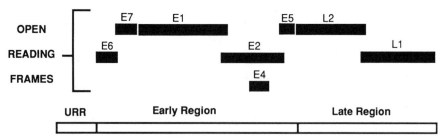

Figure 10.1. Genomic organization of HPV-16 DNA. The 8-kb circular viral DNA has been linearized at the junction between the late region and the URR. The direction of RNA transcription is from left to right. The relative location of the BPV-1 ORFs is similar.

benign, in vivo dermal fibroblast transformation that is a chracteristic of in vivo infection by this virus. In both instances the cells are nonproductively transformed, the viral DNA persists as a multicopy episome, the RNAs that encode the nonstructural (early) proteins are transcribed, but those that encode the structural viral proteins are not.

Genetic analysis of BPV-1–induced transformation has indicated that the virus contains two transforming genes, E5 and E6 [8–11]. When placed under control of a strong heterologous promoter such as a retroviral LTR, either gene by itself can induce focal transformation of mouse C127 cells. Conversely, the failure of E5⁻, E6⁻ double mutants of the full-length viral genome to induce morphological transformation implies that these two genes are the major transforming genes of the virus. Expression of these genes is required for maintenance as well as for initiation of transformation. Although cells transformed by most other papillomaviruses appear to require the continued expression of viral genes, BPV-4 has been reported to transform C127 cells via a hit-and-run mechanism [12].

For BPV-1, mutational inactivation of ORFs other than E5 and E6 has produced variable results. No effect on transformation has been seen when L1, L2, or E4 were interrupted. However, mutations in E7, which is required for a high BPV DNA copy number in some assays, have been noted to result in lower transforming activity when assayed in the context of the full-length BPV-1 genome, although no influence on transformation has been noted for these mutations when the E6E7E8 region was placed under a strong heterologous promoter [8,9,13]. E1 mutants have been found to have increased transforming activity, probably as a consequence of a faster rate of transcription from the early region [14,15]. The E2 ORF is complex; it encodes at least three protein products: a full-length E2 protein that can *trans*-activate viral transcription by binding to specific motifs that are present in several copies in the BPV URR, as well as two smaller N-terminally truncated proteins that competitively inhibit this *trans*-activation (see McBride et al Chapter 18, this volume). Mutation of the N-terminus of E2 markedly reduces the transforming activity of the full-length BPV genome, whereas C-terminal mutation results in less inhibition of transformation. These effects on transformation are probably mediated through alterations in the expression of E5 and E6, since E2 products control the expression of E5 and E6 but have no intrinsic transforming activities in these assays (McBride et al Chapter 18, this volume) [8,9,16].

Analysis of BPV-1 E5 and E6 Proteins

The E5 gene encodes a very small (44 amino-acid) hydrophobic protein that is located in the plasma membrane, where it is found as a dimer [17]. Microinjection of E5 protein can induce cellular DNA synthesis [18]. The mechanism by which the E5 protein acts has not been determined. It would seem unlikely that such a small protein would have intrinsic enzymatic activity, but rather, it seems reasonable to speculate that E5 protein may function by increasing the biological activity of one or more membrane-associated proteins.

Mutational analysis BPV-1 E5 has suggested that the major requirement of the 30 N-terminal amino acids is that the hydrophobicity of this region be maintained; sequence specificity for biological activity is apparently limited to the 15 C-terminal amino acids [17]. This region of the protein contains two cysteine (Cys) residues. Dimerization and transformation activity are apparently mediated at least in part through these residues; mutation of either Cys reduces biological activity and the efficiency of dimerization, and mutation of both Cys eliminates both properties.

It is important to note that some PVs, such as CRPV and HPV-1, do not contain an ORF with homology to BPV E5. Since CRPV can induce both benign and malignant lesions, this means that an E5 protein is not absolutely required for the generation of either type of lesion. Since the viruses that induce fibropapillomas possess similar E5 genes, it is tempting to speculate that this gene contributes significantly to the expanded host range of this class of virus. However, several HPV types, including HPV-16, which is the HPV type identified most frequently in cervical tumors, do contain an ORF with some homology to BPV E5. Its function in these HPVs has not yet been established.

Although BPV-1 E5 transforms both C127 and NIH 3T3 cells, the biological activity of BPV-1 E6 is unusual in that it transforms C127 but not NIH 3T3 cells, which is the standard cell line used to detect transforming genes. All PVs possess E6 ORFs that share some sequence homology; a striking feature of each E6 ORF is that it encodes four Cys-X-X-Cys repeats. These repeats have some features of the zinc fingers of DNA-binding proteins, although the interval between the repeats differs from that of classic zinc fingers. The BPV-1 E6 polypetide binds zinc in vitro and, under physiological salt conditions, has avidity for DNA [19]. The possible biological significance of this association remains to be established, however, since no sequence specificity has been detected. E6 encodes a 137 amino-acid protein that has been localized in transformed cells to membranous and nuclear matrix fractions [20].

The Cys residues in the repeats are important for transformation of C127 cells, since mutation of any one of them significantly reduces or abolishes this activity [21]. The extreme C terminus also appears to be functionally important, although it is not conserved at the sequence level among different E6 proteins; a premature termination mutant that encodes a protein that is only 4 amino acids shorter than full length fails to tranform, although its protein is stable. Subcellular localization of the mutant proteins suggests that nuclear matrix localization may be required for biological activity, since all biologically active E6 mutant proteins localized to this fraction, although the protein of several transformation defective mutants was not detected in this fraction. These results are consistent with the hypothesis that E6 may transform cells by influencing transcription by an undetermined mechanism.

Transformation by HPV

The analysis of HPV-induced transformation has been governed by two observations concerning the association of HPVs and cervical cancer [5]. The first is that certain types, such as 16, 18, 31, and 33, are usually found in malignant lesions, whereas other types, such as 6 and 11, are characteristically associated with benign lesions only. Second, the region of the genome containing the E6 and E7 ORFs is preferentially retained and expressed in carcinomas and in carcinoma-derived cells lines, where as the E1E2E5 region is preferentially interrupted or deleted. Several different assays have been employed to study the transforming activity of HPVs, primarily of HPV-16, since this is the type most frequently found in cervical cancers. These assays include the use of established rodent cell lines, such as NIH 3T3, as well as primary rodent cells and human keratinocytes. Distinctions between in vitro biological activities of HPV-6 and -11 versus HPV-16 and -18 DNA have been identified in that HPV-16 and -18 DNA possess greater activity than HPV-6 and -11 DNA in primary human and rodent cells [22–24]. It has not yet been determined if the functional basis for these biological differences lies principally in regulatory sequences, protein-coding sequences, or a combination of the two.

In NIH 3T3 cells, HPV-16 and -18 DNA under the control of a heterologous promoter have the unusual property of inducing anchorage independent growth more efficiently than focal transformation [25,26]. This does not appear to be the case for established rat 3Y1 cells in which focal transformation appears to be relatively efficient [27,28]. Also HPV-16 DNA can immortalize primary rodent cells and, as previously shown for *myc* and adenovirus E1a, HPV-16 DNA can cooperate with *ras* to induce transformation of primary rodent cells [22,24,29].

Analysis of HPV E6E7 Region

Genetic analysis has established that the E6E7 region, the coding sequences of the viral DNA retained in the carcinomas, is required for mediating these biological activities but that the E1, E2, and E5 ORFs are dispensable [22,25–31]. (One group [30] has found that established rodent cells transfected with the E2E4E5 region under control of a heterologous promoter formed small tumors, but even in this assay, the major biological activity resided in E6E7.) In contrast to the results obtained with BPV, E7 has been found to be responsible in HPV-16 and -18 for most, if not all, the phenotypic changes in those instances in which the genetics of the E6E7 region has been analyzed in detail, including immortalization and cell transformation [22,26–31]. HPV types associated with EV-induced cutaneous carcinomas appear, however, to resemble BPV-1 in that E6, rather than E7, is the transforming gene in the E6E7 region [32].

Analysis of the E7 protein function has provided some provacative clues that may help elucidate its mechanism of action. HPV-16 E7 can serve as a *trans*-activator of transcription, stimulating low-level activation of the E1a-inducible adenovirus E2 promoter [31,33]. Although this function may be of considerable importance to E7, cell transformation and *trans*-activation appear to be separable E7 functions in at least one in vitro system, since non-*trans*-activating mutants that still transform have been identified [33]. A small peptide within the E7 protein is homologous to two conserved segments of adenovirus E1a that are important for E1a-mediated immortalization. The corresponding peptide in E1a is required for binding of the retinoblastoma protein to E1a, and in vitro translated retinoblastoma protein has been found to bind to HPV-16 E7 protein [34]. If such an interaction occurs in vivo, it might suggest that E7 transforms, at least in part, by binding to this cell-encoded protein. This segment in E7 is required for both the *trans*-activation and transformation functions [33]. Since most proteins with *trans*-activation and/or immortalization activities are found in the nucleus, as is the retinoblastoma protein, it will be necessary to reconcile these functional observations with the immunological localization of HPV-16 E7 protein to the soluble cytoplasmic fraction, where it has also been found to be phosphorylated [35]. Another apparent similarity between E7 and E1a is that each contain two Cys-X-X-Cys repeats, which have been found to bind zinc in vitro [19,36].

HPV-16 E6 has been found to be active thus far in only one in vitro assay (immortalization of primary rat brain cells), but even in this assay, E7 is the more active gene [29]. In addition, E6 appears to be required for the tumorigenic phenotype of HPV-16–transformed established mouse $\varphi 2$ cells [30]. The HPV-16 E6 protein has been detected in cervical carcinoma derived cell lines and localized, like the BPV E6, to nuclear matrix and nonnuclear membrane fractions [37].

Keratinocyte Assays

HPV-16 and HPV-18 can also immortalize human primary foreskin keratinocyes and exocervical epithelial cells [23,38,39]. Since these are the authentic host cells for HPV, human keratinocyte assays may provide a closer approximation of the in vivo situation than those based on other cell types. It should be noted, however, that in vitro keratinocyte systems only incompletely mimic the in vivo differentiation program. Several stages of increasing transformation can be distinquished in these assays. Both oncogenic and nononcogenic HPVs can induce transient cell proliferation after transfection, but only transfectants containing the oncogenic types will give raise to immortalized cultures [23]. Most of the immortalized cells retain the capacity to undergo terminal differetiation in response to serum or calcium; at low frequency, however, colonies that are resistant to these differentiation signals occasionally arise. Unlike the carcinoma-derived cell lines, none of these lines are tumorigenic, suggesting that this further progression requires alterations in cellular genes. This phenomenon may have an in vivo correlate, since abnormalities of *ras* and *myc* have been noted in a significant proportion of cervical cancers [40]. Using an in vitro assay in which keratinocytes undergo more complete differentiation than usually obtained with keratinocyte cultures, HPV-16 DNA has been found to alter this differentiation and induce histological abnormalities [41].

The genes responsible for the induction of these phenotypes have yet to be determined, but it is likely that E6 and/or E7 play a role, since the E6E7 region of the genome continues to be expressed in the immortalized lines. This does not rule out the possibility that other genes also play a role in early events. Because of the correlation between the malignant potential of the HPVs and their ability to immortalize keratinocytes, the keratinocyte assays should also be useful in the delineation of pathologically significant differences between oncogenic and nononcogenic types.

References

1. Lancaster WD, and Olson C (1982) Animal papillomaviruses. Microbiol Rev 46:191–207
2. Broker TR, Botchan M (1986) Papillomaviruses: Retrospectives and prospectives. *In* Botchan M, Grodzicker T, Sharp PA, (eds.) Cancer Cells 4. DNA Tumor Viruses. Cold Spring Harbor Laboratory, Cold Spring Harbor, New York, pp 17–36
3. Salzman NP, Howley PM (eds.) (1987) The Papovaviridae 2: The Papillomaviruses. Plenum Publishing, New York, 387 p
4. Orth G (1987) Epidermodysplasia verruciformis. *In* Salzman NP, Howley PM (eds) The Papovaviridae 2: The Papillomaviruses. Plenum Publishing, New York, pp 199–243
5. Zur Hausen H, Schnieder A (1987) The role of papillomaviruses in human anogenital cancer. *In* Salzman NP, Howley PM (eds) The Papovaviridae 2: The Papillomaviruses. Plenum Publishing, New York, pp 245–263
6. Pettersson U, Ahola H, Stenlund A, Moreno-Lopez J (1987) Organization and expression of papillomavirus genomes. *In* Salzman NP, Howley PM (eds) The Papovaviridae 2: The Papillomaviruses. Plenum Publishing, New York, pp 67–107
7. Dvoretzki I, Shober R, Chattopadhyay SK, Lowy DR (1980) A quantitative in vitro focus forming assay for bovine papillomavirus. Virology 103:516–528
8. DiMaio D, Neary K (1989) The gentics of bovine papillomavirus type 1. *In* Pfister H (ed) Papillomaviruses and Human Cancer (in press)
9. Lambert PF, Howley PM (1988) The genetics of bovine papillomavirus type 1. Ann Rev Genet 22:235–258

10. Schiller J, Vousden K, Vass WC, Lowy DR (1986) The E5 open reading frame of bovine papillomavirus type 1 encodes a transforming gene. J Virol 57:1–6

11. Groff DE, Lancaster WD (1986) Genetics analysis of the 3′ early region transformation and replication functions of bovine papillomavirus type 1. Virology 150:221–230

12. Smith KT, Campo MS (1988) "Hit and run" transformation of mouse C127 cells by bovine papillomavirus type 4: The viral DNA is required for the initiation but not for maintenance of the transformed phenotype. Virology 164:39–47

13. Neary K, DiMaio D (1989) Open reading frames E6 and E7 of bovine papillomavirus type 1 are both required for full transformation of mouse C127 cells. J Virol 63:259–266

14. Lambert PF, Howley PM (1988) Bovine papillomavirus type 1 E1 replication-defective mutants are altered in their transcriptional regulation J Virol 62:4009–4015

15. Schiller JT, Kleiner E, Androphy EJ, Lowy DR, Pfister H (1989) Identification of bovine papillomavirus E1 mutants with increased transforming and transcriptional activity J Virol 63:1775–1720

16. Prakash SS, Horwitz BH, Zibello TZ, Settleman J, DiMaio D (1988) Bovine papillomavvirus E2 gene regulates expression of the viral E5 transforming gene. J Virol 62:3608–3613

17. Horwitz BH, Burkhardt AL, Schlegel R, DiMaio D (1988) 44-Amino-acid E5 transforming protein of bovine papillomavirus requires a hydrophobic core and specific carboxyl-terminal acino acids. Mol Cell Biol 8:4071–4078

18. Green M, Loewenstein PM (1987) Demonstration that a chemically synthesized BPV1 oncoprotein and its C-terminal domain function to induce cellular DNA synthesis. Cell 51:795–802

19. Barbosa MS, Lowy DR, Schiller JT (1989) Papillomavirus polypeptides E6 and E7 are zinc-binding proteins. J Virol 63:1404–1407

20. Androphy EJ, Schiller JT, Lowy DR (1985) Identification of the protein ncoded by the E6 transforming gene of bovine papillomavirus. Science 230:442–445

21. Vousden KH, Androphy EJ, Schiller JT, Lowy DR (1989) Mutational analyis of bovine papillomavirus E6 gene. J Virol (in press)

22. Storey A, Pim D, Murrary A, Osborn K, Banks L, Crawford L (1988) Comparison of the in vitro transforming activities of human papillomavirus types. EMBO J 7:1815–1820

23. Schlegel R, Phelps WC, Zhang YL, Barbosa M (1988) Quantitative keratinocyte assay detects two biological activities of human papillomavirus DNA and identifies viral types assocaited with cervical carcinoma. EMBO J 7:3181–3187

24. Pater MM, Hughes GA, Hyslop DE, Nakshatri H, Pater A (1988) Glucocorticoid-dependent oncogenic transformation by type 16 but not type 11 human papilloma virus DNA. Nature 335:832–835

25. Bedell MA, Jones KH, Laimins LA (1987) The E6-E7 region of human papillomavirus type 18 is sufficient for transformation of NIH 3T3 and rat-1 cells. J Virol 61:3635–3640

26. Vousden KH, Doniger J, DiPaolo JA, Lowy DR (1988) The E7 open reading frame of human papillomavirus type 16 encodes a transforming gene. Oncogene Res 3:167–175

27. Kanda T, Furuno A, Yoshiike K (1988) Human papillomavirus type 16 open

reading frame E7 encodes a transforming gene for rat 3Y1 cell. J Virol 62:610–617

28. Watanabe S, Yoskiike K (1988) Transformation of rat 3Y1 cells by human papillomavirus type-18 DNA. Intl J Cancer 41:896–900

29. Kanda T, Watanabe S, Yoshiike K (1988) Immortalization of primary rat cells by human papillomavirus type 16 subgenomic DNA fragments controlled by the SV40 promoter. Virology 165:321–325

30. Yutsudo M, Okamoto Y, Hakura A (1988) Functional dissociation of transforming genes of human papillomavirus type 16. Virology 166:594–597

31. Phelps WC, Yee CL, Münger K, Howley PM (1988) The human papillomavirus type 16 E7 gene encodes transactivation and transformation functions similar to those of adenovirus E1A. Cell 53:539–547

32. Iftner T, Bierfelder S, Csapo Z, Pfister H (1988) Involvement of human papillomavirus type 8 genes E6 and E7 in transformation and replication. J Virol 62:3655–3661

33. Edmonds C, Vousden K (1989) A point mutational analysis of HPV16 E7 protein: Transformation is separable from trans-activation in NIH 3T3 cells J Virol (in press)

34. Dyson N, Howley PM, Münger K, Harlow E (1989) The human papilloma virus-16 E7 oncoprotein is able to bind to the retinoblastoma gene product. Science 243:934–937

35. Smotkin D, Wettstein FO (1987) The major-human papillomavirus protein in cervical cancers is a cytoplasmic phosphoprotein J Virol 61:1686–1689

36. Culp JF, Webster LC, Friedman DJ, Smith CL, Huang W-J, Wu FY-H, Rosenberg M, Ricciardi RP (1988) The 289-amino acid E1a protein of adenovirus binds zinc in a region that is important for trans-activation. Proc Natl Acad Sci USA 85:6450–6465

37. Androphy EJ, Hubbert NL, Schiller JT, Lowy DR (1987) Identification of the HPV-16 E6 protein from transformed mouse cells and human cervical carcinoma cell lines. EMBO J 6:898–992

38. Pirisi L, Yasumoto S, Feller M, Doniger J, DiPaolo J (1987) Transformation of human fibroblasts and keratinocytes with human papillomavirus type 16 DNA. J Virol 61:1061–1066

39. Woodworth CD, Bowden PE, Doniger J, Pirisi L, Barnes W, Lancaster WD, DiPaolo JA (1988) Characterization of normal human exocervical epithelial cells immortalized in vitro by papillomavirus types 16 and 128 DNA. Cancer Res 48:4620–4628

40. Riou G, Barrois M, Sheng Z-M, Duvillard P, Lhomme C (1988) Somatic deletions and mutations of c-Ha-*ras* gene in human cervical cancers. Oncogene 3:329–333

41. McCance DJ, Kopan R, Fuchs E, Laimins LA (1988) Human papillomavirus type 16 alters human epitheila cells differentiation in vitro. Proc Natl Acad Sci USA 85:7169–7173

Cell Biology of Viral Infections

CHAPTER 11
Retrovirus Receptors and Cell Tropism

MAJA A. SOMMERFELT AND ROBIN A. WEISS

Retroviruses are associated with a wide spectrum of diseases affecting vertebrate hosts, ranging from bony fish to mammals, and including humans. Genetic and interference studies have demonstrated the variety of receptors recognized by the different virus strains, although the extent to which these retroviral receptors determine the pathogenesis of any particular strain is still largely unknown. During the course of vertebrate evolution, retroviral genomes have on occasion become incorporated and maintained in the host germ line [1]. Such "endogenous" viruses must give some selective advantage to the host to be preserved in this way, and indeed endogenous viral proteins are frequently expressed. Such viruses, however, are often xenotropic [2]; that is, they cannot reinfect the cells of the species that harbor them but are infectious to foreign cells. It would appear that the host allows some viral expression but blocks the cell-to-cell spread of activated virus by receptor incompatibility. In some cases the host species does not express the relevant receptors; in others the receptors are blocked by an endogenous interference phenomenon [3,4]. The nature of most receptor molecules has not been elucidated; indeed, the only receptor that has been clearly identified is the CD4 antigen utilized by the human and simian immunodeficiency viruses (HIV and SIV respectively) [5].

For simplicity, the host range of a retrovirus can be determined at two levels, the presence or absence of cell surface receptors and the postpenetration host control of viral replication. For example, mouse leukemia viruses are classified into eco-, xeno-, ampho-, and dualtropic strains according to receptor-determined host range. Each type appears to use distinct cell surface receptors [6]. The ecotropic viruses are further subdivided into N, B, and NB tropic strains, according to postpenetration host control of susceptibility. The cell tropism of a retrovirus may be viewed either in terms of the cell types susceptible to infection, which necessarily express receptors, or by the pathogenesis of the virus. In the case of HIV, however, these tropisms in the main coincide; in the case of Human T cell leukemia virus type 1 (HTLV-1), many cell types can be infected in vitro, but only T lymphocytes (usually CD4-positive T cells) are liable to cell transformation, mimicking the type of malignancy that develops in vivo. Despite the susceptibility of human cells in vitro to infection by animal retroviruses, the resistance to infection in vivo is probably related to the ability of human serum to allow complement-mediated, antibody-independent virolysis, a phenomenon that does not occur for HTLV-1, HTLV-2, or HIV [7,8].

The outer envelope glycoprotein of the retrovirus interacts with receptors, and small changes in sequence can determine recognition of distinct receptor molecules, as shown by Dorner et al. for avian leukosis viruses [9], and by Overbaugh et al. for the subgroups B and C of the feline leukemia viruses that appear to be derived from subgroup A through recombination with endogenous retroviral *env* genes to give new receptor specificity [10]. Murine dualtropic viruses are derived in a similar manner.

Most replication-competent retroviruses establish persistent infection that cannot be easily visualized as a cytopathic effect. For this reason most studies have employed the use of phenotypically mixed virus particles (pseudotypes). The envelope and therefore the host range of the pseudotype are determined by the retrovirus, and the genome is derived from another virus. This may be a defective retrovirus carrying an oncogene so that infection is visualized by focus formation, or an unrelated cytopathic plaque-forming virus, for example, vesicular stomatitis virus (VSV) [11].

Receptor interference occurs when persistently infected cells block the infection of pseudotypes that require the same receptors for binding and penetration. In this way retroviruses can be grouped according to their receptor specificity, without knowing the nature of the receptor. Host genetic determination of receptor expression has similarly been demonstrated in classic studies with avian, murine, and feline retroviruses [11]. Purified envelope glycoprotein can be used to study the kinetics and affinity of virus–receptor interactions and its interference [12,13].

With many retroviruses, giant multinucleate cells (syncytia) arise when chronically infected cells are cocultivated with uninfected cells expressing the relevant receptor. Syncytial induction has been used for many retroviruses in the development of bioassays, often termed *plaque assays,* in order to visualize and quantitate viral titers. It should be noted that all cells that induce syncytia in response to a virus have receptors for that virus but that not all receptor-positive cells are susceptible to fusion: for example, ecotopic murine leukemia virus (MLV-E) induces syncytial formation with XC and rat myoblast cells, whereas receptors are also widespread in murine cells [14]. Syncytia are comprised of both infected and uninfected cells, although the precise mechanism of cell fusion is not fully understood [15].

Conformational changes in the viral-envelope glycoproteins exposing hydrophobic domains probably triggers this process. As with influenza virus, MLV-E infection on mouse cells requires the low pH of the endosome, and therefore cell–cell fusion does not occur in MLV-E infection of most cell types. However, we have recently tested the pH dependence of pseudotypes of many retrovirus strains and those that readily form syncytia with certain cell types are pH independent (unpublished observations). Interestingly, VSV(MLV-E) pseudotype infection is pH-independent in XC cells. Syncytial induction can be inhibited by agents that will block the virus–receptor interaction, for example, both antiviral and antireceptor antibodies (Figure 11.1.) As has been shown for HIV, syncytial formation and infection can be inhibited by preincubating virus with recombinant, soluble, CD4-receptor molecules [5]. Syncytial formation has also been exploited to study receptor interference when retrovirus-infected cells are cocultivated. Cells chronically infected with different retroviruses that utilize the same call surface receptor will not fuse, as the receptors will be mutually blocked; fusion will occur if the viruses utilize distinct cell surface receptors (Figure 11.2).

Both syncytial assays and pseudotype assays have been used to determine the variety of retroviral receptors. For instance, we have recently examined some 20 strains of retrovirus that plate on human cells and have been able to classify them by syncytial and pseudotype interference into a least eight distinct receptor groups (Table 11.1)

Although the precise identity of most retroviral receptors has not been elucidated, mammalian retrovirus receptor genes have been assigned to specific chromosomes by studying the susceptibility of interspecies somatic cell hybrids segregating chromosomes. Hamster–mouse hybrids have been used to assign receptor genes for MLV-E to murine chromosome 5 [16,17],

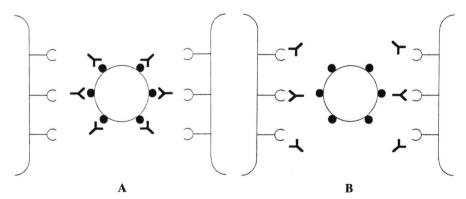

A   **B**

Figure 11.1. Syncytial inhibition by antibodies. If virion glycoproteins fail to interact with receptors, syncytial fusion does not occur. **A** Inhibition by neutralizing antiviral antibodies. **B** Inhibition by antireceptor antibodies.

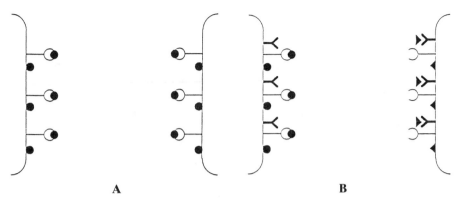

A   B

Figure 11.2. Models of syncytial induction and interference. Two apposing cell surfaces are shown. **A** Cocultivation of cells infected with viruses that utilize the same cell surface receptor inhibits fusion as the receptors are mutually blocked (syncytial interference). **B** Cocultivation of cells infected with viruses that utilize distinct cell surface receptors allows for compatible interactions between viral envelope glycoproteins on one cell type and the relevant cell surface receptors on the other, permitting fusion (syncytial formation).

amphotropic MLV to murine chromosome 8 [16], and mouse mammary tumour virus (MMTV) to chromosome 16 [18]. Human–mouse hybrids have been used to assign the gene determining the receptor shared by the cat endogenous retrovirus RD114 and baboon endogenous virus (receptor group 1) to human chromosome 19 [19,20]. We have recently localized this receptor gene, also shared by the D type retroviruses, to 19q13,1–q13,3 (M. A. Sommerfelt, unpublished observation). The single genetic locus determining cellular susceptibility for all group 1 retroviruses accords with the interference data that all these viruses utilize a common receptor. We have also assigned the receptor gene shared by HTLV-1 and HTLV-2 to human chromosome 17 [21]. Assigning a chromosome and localizing a receptor gene regionally does not identify that receptor molecule, but does permit the application of molecular techniques leading to the characterization of the receptor and its gene.

As the receptors for retroviruses become known, it will be of interest to study their expression and tissue distribution in vivo. For example, using fluorescently labeled HTLV virions, these parameters have been examined in hemotopoietic cells by flow cytometry [22], to show a greater variation of receptor expression than was apparent from pseudotype studies of cells in culture [23]. With the interest in CD4 as the HIV receptor, it has become apparent that this differentiation antigen is by no means restricted to T helper lymphocytes, and its expression in vivo goes far to explain the tropism of HIV and its consequent pathogenesis. There is evidence, however, that CD4 may not be the HIV receptor on brain and muscle cells [24].

Table 11.1. A summary of retroviral receptor groups on human cells.

Receptor group	Virus		Type	Species from which isolated	Exogenous or endogenous
1	RD114	Cat endogenous virus	C	Domestic cat	Endo
	BaEV	Baboon endogenous virus	C	Baboon	Endo
	MPMV	Mason Pfizer monkey virus	D	Rhesus macaque	Exo
	SRV-1	Simian retrovirus type 1	D	Macaque spp.	Exo
	SRV-2	Simian retrovirus type 2	D	Rhesus macque	Exo
	PO-1-Lu	Langur virus	D	Spectacled langur	Endo
	SMRV	Squirrel monkey virus	D	Squirrel monkey	Endo
2	MLV-A	Murine leukemia virus amphotropic	C	Mouse	Exo
3	MLV-X	Murine leukemia virus xenotropic	C	Mouse	Endo
4	FeLV-C	Feline leukemia virus subgroup C	C	Domestic cat	Exo
5	FeLV-B	Feline leukemia virus subgroup B	C	Domestic cat	Exo
	SSAV	Simian sarcoma associated virus	C	Woolly monkey	Exo
	GALV	Gibbon ape leukemia virus	C	Gibbon Ape	Exo
6	BLV	Bovine leukemia virus	C	Cow	Exo
7	HTLV-1	Human T cell leukemia virus type 1	C	Human	Exo
	HTLV-2	Human T cell leukemia virus type 2	C	Human	Exo
8	HIV-1	Human immunodeficiency virus type 1	Lenti	Human	Exo
	HIV-2	Human immunodeficiency virus type 2	Lenti	Human	Exo
	SIV_{mac}	Simian immunodeficiency virus	Lenti	Macaque	Exo
	SIV_{smm}	Simian immunodeficiency virus	Lenti	Mangabey	Exo
	SIV_{agm}	Simian immunodeficiency virus	Lenti	Vervet	Exo

There has been renewed interest in retrovirus receptors with the emergence of the human retroviruses associated with leukemia and AIDS. Indeed, the observations that recombinant, soluble, CD4-receptor molecules strongly inhibit HIV infection suggests that such molecules or peptides mimicking the receptor-binding site could be used as potential therapeutic agents. This underlies the importance of understanding retrovirus receptors.

References

1. Coffin J (1982) Endogenous viruses. *In* Weiss RA, Teich N, Varmus H, Coffin J (eds) RNA Tumor Viruses. Cold Spring Harbor Laboratory, Cold Spring Harbor, New York, pp 1109–1203
2. Levy JA (1973) Xenotropic viruses; Murine leukemia viruses associated with NIH Swiss NZB and other mouse strains. Science 182:1151–1153
3. Payne LN, Pani PK, Weiss RA (1971) A dominant epistatic gene which inhibits cellular susceptibility to RSV(RAV-O). J Gen Virol 13:455–462
4. Dandekar S, Rossitto P, Pickett S, Mockli G, Bradshaw H, Cardiff R, Gardner M (1987) Molecular characterisation of the *Akvr*-1 restriction gene: A defective endogenous retrovirus-borne gene identical to *Fv-4r*. J Virol 61:308–314
5. Sattentau QJ, Weiss RA (1988) The CD4 antigen: Physiological ligand and HIV receptor. Cell 52:631–633
6. Rein A (1982) Interference grouping of murine leukemia viruses: A distinct receptor for the MCF-recombinant virus in mouse cells. Virology 120:251–257
7. Cooper NR, Jensen FC, Welsh RM Jr, Oldstone MBA (1976) Lysis of RNA tumour viruses by human serum: Direct antibody-independent triggering of the classical complement pathway. J Exp Med 144:970–984
8. Hoshino H, Tanaker H, Miwa M, Okada H (1984) Human T-cell leukemia virus is not lysed by human serum. Nature 310:324–325
9. Dorner AJ, Stoye JP, Coffin JM (1985) Molecular basis of host range variation in avian retroviruses. J Virol 53:32–39
10. Overbaugh J, Riedel N, Hoover EA, Mullins JI (1988) Transduction of endogenous envelope genes by feline leukemia virus in vitro. Nature 332:731–734
11. Weiss RA (1982) Experimental biology and assay of retroviruses. *In* Weiss RA, Teich N, Varmus H, Coffin J (eds) RNA Tumor Viruses. Cold Spring Harbor Laboratory, Cold Spring Harbor, New York, pp 209–260
12. DeLarco J, Todaro GJ (1976) Membrane receptors for murine leukemia viruses: Characterisation using the purified envelope glycoprotein, gp 71. Cell 8:365–371
13. Lasky LA, Nakamura G, Smith DH, Fennie C, Shimasaki C, Patzer E, Berman P, Gregory T, Capon DJ (1987) Delineation of a region of the human immunodeficiency virus type 1 gp120 glycoprotein critical for interaction with the CD4 receptor. Cell 50:975–985
14. Klement V, Rowe WP, Hartley JW, Pugh WE (1969) Mixed culture cytopathogenicity: A new test for growth of murine leukemia viruses in tissue culture. Proc Natl Acad Sci USA 63:753–758
15. Lifson JD, Feinberg MB, Reyes GR, Rabin L, Banapour B, Chakrabarti S, Moss B, Wong-Staal F, Steimer KS, Engleman EG (1986) Induction of CD4-dependent cell fusion by the HTLV-III/LAV envelope glycoprotein. Nature 323:725–728
16. Gazdar AF, Oie P, Lallay W, Minna JD, Franck V (1977) Identification of mouse

chromosomes required for murine leukemia virus replication. Cell 11:949–956

17. Ruddle NH, Conta BS, Leinwand L, Kozak C, Ruddle F, Besmer P, Baltimore D (1978) Assignment of the receptor for ecotropic murine leukemia virus to mouse chromosome 5. J Exp Med 148:451–465

18. Hilkens J, Van der Zeijst B, Buijs F, Kroezen V, Bleumink N, Hilgers J (1983) Identification of a cellular receptor for mouse mammary tumour virus and mapping of its gene to chromosome 16. J Virol 45:140–147

19. Brown S, Oie HK, Gazdar AF, Minna JD (1979) Requirement of human chromosomes 19, 6 and possibly 3 for infection of hamster × human hybrid cells with baboon M7 type C virus. Cell 18:135–143

20. Schnitzer TJ, Weiss RA, Juricek DK, Ruddle FH (1980) Use of vesicular stomatitis virus pseudotypes to map viral receptor genes: Assignment of RD114 virus receptor gene to human chromosome 19. J Virol 35:575–580

21. Sommerfelt MA, Williams BP, Clapham PR, Solomon E, Goodfellow PN, Weiss RA (1988) Human T cell leukemia viruses use a receptor determined by human chromosome 17. Science 242:1557–1559

22. Krichbaum-Stenger K, Poiesz BJ, Keller P, Ehrlich G, Gavalchin J. Davis B, Moore J (1987) Specific adsorption of HTLV-1 to various target human and animal cells. Blood 70:1303–1311

23. Clapham PR, Nagy K, Weiss RA (1984) Pseudotypes of Human T-cell leukemia virus types 1 and 2: Neutralisation by patients' sera. Proc Natl Acad Sci USA 81:2886–2889

24. Clapham PR, Weber JN, Whitby D, McIntosh K, Dalgleish AG, Maddon PJ, Deen KC, Sweet RW, Weiss RA (1989) Soluble CD4 blocks the infectivity of diverse strains of HIV and SIV for T cells and monocytes but not for brain and muscle cells. Nature 337:368–370

CHAPTER 12
Coronavirus Receptors

KATHRYN V. HOLMES, RICHARD K. WILLIAMS,
AND CHARLES B. STEPHENSEN

Genetics of Mouse Strain Susceptibility to MHV

One of the classic examples of host resistance to virus infection is mouse hepatitis virus (MHV), a murine coronavirus. In susceptible mouse strains, various strains of MHV cause enteric, respiratory, and neurological diseases. Bang and his colleagues showed that MHV could cause a fatal infection in one strain of mice but not in another. Peritoneal macrophages isolated from these mouse strains showed susceptibility or resistance to MHV corresponding to that of the intact animal [1]. This difference in host susceptibility was due to a single autosomal gene, *hv-1*, and the allele for resistance to the fatal disease was recessive [2].

Many additional strains of mice have since been tested for susceptibility to various strains of MHV. The best-characterized host resistance system for MHV is the SJL/J mouse, which is profoundly resistant to both the JHM and A59 strains of MHV. These animals have an LD_{50} more than 1,000-fold higher than susceptible BALB/c mice [3], and macrophages and glial cells isolated from SJL/J mice are resistant to infection with MHV-JHM or MHV-A59 in vitro [4,5]. Genetic analysis of the SJL/J mouse showed that resistance to MHV is determined by a single autosomal recessive allele, *hv-2*, located on mouse chromosome 7 [4–6]. Cellular resistance to viral infection appears to be determined at a very early step in virus replication, such as adsorption, penetration, or primary translation of the genome.

Recent studies in our laboratory suggest that the molecular basis for the genetic resistance of SJL/J mice to MHV is the failure of these animals to express a specific receptor for MHV on the normal target cells for virus replication [7]. This chapter will describe the identification and characterization of the MHV receptor and demonstrate its role in determining host susceptibility to virus infection.

Identification and Characterization of the 110-K Glycoprotein Receptor for MHV

A solid-phase virus-receptor assay was developed to detect binding of MHV to plasma membranes purified from the natural target tissues for MHV [7]. Brush border membranes isolated from the small intestine of susceptible BALB/c or resistant SJL/J mice were immobilized on nitrocellulose in a dot blot apparatus and then incubated with MHV-A59 virus. Virus bound to the intestinal brush border membranes was detected immunologically with antibody directed against the peplomer glycoprotein E2 and radioiodinated staphylococcal protein A. Virus binding to brush border membranes from BALB/c mice was directly proportional to the amount of membranes used, but no binding of virus to brush border membranes from SJL/J mice was detected. Similar findings were made with hepatocyte membranes from BALB/c and SJL/J mice. These data suggest that SJL/J mice are resistant to MHV infection because they fail to express a specific receptor for MHV on the membranes of the normal target cells for this virus.

To determine the molecular weight of the MHV receptor on BALB/c membranes, a virus-overlay protein blot assay (VOPBA) was used [7]. Membrane proteins from BALB/c or SJL/J intestinal brush borders or hepatocytes were solubilized in SDS, separated by SDS-PAGE, and blotted onto nitrocellulose sheets. The sheets were then incubated sequentially with bovine serum albumin to block nonspecific binding, MHV, anti-E2 antibody, and radioiodinated staphylococcal protein A. With BALB/c brush border membranes or hepatocyte membranes, MHV bound only to a single broad band with a molecular weight of approximately 100K to 110K. No virus-binding activity was found in SJL/J membrane proteins. The 110K protein from BALB/c brush borders was specific for MHV and did not bind other viruses that can also infect murine enterocytes, such as mouse polio or rotavirus.

Additional characteristics of the MHV receptor were determined. Treatment of BALB/c brush border membranes with deoxycholate solubilized MHV–binding activity, which could then be quantitatively adsorbed to beads coated with any of several lectins including concanavalin A or ricin 120 [8]. This showed that the MHV receptor is a glycoprotein. The 110K glycoprotein was excised from SDS-PAGE gels of BALB/c brush border membrane and treated with endoglycosidase-F to release N-linked oligosaccharides. Endo-F treatment reduced the apparent molecular weight of the receptor to approximately 70K, without substantially reducing virus-binding activity [9]. Receptor eluted from the 100K to 110K region of an SDS-PAGE gel was very sensitive to proteases, and treatment with proteases destroyed virus-binding activity. However, virus-binding activity was not inhibited by treatment of the receptor with neuraminidase [8]. Taken together, the data show that MHV bound to an SDS-insensitive part of the 110K glycoprotein that was not removed by endo-F. Thus, the domain of the receptor that is recognized by the E2 glycoprotein of MHV is likely to be a linear sequence of amino acids.

Characterization of Antireceptor Antibodies

A second approach to the characterization of the MHV receptor was the development of antibodies directed against the receptor. A highly specific antireceptor antibody was developed by immunization of SJL/J mice, which do not express the MHV receptor, with extracts of intestinal brush border membranes from BALB/c mice, which do express the receptor [9]. In Western blots this polyclonal antibody recognized only the 110K receptor in BALB/c brush border membrane proteins, and the antibody did not bind to any proteins of SJL/J membranes. Pretreatment of MHV-susceptible mouse fibroblast cell lines with the polyclonal antireceptor antibody protected them from infection with MHV [9].

Hybridoma cell lines that produce monoclonal antibodies (MAbs) directed against the MHV receptor were derived from the spleens of mice producing antireceptor antibodies. More than 30 MAbs, which reacted in an enzyme-linked immunoassay with 110K receptor eluted from SDS-PAGE gels, were identified and tested for their ability to recognize the 110K receptor by immunoblot and the ability to protect L2 cells from infection with MHV-A59 [9]. Only three of the MAbs protected cells from infection. One of these MAbs, CC1, was studied extensively. It blocked infection of mouse fibroblasts with MHV-A59, MHV-JHM, MHV-3, MHV-S, and MHV-1 [9]. This indicates that, on the mouse fibroblast cell lines, the 110K glycoprotein receptor is the only receptor for all five of these prototype strains of MHV.

MAb CC1 recognized the 110K MHV receptor in Western blots of BALB/c brush border membranes, but it did not detect any protein band in SJL/J brush border membranes. This antireceptor MAb also reacted in dot immunoassays and in immunoblots with murine intestinal brush border membranes, but not with comparable brush border membrane preparations from other species [9]. Fluorescent antibody labeling of murine intestine showed that MAb CC1 specifically recognized the apical brush border of intestinal epithelial cells of BALB/c mice but failed to react with membranes of SJL/J intestine (Figure 12.1) [9]. Thus, all our results based on immuno-blotting, inhibition of biological activity, and distribution of receptor activity on cells and tissues indicate that the antireceptor MAb CC1 recognizes the 110K glycoprotein to which MHV binds. Furthermore, the ability of MAb CC1 to inhibit MHV-binding activity indicates that the domain of the 110K glycoprotein recognized by MAb CC1 may be either close to or identical with the domain to which the E2 glycoprotein of MHV binds.

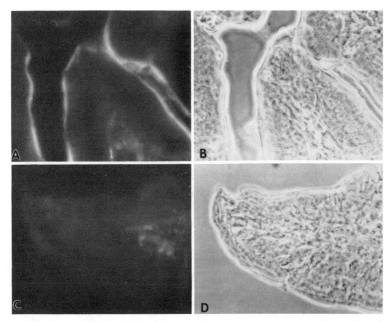

Figure 12.1. Expression of the glycoprotein receptor for mouse hepatitis virus on brush border membranes of mouse small intestine correlates with susceptibility of the mice to MHV. Monoclonal antireceptor antibody CC1 was used to detect the MHV receptor on frozen sections of mouse small intestine. BALB/c mice, which are highly susceptible to MHV infection, expressed the MHV receptor on brush border membranes **A** In contrast, SJL/J mice, which are resistant to MHV infection, did not express the receptor on their intestinal brush border membranes **C** Panels **B** and **D** are phase-contrast images of the same sections.

Affinity Purification of the Receptor Glycoprotein

MAb CC1 was used for affinity purification of MHV receptor from hepatocyte membranes from an MHV-susceptible mouse strain [9]. The amino-acid sequence of the first 15 amino acids from the N-terminus of this affinity-purified receptor was determined. A synthetic peptide corresponding to these 15 amino acids was conjugated with keyhole limpet hemocyanin and used to raise antiserum that recognized the 110K receptor from BALB/c intestinal brush border membranes in a Western blot, indicating that the derived sequence was indeed the MHV receptor [9].

The normal cellular function of the MHV receptor is not yet known. Receptors for viruses in several other groups have been identified, and these are often molecules that function as membrane receptors for such ligands as growth factors, hormones, or other cells [10]. Cloning, sequencing, and expression of the MHV receptor gene should help to identify both its normal cellular function and the domain that binds the E2 glycoprotein of MHV.

Role of Receptors in Coronavirus Species Specificity

One of the central questions about coronavirus receptors is whether they play a role in determining the marked species specificity of coronaviruses. Most coronaviruses infect only one species or a few closely related species [11]. For example, MHV causes natural infection only in mice and can be made to infect rats only by intracerebral inoculation of infant animals. Several experimental approaches were used to determine whether the species specificity of MHV infection is the consequence of the E2 glycoprotein of MHV binding selectively to a domain of the 110K glycoprotein that is only expressed on murine tissues. Immunofluorescence experiments on cell lines of different species showed that antireceptor antibody bound only to MHV-susceptible murine cell lines and not to cell lines from humans, hamsters, cats, or dogs [12]. Solid-phase, MHV-receptor assays using brush border membranes from nine species demonstrated that MHV-A59 bound only to brush border membranes from BALB/c mice and not to brush border membranes from human, pig, cat, dog, rat, or cow [12]. These experiments suggest that the species specificity of MHV is due to absence of the MHV-binding moiety of the receptor on cells of other species.

Coronavirus Hemagglutination

For some virus groups, receptors have been characterized by studying virus interactions with erythrocytes. For example, orthomyxoviruses and paramyxoviruses bind to glycolipids or glycoproteins on erythroctye membranes by an interaction of the viral hemagglutinin glycoprotein with N-acetyl neuraminic acid residues on the cellular macromolecules [10]. Although most coronaviruses do not cause hemagglutination, several coronaviruses, including bovine coronavirus (BCV), hemagglutinating encephalomyelitis virus of swine (HEV), human respiratory coronavirus (OC43), and some strains of infectious bronchitis virus (IBV), can cause hemagglutination [13]. For BCV, hemagglutination is caused by a 140K viral glycoprotein called HE, which is a disulfide-linked dimer of a 65K glycoprotein [14]. This hemagglutinin is not expressed in most strains of MHV, although an open reading frame homologous to the BCV hemagglutinin but lacking an initiator methionine has been found in the genome of MHV-A59 [15]. Surprisingly, this coronavirus hemagglutinin shares about 30% homology with the single glycoprotein of influenza C [15]. The hemagglutinating glycoproteins of both BCV and influenza C bind specifically to 9-O-acetylated neuraminic acid residues on erythrocytes and have esterase activity that inactivates 9-O-acetylated neuraminic acid–containing receptors on erythrocyte membranes [16]. This viral enzymatic activity has therefore been termed *receptor-destroying activity*. It is not yet clear, however, whether infection of susceptible cells in vitro or in vivo results from binding of BCV by means of the interaction of the HE glycoprotein with 9-O-acetylated neuraminic acid residues on the cell or from the interaction of the E2 glycoprotein with a specific glycoprotein analogous to the 110K MHV receptor. Coronavirus-induced membrane fusion is a function of the E2 glycoprotein [17,18]. Therefore, even if the BCV virion should bind to susceptible cells via the HE glycoprotein, interaction of the E2 glycoprotein with some component of the cell membrane may be required for penetration of the viral nucleocapsid.

Several lines of evidence show that MHV-A59 does not bind to susceptible cells by recognizing 9-O-acetylated neuraminic acid–containing receptors. First, MHV-A59 does not express the HE glycoprotein. Second, MHV-A59 binds to only a single glycoprotein in VOPBA, whereas OC43, a serologically related human coronavirus that expresses an HE glycoprotein, binds to multiple proteins that presumably all bear 9-O-acetylated neuraminic acid moieties [19]. Third, endo-F treatment of purified 110K MHV receptor, which should remove any N-linked oligosaccharides bearing 9-O-acetylated neuraminic acid, does not prevent binding of MHV. Finally, neuraminidase treatment of murine cells, which should remove the 9-O-acetylated neuraminic acid residues, fails to prevent binding of MHV-A59.

In some other virus groups, the virus–membrane interaction associated with hemagglutination is not necessarily identical to the interaction that leads to infection of a susceptible cell. For example, the erythrocyte-binding hemagglutinin of polyomavirus is found on noninfectious particles, whereas a different domain of the viral capsid protein binds to a component of the plasma membrane of susceptible cells that is not present on erythrocytes [20]. Possibly the hemagglutinin of some coronaviruses may, like that of polyomavirus, be irrelevant in the infectious process.

Nevertheless, it is interesting to speculate upon the biological significance of the observation that two different viral glycoproteins can mediate attachment of some coronaviruses to cell membranes. Those coronaviruses, such as MHV-A59, canine coronavirus, feline infectious peritonitis virus, or human coronavirus 229E, that do not express the HE glycoprotein are nevertheless quite capable of causing disease and being transmitted naturally, so the HE glycoprotein is not essential for infectivity of a coronavirus. It would be quite interesting to induce expression of an HE glycoprotein in such a virus in order to study what effect it may have on the pathogenesis and epidemiology of the virus. Possibly the virulence, tissue tropism, or species specificity of these viruses would be altered significantly.

The two membrane-binding activities of coronavirus glycoproteins may provide some coronaviruses with a choice of two different ways to interact with cells in order to initiate infection. Alternatively, the hemagglutinin could be essential for some coronaviruses if their E2 glycoprotein fails to recognize a glycoprotein receptor on host cells. It would be interesting to delete or mutagenize the gene for the HE glycoprotein of a coronavirus that normally expresses it, such as BCV or OC43, in order to determine how absence of this protein affects viral replication and pathogenesis.

Sensitive assays to detect binding of many different coronaviruses to glycoprotein receptors, or 9-O-acetylated neuraminic acid receptors on their normal target cells, and clones of the viral glycoprotein genes that can be sequenced and expressed either separately or in concert are becoming available. In the near future, these will be used to elucidate the mechanism(s) by which coronaviruses bind to cells and initiate infection.

Acknowledgments

This research was supported in part by research grants AI 18997 and AI 25231 from the National Institutes of Health. The opinions in this chapter are the private ones of the authors and are not to be construed as official or reflecting the views of the Department of Defense or the Uniformed Services University of the Health Sciences.

References

1. Bang FB, Warwick A (1960) Mouse macrophages as host cells for the mouse hepatitis virus and the genetic basis of their susceptibility. Proc Natl Acad Sci USA 46:1065–1075
2. Weiser W, Vellisto I, Bang FB (1976) Congenic strains of mice susceptible and resistant to mouse hepatitis virus. Proc Soc Biol Exp Med 152:499–502
3. Barthold SW, Smith AL (1984) Mouse hepatitis virus strain-related patterns of tissue tropism in suckling mice. Arch Virol 81:103–112
4. Smith MS, Click RE, Plagemann PG (1984) Control of mouse hepatitis virus replication in macrophages by a recessive gene on chromosome 7. J Immunol 133:428–432
5. Knobler RL, Haspel MV, Oldstone MB (1981) Mouse hepatitis virus type 4 (JHM strains) induced fatal central nervous system disease. I. Genetic control and murine neuron as the susceptible site of disease. J Exp Med 153:832–843
6. Knobler RL, Tunison LA, Oldstone MB (1984) Host genetic control of mouse hepatitis virus type 4 (JHM strain) replication. I. Restriction of virus amplification and spread in macrophages from resistant mice. J Gen Virol 65:1543–1548
7. Boyle JF, Weismiller DG, Holmes KV (1987) Genetic resistance to mouse hepatitis virus correlates with absence of virus-binding activity on target tissues. J Virol 61:185–189
8. Holmes KV et al (1987) Identification of a receptor for mouse hepatitis virus. Adv Exp Med Biol 218:197–202
9. Holmes KV (unpublished observations)
10. Crowell RL, Lonberg-Holm K (eds) (1985) Virus Attachment and Entry into Cells, ASM Publications, Washington, DC
11. Wege H, Siddell S, ter Meulen V (1982) The biology and pathogenesis of coronaviruses. Curr Top Microbiol Immunol 99:165–200
12. Compton SR (1988) Coronavirus attachment and replication. PhD thesis. Uniformed Services University of the Health Sciences, Bethesda, MD

13. Holmes KV (1985) Replication of coronaviruses. *In* Fields BN et al (eds) Virology. Raven Press, New York, pp 1331–1344

14. King B, Brian DA (1982) Bovine coronavirus structural proteins. J Virol 42:700–707

15. Luytjes W, Bredenbeek PJ, Noten AF, Horzinek MC, Spaan WJ (1988) Sequence of mouse hepatitis virus A59 mRNA2: Indications for RNA-recombination between coronaviruses and influenza C virus. Virology 166: 415–422

16. Vlasak R, Luytjes W, Leider J, Spaan WJ, Palese P (1988) The E3 protein of bovine coronavirus is a receptor-destroying enzyme with acetylesterase activity. J Virol 62:4686–4690

17. Holmes KV, Doller EW, Behnke JN (1981) Analysis of the functions of coronavirus glycoproteins by differential inhibition of synthesis with tunicamycin. Adv Exp Med Biol 142:133–142

18. Collins AR, Knobler RL, Powell H, Buchmeier MJ (1982) Monoclonal antibodies to murine hepatitis virus 4 (strain JHM) define the viral glycoprotein responsible for attachment and cell–cell fusion. Virology 119:358–371

19. Holmes KV, Williams RK, Stephensen CB, Compton SR, Cardellichio CB, Hay CM, Knobler RL, Weismiller DG, Boyle JF (1989) Coronavirus receptors. *In* Compans R, Helenius A, Oldstone M (eds) Cell Biology of Virus Entry, Replication and Pathogenesis. UCLA Symposium on molecular and cellular biology, New Series, vol 90, Alan R. Liss, New York, pp 85–95

20. Bolen JB, Anders DG, Trempy J, Consigli RA (1981) Differences in the subpopulations of the structural proteins of polyoma virions and capsids: Biological functions of the multiple VP_1 species. J Virol 37:80–91

CHAPTER 13
Penetration of Influenza Virus Into Host Cells

ROBERT W. DOMS, TOON STEGMANN, AND ARI HELENIUS

Influenza virus remains an important human pathogen, causing excess morbidity and mortality every year, and its impressive ability to evade immune detection through antigenic shift and drift leads to periodic epidemics and occasional pandemics [1]. As a result, considerable effort has been directed toward understanding the molecular pathogenesis of influenza to the point that it is now one of the best-characterized animal viruses. In this chapter we examine what is known about the early stages of infection, namely, the means by which influenza virus binds to and penetrates host cells.

Viral Structure

The orthomyxoviruses, consisting of the influenza A, B, and C viruses, possess a segmented, negative-stranded RNA genome. Due to this segmentation, recombination rates in doubly infected cells are exceptionally high, providing a means for the exchange of genetic material between strains and the production of novel genotypes that may be responsible for new pandemic strains [1]. The viral nucleocapsid is a complex consisting of the viral RNA and two virally encoded proteins, the nucleoprotein and the matrix protein. During infection the nucleocapsid buds through the plasma membrane of the host cell, acquiring a lipid envelope and the two viral spike glycoproteins, the *hemagglutinin* (HA) and *neuraminidase* (NA), which are expressed on the cell surface. HA is a homotrimeric molecule with a combined molecular weight of approximately 228 kd; NA is a homotetramer with a combined molecular weight of 240 kd. The ectodomains of HA and NA have been crystallized and their three-dimensional structures determined to high resolution [2,3]. The HA is initially synthesized as a precursor, HAO, which is activated to the mature HA by a posttranslational proteolytic cleavage [4]. The resulting HA consists of two polypeptide chains, HA1 and HA2, that are connected by a disulfide bond and numerous noncovalent interactions. The crystal structure of HA shows that a globular head domain composed solely of HA1 rests atop a a 75 Å long stem domain composed mostly of HA2. The head domain contains all the major antigenic epitopes as well as the receptor binding site [5]. Interactions between adjoining head and stem domains are responsible for stabilizing the trimeric structure. The structure and function of influenza HA have been recently reviewed [5].

Binding to the Cell Surface

The first step in *viral penetration* is binding to a cell surface *receptor*. The receptors for influenza have long been known to be the terminal sialic acid residues present on membrane glycoproteins and glycolipids [5]. Given the ubiquitous nature of sialic acids, it is not surprising that influenza can bind to most tissue culture cells, even though in nature it is largely confined to the epithelium of the upper respiratory tract. The region of HA responsible for binding is a pocket of highly conserved residues at the distal end of the trimer. Indirect proof that this is indeed the sialic acid binding site came from the observation that although some influenza strains preferentially bind to sialic residues with an $\alpha2,6$ linkage, others bind the $\alpha2,3$ linkage with higher affinity [5], thus making it possible to select for binding variants. When variants were selected, each had a single amino-acid change in the proposed binding site [6]. Direct evidence has recently been obtained by Weiss et al. [7], who infused HA crystals with $\alpha2,3$ sialyl lactose and were able to visualize the interactions between sialic acid and the HA directly, thus raising the possibility of designing molecules that can block this important interaction.

Entry Into the Host Cell

All enveloped animal viruses face a common problem once they are bound to the surface of the host cell. Namely, the large hydrophilic viral nucleocapsid must gain entry to the host cell cytoplasm by passing through two hydrophobic barriers, the viral membrane and the cellular membrane. This is accomplished by *membrane fusion,* which is invariably catalyzed by specific viral envelope proteins. *fusion* may occur at the cell surface, as for Sendai virus and HIV, or from within intracellular organelles, as for influenza and many other enveloped viruses, including togaviruses, bunyaviruses, and rhabdoviruses [8]. Early morphological studies on the entry of influenza into cells showed virus particles in small cytoplasmic vacuoles [9]. Although coated pits play an important role in the *endocytosis* of influenza virus, the virus may also enter by smooth vesicles.

Once bound to the cell surface, the influenza virus may be released back into the extracellular space by the action of NA, which cleaves sialic acid, or it may be endocytosed. The presence of a receptor-destroying compound (NA) on the viral envelope probably serves several functions. NA may play a role in releasing newly made virions from the surface of the infected cell to which they might otherwise bind. In addition, NA may release the virus from mucopolysaccharides and cell debris in the upper respiratory tract, enabling the virus to find a susceptible host cell more readily.

Morphological and biochemical studies reveal that virus is internalized with a $t_{\frac{1}{2}}$ of approximately 10 minutes [10]. Once internalized, virions are delivered to *endosomes*, acidic prelysosomal vacuoles that play an important role in receptor–ligand disassembly and membrane recycling [11]. That influenza virus penetrates into the host cell cytoplasm from the endosomal compartment was first shown in an elegant experiment by Yoshimura and Ohnishi [12]. Through the use of fluorescent probes, they showed that influenza reaches acidic endosomes. Low pH induces the fusion reaction (see below), with kinetics that correspond to those of penetration and the initiation of replication. More recently it has been shown that the virus is internalized as an intact entity [13], and fusion from within the endosomal compartment has been shown directly [14,15]. These studies, plus observations that agents that elevate endosomal pH inhibit infection, demonstrate that influenza virus infects cell via endocytosis followed by a low pH-induced, membrane-fusion reaction.

Characteristics of the Fusion Reaction

The first indication that influenza virus is capable of membrane fusion came from observations by several groups that the virus induces hemolysis and cell-to-cell fusion at low pH [8]. Quantitative methods have been developed to monitor fusion of the virus with either biological or artificial target membranes. In one, virus is allowed to fuse with liposomes that contain proteases. The extent of fusion can then be assayed by determining the fraction of viral nucleocapsids that become degraded. The extent of leaky versus nonleaky fusion can be determined by including protease inhibitors in the extravesicular space [8]. Other assays employ either a self-quenching fluorescent probe incorporated in the viral membrane or measure resonance energy transfer between fluorescent lipids in the target membrane [16]. Upon fusion with the target membrane, energy transfer or quenching is relieved as the probes diffuse into the larger membrane area. The results of these assays are in good agreement and show that fusion occurs only at mildly acidic pH, such as that found in endosomes. Fusion occurs generally within a minute, does not require divalent cations, does not depend on the presence of specific lipids in the target membranes, and is strongly temperature dependent [8,17,18].

Mechanism of Fusion

That HA plays a role in infectivity was first predicted from the observation that only viruses that contain HA that has undergone the cleavage into the HA1 and HA2 subunits are infectious [4]. Viruses that contain the precursor HAO are not infectious. The new amino terminus generated by the cleavage of HAO is markedly hydrophobic and the most highly conserved region of the molecule. It also bears sequence homology with the corresponding amino terminus on the F1 subunit of Sendai virus, a region implicated in its membrane fusion activity [19]. That HA alone is both sufficient and necessary for fusion was shown by White et al. [20], who expressed HA in the absence of other viral components in tissue culture cells. Cells expressing HA on the plasma membrane could be fused to form giant syncytia provided that the cell surface HAO was converted to the mature HA by trypsin treatment and that the cells were briefly incubated at mildly acid pH. More recently the membrane glycoproteins of influenza have been successfully reconstituted into liposomes in the absence of other viral proteins [18,21]. Reconstituted virosomes fuse with parameters that closely match those of the intact virus.

Because the structure of influenza HA is known from X-ray crystallographic studies, it represents an ideal model for protein-induced membrane fusion. The HA trimer undergoes an irreversible *conformational change* at low pH [22,23]. The conditions under which the alteration occurs closely resemble those needed for fusion, strongly suggesting that the conformational change is involved in the membrane fusion activity of the virus. The requirement for low pH seems to be due at least in part for the need to protonate acidic residues involved in intersubunit salt bridges, thereby destabilizing the trimer interface. This is suggested by the fact that single amino-acid changes that abolish any one of a number of interchain salt bridges, or other noncovalent interactions, allow the HA to undergo the conformational change and catalyze fusion at higher pH [24,25]. That the trimer may largely dissociate at low pH has been suggested by a number of studies. Epitopes in the trimer interface are exposed only after acid treatment, trypsin cleavage sites become accessible, and the HA1 subunits can be released from the acid conformation of the molecule in monomeric form. One consequence of trimer destabilization may be exposure of the hydrophobic amino terminus of the HA2 subunit, which is normally sequestered in the trimer interface near the base of the spike. Antibodies to this region indicate that the fusion peptide does in fact become exposed at low pH [26], and the replacement of hydrophobic residues by charged residues has deleterious effects on fusion activity [27]. Studies with hydrophobic photoactivatable cross linkers also show that the HA2 subunit mediates the interaction with the target membrane [28,29]. The exact nature of this interaction has yet to be determined, so the possible involvement of other regions of the HA cannot yet be exluded.

HA-induced fusion is a cooperative event at several levels. First, a majority of the viral spikes must convert to the low pH form before fusion can begin [23]. Second, the conformational change itself is a cooperative event, with all three HA subunits undergoing the conformational change under the same conditions [30]. Finally, recent studies appear to indicate that the fusion active unit is likely to be a trimer or higher-order oligomer since trimers containing a mixture of *wt* and fusion mutant subunits do not cause fusion (T. Stegmann and A. Helenius, unpublished observations).

Summary

Influenza virus and many other enveloped animal viruses have evolved to take advantage of the host cell's endocytic machinery. There are several advantages for the virus that enters cells by this route, rather than through fusion at the plasma membrane. First, by entering the cell by receptor-mediated endocytosis, the virus effectively covers its tracks, leaving no trace of it behind on the plasma membrane. This could help the virus-infected cell evade immune detection until newly synthesized viral spikes are expressed on the cell surface several hours later. Second, by fusing only at low pH, the virus ensures itself that it has entered a viable, metabolically active cell. This may be of particular importance to viruses that infect the upper respiratory tract where they could conceivably bind to mucopolysaccharides or dead cells. Finally, internalized viruses are carried to the perinuclear region, an area rich in the cellular machinery required for viral replication and near the nucleus, in which viruses such as influenza replicate. Fusion at the plasma membrane can leave the capsid in an area destitute of cellular organelles or trapped in an underlying meshwork of cytoskeletal components. Such an area, the terminal web, is particularly prominent in the pseudostratified columnar epithelium of the upper respiratory tract. Indeed, in the case of Semliki Forest Virus, it has long been known that fusion of virus bound to the cell surface by low pH treatment results in a drastic decrease in infectivity, even though the fusion reaction itself is highly efficient. It has recently been shown that the addition of microfilament depolymerizing agents prior to such plasma membrane fusion increases the efficiency of infection to levels approaching that of infection by receptor-mediated endocytosis (M. Marsh, personal communication).

In conclusion, the *infectious entry pathway of influenza* and of certain other acid-activated viruses has been relatively well worked out. These viruses also serve as useful models that have revealed much about the endocytic pathway itself. Further studies directed at several points of the entry pathway should also help illuminate other important cellular functions.

For example, the HA is by far the best-characterized fusion protein. It is hoped that what is learned about the mechanism of HA-induced membrane fusion can eventually be applied to the identification and understanding of cellular fusion factors. Finally, influenza replication occurs in the nucleus, but little is known about the events between penetration from the endosome and initiation of viral RNA replication. The means by which the viral RNA and accessory proteins are transported to the nucleus may very well utilize the cell's transport machinery and represents an important area for further study.

Acknowledgments

This work was supported by NIH Grant AI18599 and by the Medical Scientist Training Program.

References

1. Webster RG, Laver WG, Air GM, Schild GC (1982) Molecular mechanisms of variation of influenza viruses. Nature 296:115–121
2. Wilson IA, Skehel JJ, Wiley DC (1981) Structure of the hemagglutinin membrane glycoprotein of influenza virus at 3 Å resolution. Nature 289:366–373
3. Varghese JN, Laver WG, Colman PM (1983) Structure of the influenza virus glycoprotein antigen neuraminidase at 2.9 Å resolution. Nature 303:35–40
4. Klenk H-D, Rott R, Orlich M, Blodorn J (1975) Activation of influenza A viruses by trypsin treatment. Virology 68:426–439
5. Wiley DC, Skehel JJ (1987) The structure and function of the hemagglutinin membrane glycoprotein of influenza virus. Ann Rev Biochem 56:365–394
6. Rogers GN, Paulson JC, Daniels RS, Skehel JJ, Wiley DC (1983) Single amino acid substitutions in influenza hemagglutinin change receptor binding specificity. Nature 304:76–78
7. Weis W, Brown JH, Cusack S, Paulson JC, Skehel JJ, Wiley DC (1988) Structure of the influenza virus haemagglutinin complexed with its receptor, sialic acid. Nature 333:426–431
8. White J, Kielian M, Helenius A (1983) Membrane fusion proteins of enveloped animal viruses. Q Rev Biophys 16:151–195
9. Dourmashkin RR, Tyrrell DAJ (1974) Electron microscopic observations on the entry of influenza virus into susceptible cells. J Gen Virol 24:129–141
10. Matlin KS, Reggio H, Helenius A, Simons K (1981) Infectious entry pathway of influenza virus in a canine kidney cell line. J Cell Biol 91:601–613
11. Helenius A, Mellman I, Wall D, Hubbard A (1983) Endosomes. Trends Biochem Sci 8:245–250
12. Yoshimura A, Ohnishi S-I (1984) Uncoating of influenza virus in endosomes. J Virol 51:497–504
13. Richman DG, Hostetler KY, Yazaki PJ, Clark S (1986) Fate of influenza A virion proteins after entry into subcellular fractions of LLC cells and the effect of amantadine. Virology 151:200–210
14. Stegmann T, Morselt HWM, Scholma J, Wilschut J (1987) Fusion of influenza virus in an intracellular acidic compartment measured by fluorescence dequenching. Biochim Biophys Acta 904:165–170

15. Nussbaum O, Loyter A (1987) Quantitative determination of virus-membrane fusion events. Fusion of influenza virions with plasma membranes and membranes of endocytic vesicles in living culture cells. FEBS Lett 221:61–67

16. Hoekstra D, DeBoer T, Klappe K, Wilschut J (1984) Fluorescence method for measuring the kinetics of fusion between biological membranes. Biochemistry 23:5675–5681

17. Stegmann T, Hoekstra D, Wilschut J (1986) Fusion activity of influenza virus. A comparison between artificial and biological target membrane vesicles. J Biol Chem 261:10966–10969

18. Nussbaum O, Lapidot M, Loyter A (1987) Reconstitution of functional influenza virus envelopes and fusion with membranes and liposomes lacking virus receptors. J Virol 61:2245–2252

19. Gething M-J, White J, Waterfield M (1978) Purification of the fusion protein of Sendai virus: Analysis of the NH_2-terminal sequence generated during precursor activation. Proc Natl Acad Sci USA 75:2737–2740

20. White J, Helenius A, Gething M-J (1982) Hemagglutinin of influenza virus expressed from a cloned gene promotes membrane fusion. Nature 300:658–659

21. Stegmann T, Morselt HWM, Booy FP, van Breemen JFL, Scherphof G, Wilschut J (1987) Functional reconstitution of influenza virus envelopes. EMBO J 6:2651-2659

22. Skehel JJ, Bayley PM, Brown EB, Martin SR, Waterfield MD, White JM, Wilson IA, Wiley DC (1982) Changes in the conformation of influenza hemagglutinin at the pH optimum of virus mediated membrane fusion. Proc Natl Acad Sci USA 79:968–972

23. Doms RW, Helenius AH, White J (1985) Membrane fusion activity of the influenza virus hemagglutinin. The low pH induced conformational change. J Biol Chem 260:2973–2981

24. Daniels RS, Downie JC, Hay AJ, Knossow M, Skehel JJ, Wang ML, Wiley DC (1985) Fusion mutants of the influenza virus hemagglutinin glycoprotein. Cell 40:431–439

25. Doms RW, Gething M-J, Henneberry J, White J, Helenius A (1986) Variant influenza virus hemagglutinin that induces fusion at elevated pH. J Virol 57:603–613

26. White J, Wilson IA (1987) Anti-peptide antibodies detect steps in a protein conformational change: Low-pH activation of the influenza virus hemagglutinin. J Cell Biol 105:2887–2897

27. Gething M-J, Doms RW, York D, White J (1986) Studies on the mechanism of membrane fusion: Site-specific mutagenesis of the hemagglutinin of influenza virus. J Cell Biol 102:11–23

28. Boulay F, Doms RW, Helenius A (1986) Photolabeling of the influenza hemagglutinin with a hydrophobic probe. *In* Brinton MA, Rueckert RR (eds) Positive Strand RNA Viruses. Alan R Liss, New York, pp 103–112

29. Harter C, Bächi T, Semenza G, Brunner J (1988) Hydrophobic photolabeling identifies BHA2 as the subunit mediating the interaction of bromelain-solubilized influenza virus hemagglutinin with liposomes at low pH. Biochemistry 27:1856–1864

30. Boulay F, Doms RW, Helenius A (1988) Post-translational oligomerization and cooperative acid activation of mixed influenza hemagglutinin trimers. J Cell Biol 106:629–639

CHAPTER 14
Adenoviral Immune Suppression

PER A. PETERSON

A pathogenic virus has to overcome the immune system in order to establish a successful infection. Most viruses seem to succeed by having a temporal advantage over the specific immune defense mechanisms that are elicited only when the organism encounters the virus. Eventually the immune system catches up with the replication of the virus and the pathogen is eliminated. Although several separate immune mechanisms are responsible for containing and eliminating the infectious agent, cytotoxic T lymphocytes (CTL) have a central role in destroying virally infected cells such that viral replication is stopped. The virus-specific CTLs recognize not only specific viral products but also class I major histocompatibility complex (MHC) antigens [1]. Recent data demonstrate that the class I antigens, which contain a binding groove for peptides [2,3], associate with peptides intracellularly [4,5]. The class I antigen–peptide complexes are brought to the cell surface such that CTLs with specificity both for the foreign peptide and the particular class I MHC antigen can recognize and lyse the infected cell.

Class I MHC antigens are found on the surface of most if not all nucleated, somatic cells. These molecules will serve as the scaffold for a variety of peptides derived from viruses with widely different tropism, and they will expose the peptides on the cell surface such that CTLs of the immune system are alerted even when the virus has adopted an entirely intracellular life cycle. Thus, it seems reasonable to suggest that newly synthesized class I antigens associate with peptides at some point during their intracellular transport to the cell surface. Given the exquisite sensitivity of the CTLs, it can be assumed that only a limited number of virally derived peptides are needed to elicit a CTL response.

Although many viruses do not seem to have adopted any specific strategies to evade immune surveillance, it seems likely that most viruses that can give rise to persistent infections have evolved such strategies. One group of viruses of the latter type is the adenoviruses. Different subgroups of these viruses have acquired two genetically separate mechanisms by which they can modulate the expression of class I MHC antigens such that the phenotypes of the infected cells become similar. This review summarizes the present knowledge of these mechanisms.

Adenoviruses

More than 40 serotypes of the human adenoviruses have been distinquished following the discovery of the first member of this group of viruses [6]. It has been shown that the adenoviruses are prevalent in many mammalian and avian species. The human adenoviruses have been classified into several subgenera based on their oncogenicity in newborn hamsters and on their antigenic properties and DNA homologies [7]. Viruses of subgeneus A are highly oncogenic, whereas subgenus C viruses, for example, have little oncogenic potential. It is noteworthy, however, that all adenoviruses can transform rodent cells in vitro. Viruses of the different subgenera usually have different clinical effects, which are usually confined to the epithelial portions of the affected organs. Several of the viruses may induce latency in the infected host.

The genomes of all human adenoviruses are very similar in their organization. Five different transcription units can be temporally resolved. The adenoviral infection is initiated by the activation of the pre-early transcription unit E1A [8]. Gene products derived from this region induce the transcription from the other early transcription units [9]. The gene products derived from E1A and E1B are also responsible for the oncogenicity caused by certain adenoviruses [10]. The early and late phases of the infection are demarcated by the onset of viral DNA replication, with all the late genes expressed under the control of a single promoter.

The various adenoviral transcription units seem to encode proteins with related functions. This is, of course, obvious during the late phase of the infection when mainly structural proteins are manufactured but it also seems to occur during the various stages of the early infection. Thus, the E2 region expresses genes important for the replication of the virus, such as a DNA-binding protein [11], the terminal protein [12,13], and a DNA polymerase [13]. The E3 region expresses at least four different proteins but contains additional open reading frames [14–16]. Five of the putative gene products contain hydrophobic sequences typical of membrane-integrated proteins [17,18]. The E3 region is dispensable when the virus is grown in vitro. This observation raised the suggestion that the E3 transcription unit is not involved in any essential viral function, but at least two of the proteins encoded in this region have immunoregulatory functions (see below).

Although the function of the E4 region has not yet been completely resolved, mutants lacking this region display reduced replication, diminished late protein synthesis, and only partial reduction of host macromolecular synthesis [19,20].

Adenoviral Gene Products That Modulate Expression of Class I MHC Antigens

The E1A region encodes two proteins that reside in the nucleus [21]. Both proteins have identical amino-terminal and carboxy-terminal sequences, but an internal stretch of 46 amino-acid residues is missing from the smaller one. A synthetic fragment corresponding to the unique 46 amino-acid stretch is able to induce transcription from the E2 unit [22]. The gene products of the E1A region serve as transcriptional activators of a variety of viral and cellular promoters, but they also down-regulate the expression of other genes by inhibiting viral and cellular enhancers [22,24]. In the latter case, the effect has been mapped to a region of the two gene products that they hold in common, which is also required for adenoviral transformation [25]. Thus, it appears as if the longer protein acts primarily as a transcriptional activator, whereas the smaller variant serves the function of a repressor of enhancer-dependent transcription.

The role of the E1A region discussed above is common to adenoviruses of all serotypes. The E1A region of viruses of subgenus A, however, also has the unique property of down-regulating the class I MHC antigens in transformed rat cells [26]. This trait is contained in the longer of the two E1A proteins, but the activity can be separated from the activating function of the 46 amino-acid stretch [27]. The expression of the E1A protein greatly reduces the steady-state levels of mRNA-encoding, class I MHC antigen heavy chains in rat, mouse, and human cells [28,29], but the mRNA level of the small class I antigen subunit, β-2-microglobulin, is unaffected. The E1A region of other adenoviruses, for example, from subgenus C, does not affect the class I MHC antigen expression in transformed cells. In fact, coexpression of E1A proteins from adenoviruses of subgenera A and C gives rise to the subgenus C phenotype. The reduced class I antigen expression induced by E1A of the subgenus A viruses can be reversed by the addition of gamma interferon. This demonstrates that the class I gene expression is actively but not irreversibly suppressed. Exactly how the E1A protein works to diminish the steady-state levels of the class I MHC antigen mRNA is not yet fully elucidated.

Although many of the cells transformed or infected by adenoviruses of various subgenera display reduced cell surface expression of class I MHC antigens, only the E1A region of the subgenus A viruses induces a reduction in the class I antigen mRNA level. Accordingly, other adenoviruses must have adopted another strategy to affect the class I MHC antigen cell surface expression. This strategy is post- rather than pretranslational and depends on a single protein encoded in early region 3. This protein is a resident of the endoplasmic reticulum (ER) and is composed of 104 intralumenal amino-acid residues, 23 residues spanning the membrane, and 15 residues exposed on the cytoplasmic side of the membrane [18]. The molecular weight of the mature protein is 25,000, including two asparagine-linked carbohydrate moieties [30], which always occur in the high mannose form due to the ER residency. Since the 142 amino-acid residues of the mature protein and the signal sequence have a combined molecular weight of about 19,000, the name E19 (also designated E3/19K and E3-gp19) has been adopted.

Immunoprecipitates of the E19 protein from transformed rat cells and from productively infected cells contain class I MHC antigens [31–33]. Such complexes contain no other viral or cellular constituents, and complex formation can be generated in vitro with purified components [34,35]. The association between the adenoviral protein and the class I MHC antigens occurs via the heavy chain, inasmuch as E19-class I antigen, heavy-chain complexes have been generated in the absence of β-2-microglobulin [36]. The class I antigen heavy chains bound to the E19 protein have their asparagine-linked carbohydrate moiety in the high mannose form, which indicates that complex formation in the ER leads to the retention of the nascent class I antigens in this locality. In fact, biosynthetic experiments have indicated that, only 4 hours after the onset of an adenovirus infection, newly synthesized class I antigens are quantitatively retained in the ER by association with the E19 protein [37,38]. Other newly synthesized, unrelated proteins exit the ER without measurable impediment, so the complex formation between the class I antigens and the viral polypeptide does not seem to affect transport out of the ER generally. The abrogated intracellular transport of the class I antigens is not accompanied by enhanced degradation of cell surface-expressed, class I molecules synthesized prior to the virus infection, so it takes considerable time before a reduction in the cell surface-expressed amounts is noticeable. Cell surface levels of class I antigens, however, are reduced to half in about 10 hours on adenovirus-2–infected HeLa cells [39].

The E19 protein consists of two discrete and separate functional domains. The intralumenal portion of the viral protein interacts with the class I antigens [35], while the 15-membered, COOH-terminal cytoplasmic tail confers ER residency to the protein [35,40]. In fact, truncation of the last eight amino-acid residues generates a cell surface-expressed form of the E19 protein. The lumenal portion of the polypeptide seems to interact with the peptide-binding domain of the class I antigen heavy chain [41], but it is not yet known whether any portion of the E19 protein actually occupies the peptide-binding groove. It is noteworthy, however, that two human class I antigens bind the E19 protein with differing affinities [42].

Consequences of Adenoviral Modulation of Class I MHC Antigen Expression

A protein that can retain class I antigens in the ER is present in cells infected by adenoviruses of the subgenera B, C, D, E, and possibly F [43]. Although the molecular characteristics of the protein vary for different viruses, there is little doubt that it is the equivalent of the E19 protein of adenoviruses 2 and 5. The subgenus A viruses do not seem to have a functional equivalent of the E19 protein, but these viruses have instead acquired a pretranslational mechanism that generates the same phenotype as the E19 protein. Consequently, all adenoviruses seem to abrogate the cell surface expression of class I MHC antigens in infected cells. Since the reactivity of CTLs is dependent on the relative concentrations of the interacting molecules, it is reasonable to assume that adenovirus-infected cells should be less sensitive to specific CTLs than are noninfected cells. This has been shown to be the case with regard to allo-specific CTLs. Thus, higher numbers of class I allele-specific CTLs, maintained as lines and clones, are needed to lyse adenovirus-2–infected cells than their noninfected counterparts [39,41,44]. Similarly, alloantigen-specific CTLs display a drastically reduced reactivity against adenovirus-12–transformed cells [27].

The in vivo consequences of the adenoviral expression of the E19 protein has only recently begun to be unraveled. A recently developed animal model for the study of the pathophysiology of human adenovirus infections has been described. In this experimental model system, cotton rats infected with adenovirus-5 get pneumonia, which appears to be similar to that observed in humans. A viral deletion mutant that lacks a substantial portion of the E3 region produced a remarkable increase in pulmonary infiltration of monocytic cells [45]. A reasonable interpretation of these data is that, in the absence of the E19 protein, the CTL-mediated lysis of adenovirus-infected cells progressed more efficiently than in the presence of the E19 expression. Consistent with this view is the observation that a recombinant adenovirus-5, containing the hepatitis B virus antigen gene in the E3 region, produced lower titers in syrian hamsters than did the wild-type virus [46]. These data, however, may have alternative interpretations, inasmuch as the E3 region encodes a 14.7-kd protein that protects infected cells against lysis by tumor necrosis factor [16]. Thus, the precise role of the E19 protein in viral pathogenesis will only be defined once viral mutants restricted exclusively to the E19 gene have been used.

The pathophysiological effects of the E1A region of the subgenus A adenoviruses is less clear than the effects of the E19 protein. The pretranslational down-regulation of the class I MHC antigen mRNA levels has only been deomonstrated in virally transformed cells [26,28,29]. In fact, acutely adenovirus-12–infected cells have been shown to contain increased amounts of class I mRNA [47]. It is conceivable, however, that the latter observation may be caused by a viral product that overrides the effect of the E1A gene product. If so, it may be suggested that the E1A function is important only in persistent infections, which may resemble the transformed state in that early gene functions are expressed in the absence of late gene expression.

Conclusions

Adenoviruses have evolved two different strategies by which they can evade immune surveillance. Most adenoviruses express a protein, E19, that prevents nascent class I molecules from becoming expressed on the cell surface. Cell surface-expressed, class I MHC antigens that are already present are not affected by the viral protein. Since the virally infected cells are quite resistant to CTLs, it seems reasonable to suggest that class I antigens only acquire viral peptides in the ER or in the Golgi en route to the cell surface. Consequently, cell surface-expressed, class I antigens would be committed to present whatever peptide they picked up intracellularly, thereby reflecting the recent history of the cell, but they would not associate with peptides generated after their appearance on the cell surface.

The importance of the E1A region of subgenus A adenoviruses in reducing the steady-state levels of class I, heavy-chain mRNA is less obvious. This effect would hardly be noticeable in acutely infected cells, if indeed it exists in such cells, due to the relatively slow half-life of the mRNA, but it could serve as a means by which persistent infections are induced inasmuch as consistently reduced levels of class I antigens should make cells less susceptible to eliminate by CTLs.

Persistent adenoviral infections are probably not exclusively due to the E19 protein or the E1A region, but other viral components may also be of importance. The resistance toward tumor necrosis factor conferred onto infected cells by the 14.7-kd E3 protein may be only one of several contributing mechanisms.

References

1. Zinkernagel RM, Doherty PC (1975) H-2 compatibility requirement for T-cell-mediated lysis of target cells infected with lymphocyte choriomeningitis virus. J Exp Med 141:1427–1436
2. Bjorkman PJ, Saper MA, Samraoui B, Bennet WS, Strominger JL, Wiley, DC (1987a) Structure of the human class I histocompatibility antigen, HLA-A2. Nature 329:506–512
3. Bjorkman PJ, Saper MA, Samraoui B, Bennet WS, Strominger JL, Wiley DC (1987b) The foreign antigen binding site and T cell recognition regions of class I histocompatibility antigens. Nature 329:512–518
4. Townsend A, Rothbard J, Gotch J, Bahadur FM, Wraith D, McMichael AJ

(1986) The epitopes of influenza nucleoprotein recognition by cytotoxic T lymphocytes can be defined with short synthetic peptides. Cell 44:959–968

5. Moore M, Carbone F, Bevan M (1988) Introduction of soluble protein into the class I pathway of antigen processing and presentation. Cell 54:777–785

6. Rowe WP, Huebner RJ, Gilmore LK, Parrott RH, Ward TRJ (1953) Isolation of a cytopathogenic agent from human adenoids undergoing spontaneous degeneration in tissue culture. Proc Soc Exp Biol Med 84:570–573

7. Wadell G (1984) Molecular epidemiology of human adenoviruses. Curr Top Microbiol 110:191–220

8. Nevins JR (1981) Mechanism of activation of early viral transcription by the adenovirus E1A gene product. Cell 26:213–220

9. Berk AJ (1986) Adenovirus promoters and E1A transactivation. Ann Rev Genet 20:45–79

10. Graham FL (1987) In Ginsberg H (ed) The Adenoviruses. Plenum Press, New York, pp 339–398

11. Kruijer W, Van Schaik FMA, Sussenbach JS (1981) Structure and organization of the gene coding for the DNA binding protein of adenovirus type 5. Nucleic Acids Res 9:4439–4457

12. Smart JE, Stillman BW (1982) Adenovirus terminal protein precursor. J Biol Chem 257:13499–13506

13. Stillman B, Tamanoi F, Mattews M-B (1982) Purification of an adenovirus-coded DNA polymerase that is required for initiation of DNA replication. Cell 31:613–620

14. Wold WSM, Deutscher SL, Takemori N, Bhat BM, Magi SC (1986) Evidence that AGUAUAUGA and CCAAGAUGA initiate translation in the same mRNA in region E3 of adenovirus. Virology 148:168–180

15. Tollefson AE, Wold WSM (1988) Identification and gene mapping of a 14,700-molecular-weight protein encoded by region E3 of group C adenoviruses. J Virol 62:33–39

16. Gooding LR, Elmore LW, Tollefson AE, Brady HA, Wold WSM (1988) A 14,700 MW protein from the E3 region of adenovirus inhibits cytolysis by tumor necrosis factor. Cell 53:341–346

17. Cladaras C, Wold WSM (1985) DNA sequence of the early E3 transcription unit of adenovirus 5. Virology 140:28–43

18. Wold WSM, Cladaras C, Deutscher SL, Kapoor QS (1985) The 19-kDa glycoprotein coded by region E3 of adenovirus. J Biol Chem 260:2424–2431

19. Halbert DN, Cutt JR, Shenk T (1985) Adenovirus early region 4 encodes functions required for efficient DNA replication, late gene expression, and host cell shutoff. J Virol 56:250–257

20. Weinberg DH, Ketner G (1986) Adenoviral early region 4 is required for efficient viral DNA replication and for late gene expression. J Virol 57:833–838

21. Berk AJ, Sharp PA (1987) Structure of the adenovirus 2 early mRNAs. Cell 14:695–711

22. Lillie JW, Loewenstein PM, Green MR, Green M (1987) Functional domains of adenovirus type 5 E1A proteins. Cell 50:1091–1100

23. Borrelli E, Hen R, Chambon P (1984) Adenovirus-2 E1A products repress enhancer-induced stimulation of transcription. Nature 312:608–612

24. Hen R, Borrelli E, Chambon P (1985) Repression of the immunoglobulin heavy chain enhancer by the adenovirus-2 E1A products. Science 230:1391–1394

25. Lillie JW, Green M, Green MR (1986) An adenovirus E1A protein region required for transformation and transcriptional repression. Cell 46:1043–1051

26. Schrier PL, Bernards R, Vaessen RTMJ, Houweling A, van der Eb AJ (1983) Expression of class I major histocompatibility antigens switched off by highly oncogenic adenovirus 12 in transformed rat cells. Nature 305:771–775

27. Bernards RP, Schrier A, Houweling JL, Bos AJ, van der Eb AJ, Zijlstra M, Melief CMJ (1983) Tumorigenicity of cells transformed by adenovirus type 12 by evasion of T-cell immunity. Nature 305:776–779

28. Eager KB, Williams J, Breiding D, Pan S, Knowles B, Appella E, Ricciardi RP (1985) Expression of histocompatibility antigens H-2K, -D, and -L is reduced in adenovirus-12-transformed mouse cells and is restored by interferon γ. Proc Natl Acad Sci USA 82:5525–5529

29. Vaessen RTMJ, Houweling A, Israel A, Korilsky P, van der Eb AJ (1986) Adenovirus E1A-mediated regulation of class I MHC expression. EMBO J 5:335–341

30. Kornfeld R, Wold WSM (1981) Structures of the oligosaccharides of the glycoprotein coded by early region 3 of adenovirus 2. J Virol 40:440–449

31. Kvist S, Ostberg L, Persson H, Philipson L, Peterson PA (1978) Molecular association between transplantation antigens and cell surface antigen in adenovirus-transformed cell line. Proc Natl Acad Sci USA 75:5674–5678

32. Signas C, Katze MG, Persson H, Philipson L (1982) An adenovirus glycoprotein binds heavy chains of class I transplantation antigens from man and mouse. Nature 299:175–178

33. Kampe O, Bellgrau D, Hammerling U, Lind P, Paabo S, Severinsson L, Peterson PA (1983) Complex formation of class I transplantation antigens and a viral glycoprotein. J Biol Chem 258:10594–10598

34. Paabo S, Weber F, Kampe O, Schaffner W, Peterson PA (1983) Association between transplantation antigens and viral membrane protein synthesized from a mammalian expression vector. Cell 33:445–453

35. Paabo S, Weber F, Nilsson T, Schaffner W, Peterson PA (1986a) Structural and functional dissection of an MHC class I antigen-binding adenovirus glycoprotein. EMBO J 5:1921–1927

36. Severinsson L, Peterson PA (1985) Abrogation of cell surface expression of human class I transplantation antigens by an adenovirus protein in Xenopus laevis oocytes. J Cell Biol 101:540–547

37. Burgert H-G, Kvist S (1985) An adenovirus type 2 glycoprotein blocks cell surface expression of human histocompatibility class I antigens. Cell 41:987–997

38. Andersson M, Paabo S, Nilsson T, Peterson PA (1985) Impaired intracellular transport of class I MHC antigens as a possible means for adenovirus to evade immune surveillance. Cell 43:215–222

39. Andersson M, McMichael A, Peterson PA (1987) Reduced allorecognition of adenovirus-2 infected cells. J Immunol 138:3960–3966

40. Paabo S, Bhat BM, Wold WSM, Peterson PA (1987) A short sequence in the COOH-terminus makes an adenovirus membrane glycoprotein a resident of the endoplasmic reticulum. Cell 50:311–317

41. Burgert H-G, Kvist S (1987) The E3/19K protein of adenovirus type 2 binds to the domain of histocompatibility antigens required for CTL recognition. EMBO J 6:2019–2026

42. Severinsson L, Martens I, Peterson PA (1986) Differential association between two human MHC class I antigens and an adenoviral glycoprotein. J Immunol 137:1003–1009
43. Paabo S, Nilsson T, Peterson PA (1986b) Adenoviruses of subgenera B, C, D and E modulate cell-suface expression of major histocompatibility complex class I antigens. Proc Natl Acad Sci USA 83:9665–9669
44. Burgert H-G, Maryanski JL, Kvist S (1987) "E3/19K" protein of adenovirus type 2 inhibits lysis of cytolytic T lymphocytes by blocking cell-surface expression of histocompatibility class I antigens. Proc Natl Acad Sci USA 84:1356–1360
45. Ginsberg HS, Valesuso J, Horswood R, Chanock RM, Prince G (1987) Vaccines 87. Cold Spring Harbor Laboratory, Cold Spring Harbor, New York, pp 322–326
46. Morin JE, Lubeck MD, Barton JE, Conley AJ, Davis AR, Hung PP (1987) Recombinant adenovirus induces antibody response to hepatitis B virus surface antigen in hamsters. Proc Natl Acad Sci USA 84:4626–4630
47. Rosenthal A, Wright S, Quade K, Gallimore P, Cedar H, Grosveld F (1985) Increased MHC H-2K gene transcription in cultured mouse embryo cells after adenovirus infection. Nature 315:579–581

CHAPTER 15
Antigen Recognition by Cytotoxic T Cells

J. LINDSAY WHITTON AND MICHAEL B.A. OLDSTONE

The importance of cell-mediated immunity in combating virus infection has been recognized for more than a decade. In 1973 the effector immune cells were shown to bear the Thyl.2 marker [1]; the following year it was demonstrated that the function of these cytotoxic T cells (CTL) depended on recognition not of viral antigen alone, but rather of viral antigen in association with a host glycoprotein encoded by the major histocompatibility complex (MHC) class I genes [2]. For several years the precise nature of the antigen/MHC interaction remained refractory to detailed analysis, but in the past three years, the interaction has been clarified. This chapter reviewing the nature of the complex recognized by CTL addresses four points. First, why recent and current experiments are important; second, how analyses in the experiments have been carried out; third, the present status of such experiments; and finally, a projection of benefits that might accrue in the next few years.

Why Map Virus Epitopes Recognized by Cytotoxic T Cells?

As stated above, these cells play a cardinal role in the control or eradication of virus infection. They are important also in limiting the development and spread of certain tumors [3]; and in fulfilling these functions, they are dependent on expression of MHC molecules by the target cells. Furthermore, they represent in part the immunological effector mechanism involved in graft rejection, the process in which the MHC molecules were originally indentified [4]. Data from studies of virus antigen interaction with MHC, therefore, pertain to a wide variety of biological phenomena. In the approaches used, precise virus sequences involved in recognition can be identified, and this identification has underlined the profound influence exerted by the MHC molecules themselves. The findings elucidate the interactions of individual virus epitopes with individual MHC molecules and as such are relevant to the field of vaccine development. Knowledge of the exact protein sequences required for recognition allows three-dimensional modeling of the recognition moiety and may lead to the development of a predictive scheme with which CTL epitopes can be identified upon primary sequence scanning.

Techniques Used to Map Cytotoxic T Cell Epitopes

Two general methods have been employed: expression within the target cell of nucleic acid sequences encoding the virus protein (endogenous synthesis) and application from outside the target cell of synthetic oligopeptides corresponding to fragments of the virus protein (exogenous administration).

It is well known that intact protein introduced to antigen-presenting cells (such as macrophages) can be processed and presented on the cell surface by class II MHC molecules [5,6]. In marked contrast, activation of class I MHC-restricted functions (such as CTL induction and recognition) simply by applying intact virus protein exogenously to target cells has proven difficult; successful sensitization by most viral proteins requires their expression within the target cell [7]. Thus, much of the work on class I–restricted epitopes relies on expression systems in cultured cells. Early work in the influenza system [8] utilized cell lines stably transformed with incomplete copies of genes encoding influenza virus proteins. More recently, many laboratories have used expression vectors, most commonly vaccinia virus, to achieve the goal. In the first method in which lymphocytic choriomeningitis virus (LCMV) was used as a model system, we made a large number of recombinant vaccinia viruses that expressed various fragments of the LCMV proteins [9]. Some advantages of the vaccinia system are (1) it is well characterized; (2) the virus used is stable and easy to manipulate; (3) it infects most cell lines efficiently, allowing the expression of antigen in 100% of cells; (4) its broad host range in vivo facilitates dissection of requirements for CTL induction; and (5) its life cycle is entirely extranuclear, so sequences from other viruses with "cytoplasmic" life cycles (e.g., LCMV) can be expressed without the risk involved in exposing them to the nuclear environment. Disadvantages of the system are twofold: (1) The DNA to be expressed from vaccinia virus must encode an unspliced mRNA, thus rendering impossible the direct expression of many genomic DNA sequences; and (2) the construction of recombinants is relatively time consuming and labor intensive, which discourages their use in assessing their effects on CTL recognition and lysis of multiple single amino-acid substitutions.

In the second method, although intact protein cannot, as a rule, sensitize target cells to CTL lysis, short synthetic peptides can [10]. Such peptides are now widely used to advantage. Being short, they may identify with precision a CTL epitope; they are readily synthesized; and they can be made with a variety of single residue changes to dissect those residues critical for CTL recognition. There are also several disadvantages. A series of peptides, even if they overlap, might miss an epitope if some peptides straddled, but did not encompass it; and since they are applied both exogenously and at huge molar excess (10^{10} peptide molecules per target cell), this mode of sensitization may be particularly open to the formation of artifacts. Furthermore, there have been few reports of successful CTL induction in vivo by peptides [11] or of protection [12].

Comparison of the two methods shows them to be complementary; the optimal approach at present is to use a judicious combination of both methods. For example, we have identified the general location of epitopes by using nested sets of LCMV protein deletions, each of which has been expressed in recombinant vaccinia, and have precisely delimited the virus epitope by using synthetic peptides [9,13]. Other laboratories have successfully used a similar approach [14].

Current Status of Research

The above methods have yielded a wealth of information about CTL recognition and MHC function. In summary:

1. *CTL can recognize essentially any virus protein.* Potential CTL target antigens are not restricted to viral glycoproteins; "internal" moieties such as polymerases and nucleoproteins are frequent and are sometimes the major antigens [15,16]. Such observations were the first to indicate that class I MHC-restricted T cells might see something other than native virus proteins oriented alongside the MHC glycoproteins in the cell membrane. This arrangement is of clear advantage to the host, since it lays open to the immune system the entire coding capacity of the invading genome, instead of just a few membrane proteins. Additionally, it allows an immune response to proteins expressed early in the virus life cycle (often nonmembrane proteins), when CTL-mediated lysis of the infected cell might be most beneficial.

2. *CTL probably see virus antigen as a short peptide.* The fact that peptides can sensitize cells to lysis is often cited as evidence that virus antigens are normally recognized in peptidic form, but this is an unwarranted extrapolation of data. The observation confirms that CTL can recognize peptides, but does not show that this is the form in which virus antigens are normally seen, that is, after endogenous systhesis in an infected cell. The following evidence, however, does indicate that, after endogenous synthesis, fragments of protein can be recognized on the cell surface. First, as stated above, many viral proteins are undetectable in intact form

at the cell membrane, but nevertheless can act as efficient CTL target antigens. Second, cell lines stably transformed by sequences encoding incomplete influenza virus nucleoprotein are functional as CTL targets [8], which proves that intact virus protein is not a prerequisite for association with MHC and recognition by CTL. Third, we have shown that expression from recombinant vaccinia of the truncated copies of a virus protein renders target cells fully susceptible to CTL killing and that epitopes can be precisely identified using serially deleted proteins. The latter suggests that the presentation of recognizable motifs is not dependent on tertiary structure [9]. Fourth, we have found that only 22 residues of a LCMV glycoprotein sequence expressed from vaccinia virus are sufficient to allow virus-specific, MHC-restricted killing (manuscript in preparation). Finally, influenza virus hemagglutinin has a strongly hydrophobic region (which is *trans*-membrane and therefore not visible on the cell surface in the intact molecule) that is a major CTL epitope. Together these data provide compelling evidence that class I MHC-restricted virus antigens (and, implicitly, class I-restricted antigens involved in tumor and graft rejection) are presented as short peptides. A possible physical confirmation of the phenomenon is available; resolution of the crystallographic structure of a human class I MHC molecule showed a cleft potentially capable of binding a peptide of up to 20 residues, and contained within it was an unresolved density possibly corresponding to bound peptides [17,18].

3. *MHC selects the virus epitope to be presented.* We [19] and others [20,21] have found that the class I MHC molecules themselves determine which virus protein will induce CTL. Because class I glycoproteins are highly polymorphic, different class I alleles present different regions of virus protein. For example, in mice of the H2bb haplotype, there are approximately equivalent responses to the LCMV glycoprotein and the nucleoprotein. In contrast, in an H2dd mouse strain [19], or in a congenic strain differing only at the MHC locus, no response to the glycoprotein is detectable. Such genetically determined nonresponsiveness has been well documented for class II MHC-restricted responses, and we now show a class I counterpart. Furthermore, such an outcome may not be unusual. We have mapped a variety of CTL epitopes on several MHC backgrounds and have found that the number of CTL sites is surprisingly limited. On average, for any one MHC molecule, there is one epitope per 1,500 amino acids of virus protein (i.e., for an inbred mouse, with three class I alleles, there would be one epitope on a 500 residue virus protein). These observations should be borne in mind when administration of subunit vaccines is advocated. Mechanistically, how do class I molecules exert their effects? There is excellent correlation between the host's ability to mount a humoral response to a peptide molecule and the peptide's affinity for a class II molecule expressed by the host. That is, if a peptide cannot bind to any of the host's class II molecules, then it will

not be immunogenic in that host. Such studies are now under way for class I molecules; it is likely that the general mechanism will be similar.

4. *CTL epitopes can confer protection in vivo.* The precise identification of CTL epitopes permits the assessment of their biological function when epitopes are expressed individually. In general, inoculation of peptides has failed to confer protection. Administration of recombinant vaccinia viruses, however, has often protected the recipient. In several cases, immunization has been achieved using "internal" virus proteins expressed from vaccinia, suggesting that the protection is unlikely to be antibody mediated. In one case the protection proved to be exerted by $CD8^+$ T cells, that is, a CTL phenotype [22]. We have found that a single inoculation of a recombinant vaccinia encoding an LCMV nucleoprotein fragment that appears to contain a single epitope is sufficient to protect mice from challenge with lethal doses of LCMV (Klavinskis, L., Whitton, J.L., and Oldstone M.B.A., manuscript in preparation).

Future Directions

The recent advances detailed above have led to new questions: How are the viral proteins processed for presentation by class I MHC? Are there two discrete pathways of antigen processing/presentation, one for class I–restricted antigens, the other for class II–restricted antigens? What determines the selection of a particular viral sequence by the MHC molecules? By accumulating a large enough database of epitopes mapped to individual MHC alleles, can we discern a pattern that would allow us to predict epitopes from primary sequence? So far all of the work has revolved around requirements for recognition by CTL; at least as important are the viral sequence requirements for CTL induction. What are these requirements? Does the need for "T-help" vary among different epitopes? Can we learn to manipulate peptides to induce an in vivo response? Judging by the progress of the past 3 years, the answers to some or all of the above will not be long delayed.

Acknowledgments

This is Publication no. 5473-IMM from the Department of Immunology, Scripps Clinic and Research Foundation, La Jolla, CA 92037. This work was supported in part by USPHS Grants NS-12428 and AG-04342. We thank Gay Schilling for expert secretarial assistance.

References

1. Cole GA, Prendergast RA, Henney CS (1973) Fed Proc 32:964
2. Zinkernagle RM, Doherty PC (1974) Restriction of *in vitro* T cell mediated cytotoxicity in lymphocytic choriomeningitis within a syngeneic or semi-allogeneic system. Nature 248:701–702
3. Anichini A, Fossati G, Parmiani G (1987) Clonal analysis of the cytotoxic T cell response to human tumors. Immunol Today 8:385–389
4. Gorer PA (1936) J Genet 32:17
5. Sette A, Buus S, Colon S, Smith JA, Miles C, Grey HM (1987) Structural characteristics of an antigen required for its interaction with Ia and recognition by T cells. Nature 328:395–399
6. Unanue ER, Allen PM (1987) The basis for the immunoregulatory role of macrophages and other accessory cells. Science 236:551–557
7. Morrison LA, Lukacher AE, Braciale VL, Farr DP, Braciale TJ (1986) Differences in antigen presentation to MHC class I and class II restricted influenza virus-specific cytotoxic T lymphocyte clones. J Exp Med 163:903–921
8 Townsend ARM, Gotch FM, Davey J (1985) Cytotoxic T cells recognize fragments of the influenza nucleoprotein. Cell 42:457–467
9. Whitton JL, Gebhard JR, Lewicki H, Tishon A, Oldstone MBA (1988) Molecular definition of a major cytotoxic T lymphocyte epitope in the glycoprotein of lymphocytic choriomeningitis virus. J Virol 62:687–695
10. Townsend ARM, Rothbard J, Gotch F, Bahadur G, Wraith DC, McMichael AJ (1986) The epitopes of influenza nucleoprotein recognized by cytotoxic T lymphocytes can be defined with short synthetic peptides. Cell 44:959–968
11. Staerz UD, Karasuyama H, Garner AM (1987) Cytotoxic T lymphocytes against a soluble protein. Nature 329:449–451
12. Watari E, Dietzschold B, Szokan G, Heber-Katz E (1987) A synthetic peptide induces long-term protection from lethal infection with herpes simplex virus 2. J Exp Med 165:459–470
13. Oldstone MBA, Whitton JL, Lewicki H, Tishon A (1988) Fine dissection of nine amino acid glycoprotein epitope, a major determinant recognized by lymphocytic choriomeningitis virus specific class I restricted H-2D^b cytotoxic T lymphocytes. J Exp Med 168:559–570
14. Del Val M, Volkmer H, Rothbard JB, Jonjic S, Messerle M, Schickedanz J, Reddehasse MJ, Koszinowski UH (1989) Molecular basis for cytolytic T-lymphoctye recognition of the MCMV IE protein pp89. J Virol (in press)
15. Townsend AR, Skehel JJ (1984) The influenza virus nucleoprotein gene controls the induction of both subtype specific and corss-reactive T cells. J Exp Med 160:552–563
16. Bennink Jr, Yewdell JW, Smith GL, Moss B (1987) Anti-influenza virus cytotoxic T lymphocytes recognize the three viral polymerases and a nonstructural protein responsiveness to individual viral antigens is major histocompatibility complex controlled. J Virol 61:1098–1102
17. Bjorkman PJ, Saper MA, Samraoui B, Bennett WS, Strominger TL, Wiley DC (1987a) Structure of the human class I histocompatibility antigen, HLA-A2. Nature 329:506–511

18. Bjorkman PJ, Saper MA, Samraoui B, Bennett WS, Strominger TL, Wiley DC (1987b) The foreign antigen binding site and T cell recognition regions of class I histocompatibility antigens. Nature 329:512–518
19. Whitton JL, Southern PJ, Oldstone MBA (1988) Analyses of the cytotoxic T lymphocyte responses to glycoproteins and nucleoprotein components of lymphocytic choriomeningitis virus. Virology 162:321–327
20. Vitiello A, Sherman LA (1983) Recognition of influenza infected cells by cytolytic T lymphocyte clones: Determinant selection by class I recognition elements. J Immunol 131:1635–1640
21. Pala P, Askonas BA (1986) Low responder MHC alleles for Tc recognition of influenza nucleoprotein. Immunogenetics 23:379–384
22. Koszinowski UH, Volkmer H, Messerle M, Jonjic S, Witlek R (1988) Cytotoxic T lymphocytes induced by a vaccinia recombinant expressing a herpesvirus nonstructural IE protein product protect against CMV infection. *In* Ginsberg H, Brown F, Lerner R, Chanock RM (eds) Vaccines 88. Cold Spring Harbor Laboratory, Cold Spring Harbor, New York, pp 41–45

CHAPTER 16

The Use of the Epstein–Barr Virus to Probe the Human B Cell Repertoire and to Generate Monoclonal Antibodies

PAOLO CASALI AND ABNER LOUIS NOTKINS

Most studies on the interaction of viruses with the host have been concerned with the analysis of disease pathogenesis and forms of treatment. It is now becoming clear that viruses also can serve as powerful tools to explore cellular functions or as vectors to carry genetic information for the expression of a variety of biological and clinically important products. In this minireview, we summarize how the Epstein–Barr virus (EBV) is being used to probe the human B lymphocyte repertoire and to make human monoclonal antibodies.

Epstein–Barr Virus and Human B Lymphocytes

The main boilogical activity of EBV is the infection of B lymphocytes, which results in blast formation (cell activation), immunoglobulin (Ig) secretion, and, ultimately, indefinite cell proliferation, a process termed *immortalization* [1]. EBV infection of human B lymphocytes is a one-hit phenomenon [2] and involved two sequential, discrete events. The virus first attaches to the surface receptor CR2, which also mediates the binding of the complement C3d split product and is selectively present on all mature B cells and late pre-B cells [1,3]. Subsequent virus internalization leads, in most cases, to cell activation and cycling. Actual cell transformation may later ensue, yielding cell blasts capable of continuous proliferation and steady Ig secretion [4]. EBV is an ideal polyclonal B cell activator in that it acts directly on B cells independent of any T cell help [5]. Moreover, as we recently showed, EBV is equally efficient in binding to and transforming B cells bearing surface μ, γ, and α Ig H chains to produce IgM, IgG, and IgA, respectively, in the absence of any significant Ig class switch [2,6]. Although no restriction exists in EBV transformation of different B cell subsets, different phases of the cell cycle do affect the ability of the virus to transform B cells. Our studies demonstrated that, whereas EBV does bind to cells in G_1 phase almost as efficiently as it does to cells in G_0, it does not bind to and infect cells in S phase [3]. The lack of transformation of proliferating B cells is due, in the first instance, to the disappearence of the cell surface CR2 and the consequent loss of virus attachment [3]. Similarly, antibody (Ab)-producing cells in the terminal differentiation stage lose surface CR2 [7]. This might limit the rate of EBV-induced immortalization in cells spontaneously producing Ab in vivo, as found in subjects with an ongoing immune response.

Construction of Human Monoclonal Antibody-Secreting Cell Lines

Unlike mouse monoclonal antibodies that can be easily generated using spleen cells from a hyperimmunized animal, human monoclonal Abs of selected specificity are difficult to prepare because of the relative low frequency of the relevant circulating B cells in subjects with an ongoing immune or autoimmune response. Recently we overcame this obstacle by enriching for B cells of a desired specificity using flow cytometry techniques and EBV transformation in serial limiting dilution microcultures [8,9]. Even though the selected, EBV-transformed B cells obtained by these procedures are a useful source of monoclonal antibodies (mAbs), their growth and antibody production are erratic when cultured at low density and beyond a short period (3 to 4 weeks) [8–10]. This has been obviated by us and others by using somatic hybridization techniques involving fusion of selected EBV-transformed B cells with a preconstructed human–mouse hybrid partner [11–12]. The resulting human–human–mouse hybrid clones have a much higher plating efficiency, stability, and Ab secretion rate than their parental, EBV-transformed B cells in long-term culture [9]. By sequential use of cell selection, EBV-transformation, and somatic hybridization techniques, we have been able to generate a number of human mAbs (IgM, IgG, and IgA) to both exogenous antigens (Ags) and self-Ags. The data derived from the binding and inhibition studies of these Abs have been invaluable in complementing the information collected on the frequency of B lymphocytes producing the Abs with the same specificities in normal humans and patients, as described below.

Frequency of B Lymphocytes Producing Antibodies to Self-Antigens and Exogenous Antigens in Healthy Subjects

Because EBV is equally efficient in inducing B lymphocytes committed to the production of Ig of different classes, it is ideally suited for a balanced activation of the human B cell repertoire in limiting dilution assay, in which only one cell (EBV-activated B lymphocyte) is sufficient for a positive response (Ab production) [13]. The procedure adopted by us and others involves infection with EBV of fresh B lymphocytes and their immediate distribution in microcultures in the presence of irradiated peripheral blood mononuclear cells as feeders [6,8]. After a 4-week culture, the supernatant fluids are analyzed for their content of total and Ag-specific IgM-, IgG-, or IgA-producing cells; the frequencies of occurrence of these lymphocytes are calculated by statistical analysis according to a Poisson distribution [2,6,8,13]. In this approach, considerable numbers (3% to 13%) of circulating cells capable of producing Abs to self-Ags (e.g., IgG Fc fragment, ssDNA, thyroglobulin, and insulin) and exogenous Ags (e.g., tetanus toxoid) are readily detectable in healthy subjects [6,8]. The vast majority of these Abs are of the M class, but a few IgG and IgA with similar binding activities can also be detected. This would argue against any hypothesis of clonal abortion of self-reactive B cell clones [14], and provides an explanation for the relative ease with which cell hybrids producing "autoantibodies" are generated when lymphocytes from normal mice or humans are used to construct mAb-producing cell lines [15–17].

Polyreactive Antibodies

The observed high frequency of B cells committed to the production of Abs, particularly IgM, binding to self-Ags was likely due to the presence of a population of Abs that can bind to more than one Ag. Indeed, in the last few years, a number of mAbs binding multiple Ags, particularly self-Ags, including soluble hormones, nucleic acids, structural cellular constituents, and tissue components, have been generated in this and other laboratories [15–17]. Although at first these polyreactive mAbs had been found to be produced by B lymphocytes from autoimmune patients or chronically infected mice [17], it soon became apparent that similar autoantibodies could in fact be derived from healthy subjects [18]. It was also clear that, although in a few cases the detected multiple Ag-binding activity was due to the binding of one single Ab to an identical epitope present in two different molecules, in most cases the Ab recognition involved two different epitopes [18]. Moreover, it was recognized that these polyreactive autoantibodies react efficiently also with exogenous Ags. Thus, even though B cells committed to the production of polyreactive mAbs recognizing multiple self-Ags and exogenous Ags are known to be common components of the normal B cell repertoire, their physiological or pathological roles are still unknown.

Identification of the Cell Type (CD5+) Producing Polyreactive Antibodies

We recently identified the cell type that makes these polyreactive Abs. In mice it has been reported that Ly-1$^+$B lymphocytes make autoantibodies [19]. In humans the equivalent of the mouse Ly-1$^+$B cell is the CD5$^+$B cell, which constitutes a discrete subset in the normal B cell repertoire, accounting for up to 25% of total ciruclating and splenic B lymphocytes [19–22]. Using sophisticated cell sorting techniques, we determined that CD5$^+$B cells from healthy subjects include the vast majority of lymphocytes committed to the production of IgM, IgG, and IgA to self-Ags, such as IgG Fc fragment (rheumatoid factor, RF), ssDNA, and thyroglobulin, and, also, IgM to tetanus toxoid [20]. These Abs are inherently polyreactive as revealed by the study of the mAbs produced by hybrid EBV-cell lines constructed using CD5$^+$B cells from healthy subjects. They bind in a dose-dependent fashion, although to different degrees, to a number of self-Ags and exogenous Ags, including bacterial constituents and products (Figure 16.1 **A** and **B**) [11]. Moreover, their binding to different solid-phase Ags can always be cross-inhibited in a dose-dependent fashion and with different efficiency by soluble heterologous Ags (Figure 16.1 **C**) [10]. These mAbs consistently display relative low affinity for IgG Fc fragment (K_d, about 10^{-5} mole/liter) and variable affinities for other Ags (k_d, 10^{-5} to 10^{-7} mole/liter). Their high k_d values for IgG Fc fragment are compatible with the low-binding affinities attributed to human RFs, in general from patients with monoclonal gammopathies [22]. In addition, their functional features resemble those of some monoclonal anti-DNA autoantibodies and "natural autoantibodies of natural antibodies" of humans and mice [18]. Although the natural Ab functions have been traditionally associated with molecules of the IgM class, polyreactive Abs can also be IgG and IgA [6,9]. The molecular basis underlying the behavior of these Abs are still unknown, although it is possible that they utilize only selected variable (V) regions with conserved framework segments and endowed with a high degree of reactivity for different components. Along these lines we found that polyreactive antibodies from CD5$^+$ B cells do seem to utilize only selected V$_H$ gene segments, including V$_H$III, V$_H$IV, and V$_H$V, most often in unmutated configuration [24]. Utilization of selected, possibly 3′, V$_H$ segments, a feature characteristic of lymphocytes of the early B cell repertoire [25,26] and chronic lymphocytic leukemia (CLL) cells [27], suggests that CD5$^+$B lymphocytes may represent an early stage in the ontogenesis of B cell.

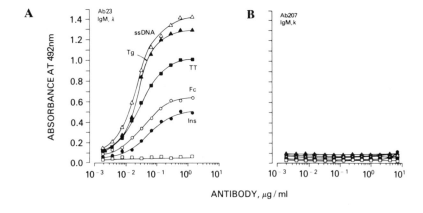

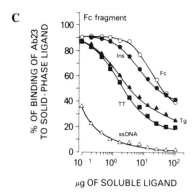

Figure 16.1. **A** Dose-dependent binding of monoclonal Ab23 (produced by a mono-clonal cell line derived from CD5$^+$B cells of a healthy subject) to solid-phase IgG Fc fragment (o), ssDNA($\triangle$), thyroglobulin ($\blacktriangle$), insulin ($\bullet$), and tetanus toxoid ($\blacksquare$). The Ag-binding activity for each molecule is expressed as optical absorbance (at 492 nm). The empty squares indicate the binding to bovin serum albumin, which was used in the dilution buffer. **B** Failure of monoclonal Ab 207 (produced by a monoclonal cell line derived from CD5$^-$B cells of the same donor used for the generation of Ab23) to bind to solid-phase self-Ags and exogenous Ags (as used in the experiments of **A**). **C** Dose-dependent inhibition of monoclonal Ab23 mAb binding to solid-phase IgG Fc fragment by soluble homologous (IgG Fc fragment) and heterologous (ssDNA, thyroglobulin, insulin, and tetanus toxoid) ligands. Samples of the Ab (0.2 μg) were incubated with increased amounts (from 0.1 to 200 μg) of soluble Fc fragment, ssDNA, thyroglobulin, insulin, or tetanus toxoid. After 18 hours, the mixtures were transferred into ELISA plates precoated with Fc fragment, and the amount of antibody bound to the solid-phase IgG Fc fragment was measured as optical absorbance (at 492 nm). The binding of each Ab observed in the presence of soluble ligand is expressed as percentage of the binding measured after incubation of the Ab under identical conditions but in the absence of any soluble ligand (100% binding activity). Modified from [11].

The B Cell Repertoire Profile in Human Autoimmune Diseases: Analysis of Polyreactive and Monoreactive Antibody-Producing Cells

Given the antiself-Ag reactivity of the polyreactive (natural) autoantibodies, it has been tempting to speculate that these molecules may play some role in autoimmune diseases. In fact, using EBV technology, we have found that in patients with a variety of systemic as well as organ-specific autoimmune diseases, such as Hashimoto's disease (HD), systemic lupus erythematosus (SLE), and insulin-dependent diabetes mellitus, the frequencies of circulating cells (CD5$^+$) capable of producing polyreactive IgM or IgA autoantibodies are similar to those of cell precursors producing Abs of the same classes and to the same Ags found in healthy subjects [6]. In contrast, a higher frequency of B cell precursors producing IgG autoantibodies monoreactive and highly specific for the relevant self-Ags, that is, thyroglobulin and ssDNA, can be detected in HD and SLE patients, respectively [6]. Similar to anti-tetanus toxoid IgG in vaccinated subjects, these monoreactive, highly specific autoantibodies are produced by CD5$^-$ B cells [21], and their affinity is always much higher (K_d, 10^{-7} to 10^{-11} mole/liter) than that displayed for the same self-Ag by polyreactive autoantibodies from healthy subjects or patients (K_d, 10^{-3} to 10^{-7} mole/liter) [6]. Taken as a whole, these findings suggest that polyreactive autoantibodies are unlikely to play any significant pathogenetic role in autoimmune diseases. Rather, an effective autoimmune attack would be initiated by the high-affinity autoantibodies, possibly the products of B cells that underwent a process of Ag-driven clonal selection and somatic mutation [9,28].

Summary and Concluding Comments

The balanced activation and amplification of discrete B cell clones and subsets and the generation of continous mAb-producing cell lines have proved that EBV is a powerful tool to investigate the diversity of the human B cell repertoire. Using EBV technology in conjunction with fine cell sorting techniques, we established that a major proportion of circulating cell precursors produce polyreactive Abs to both self-Ags and exogenous Ags in the normal B cell repertoire; we also identified a discrete CD5$^+$ B lymphocyte subset responsible for the production of these polyreactive Abs. The label "autoantibody-producing B lymphocytes" that one may be tempted to attach to these cells and to mouse Ly-1$^+$ B cells is, perhaps, not appropriate. In fact, the Ig produced by CD5$^+$B cells constitute a class of Abs that are more appropriately described as polyreactive or natural Abs. These autoantibodies are easily detected in the ciruclation of healthy humans and mice during the physiological response to an exogenous Ags, such as tetanus toxoid or bacterial lipopolysaccharides, or during the course of viral, bacterial, or parasitic infections [29–32]. Such Abs likely play a role in the first line of defense against invading microorganisms (by enhanced phagocytosis, complement-mediated lysis, etc.) and in the enhancement of an ongoing specific immune response. Indeed, polyreactive antibodies derived from CD5$^+$B cells bind tetanus toxoid and lipolysaccharides or polysaccharides from different bacterial strains [6].

The combined utilization of EBV technology and cell hybridization techniques has allowed us to generate human monoclonal IgM, IgG, and IgA Abs with a number of different specificities. These include human pathogen viruses, such as rabies virus and HIV; bacterial constituents or products, such as lipolysaccharides, saccharides, or tetanus toxoid; as well as self-Ags, such as cell components (e.g., DNA) and cellular products (e.g., insulin, thyroglobulin). The articulate technology we developed for the generation of human monoclonal Abs is necessarily far more sophisticated than the one previously devised for the production of mouse mAbs. Again, the major difference between the two systems and the limiting factor in the human one is the impossibility of intentionally inducing in humans a vigorous immune response to a given Ag in order to increase the frequency and affinity of the relevant cirulating, Ab-producing cell precursors. Our studies have clearly shown that this obstacle can be at least partially overcome by the use of precise cell selection and enriching procedures. Whereas this allows for the preparation of useful numbers of cells homogeneous for a given cell surface receptor, and therefore specificity, the generation of high-affinity, clinically useful, human mAbs is absolutely dependent on the availability of B lymphocytes with high Ag specificity, as found in subjects with an ongoing specific immune response.

In conclusion, we have shown here that the systematic exploitation of defined biological properties of a human pathogen virus (EBV) led to the definition of a systematic approach for the study of a major cellular compartment (B cell repertoire) of the immune system in healthy subjects and in patients. In addition, the use of the same virus has allowed us to generate consistently , for the first time, invaluable human immune reagents (mAbs). It is possible that better knowledge of the properties of other, still poorly characterized viruses will result in the acquisition of a new generation of biologically important intermediates.

References

1. Miller G (1985) Epstein–Barr virus. *In* Fields BN (ed) Virology 27, Raven Press, New York pp 563–589
2. Yarchoan R, Tosato G, Blaese RM, Simon RM, Nelson DL (1983) Limiting dilution analysis of Epstein–Barr virus-induced immunoglobulin production by human B cells. J Exp Med 157:1–14
3. Inghirami G, Nakamura M, Balow JE, Notkins AL, Casali P (1988) A model for

virus attachment: Identification and quantitation of EBV-binding cells using biotinylated virus in flow cytometry. J Virol 62:2453–2463

4. Bird AG, Britton S, Ernbert I, Nilsson K (1981) Characteristics of Epstein–Barr virus activation of human B lymphocytes. J Exp Med 154:832–839

5. Waldmann TA, Broder S (1982) Polyclonal B cell activators in the study of the regulation of immunoglobulin synthesis in the human system. Adv Immunol 32:1–63

6. Nakamura M, Burastero SE, Ueki Y, Larrick JW, Notkins AL, Casali P (1988) Probing the human B cell repertoire with EBV. Frequency of B cells producing monoreactive high affinity autoantibodies in patients with Hashimoto's disease and SLE. J Immunol 141:4165–4172

7. Boyd A, Anderson K, Freedman A, Fisher D, Slaughenhoupt B, Schlossman SF, Nadler LM (1985) Studies on in vitro activation and differentiation of human B lymphocytes. I. Phenotypical and functional characterization of the B cell population responding to anti-Ig antibodies. J Immunol 134:1516–1523

8. Casali P, Inghirami G, Nakamura M, Davies TF, Notkins AL (1986) Human monoclonals from antigen-specific selection of B lymphocytes and transformation by EBV. Science 234:476–479

9. Casali P, Notkins AL (1989) Probing the human B cell repertoire with EBV: Polyreactive antibodies and CD5$^+$B lymphocytes. Ann Rev Immunol 7:513–535

10. James K, Bell GT (1987) Human monoclonal antibody production. Current status and future prospects. J Immunol Meth 100:5–40

11. Nakamura M, Burastero SE, Notkins AL, Casali P (1988) Human monoclonal rheumatoid factor-like antibodies from CD5 (Leu-1)$^+$ B cells are polyreactive. J Immunol 140:4180–4186

12. Pollack M, Raubitschek AA, Larrick JW (1987) Human monoclonal antibodies that recognized conserved epitopes in the core-lipid A region of lipopolysaccharides. J Clin Invest 79:1421–1430

13. Lefkovits I, Waldmann H (1984) Limiting dilution analysis of the cells of immune system I. The clonal basis of the immune response. Immunol Today 5:265–268

14. Burnett FM (1959) The Clonal Selection Theory of Acquired Immunity. Cambridge University Press, London

15. Dighiero G, Lymberi P, Mazie JC, Rouyre S, Butler-Brown GS, Whalen RC, Avrameas S (1983) Murine hybridomas secreting natural monoclonal antibodies reacting with self-antigens. J Immunol 131:2267–2272

16. Prabhakar BS, Saegusa J, Onodera T, Notkins AL (1984) Lymphocytes capable of making monoclonal autoantibody that react with multiple organis are common feature of the normal B cell reperotire. J Immunol 133:2815–2817

17. Casali P, Prabhakar BS, Notkins AL (1988) Characterization of multireactive autoantibodies and identification of Leu-1$^+$B lymphocytes as cells making antibodies binding multiple self and exogenous molecules. Int Rev Immunol 3:17–45

18. Ternynck T, Avrameas S (1986) Murine natural monoclonal autoantibodies: A study of their polyspecificities and their affinities. Immunol Rev 94:99–112

19. Hayakawa K, Hardy RR, Honda M, Herzenberg LA, Steinberg AD, Herzenberg LA (1984) Ly-1 B cells: Functionally distinct lymphocytes that secrete IgM autoantibodies. Proc Natl Acad Sci USA 81:2494–2498

20. Casali P, Burastero SE, Nakamura M, Inghirami G, Notkins AL (1987) Human

lymphocytes making rheumatoid factors and antibodies to single stranded DNA belong to the Leu-1$^+$B cell subset. Science 236:77–81

21. Hardy RR, Hayakawa K, Shimizu M, Yamasaki K, Kishimoto T (1987) Rheumatoid factor secretion from human Leu-1$^+$B cells. Science 236:81–83

22. Burastero SE, Casali P (1989) Characterization of human CD5 (Leu-1 OKT1)$^+$B lymphocytes and the antibodies they produce. *In* del Guercio P (ed) B cells and B cell Products. S. Karger AG, Basel (in press)

23. Carson DA, Chen PP, Fox RI, Kipps TJ, Jirik F, Goldfien RD, Silverman G, Redoux V, Fong S (1987) Rheumatoid factors and immune networks. Ann Rev Immunol 5:109–126

24. Sanz I, Casali P, Thomas JW, Notkins AL, Capra JD (1989) Genetic basis of human polyreactive antibodies: Organization, complexity and mechanisms of diversity of the human B cell repertoire. J Immunol (in press)

25. Yancopoulos GD, Desiderio SV, Paskind M, Kearney JF, Baltimore D, Alt FW (1984) Preferential utilization of the most J_H-proximal V_H gene segments in pre-B cell lines. Nature 311:727–733

26. Schroeder HW Jr, Hillson JL, Perlmutter RM (1987) Early restriction of the human antibody repertoire. Science 238:791–793

27. Humpries GC, Shen A, Kuziel WA, Capra JD, Blattner FR, Tucker PW (1988) A new immunoglobulin V_H family preferentially rearrenged in immature B-cell tumors. Nature 331:446–449

28. Burastero SE, Casali P, Wilder RL, Notkins AL (1988) Monoreactive high affinity and polyreactive low affinity rheumatoid factors are produced by CD5$^+$B cells from patients with rheumatoid arthritis. J Exp Med 168:1979–1992

29. Welch MJ, Fong S, Vaughan J, Carson DA (1983) Increased frequency of rheumatoid factor precursor B lymphocytes after immunization of normal adults with tetanus toxoid. Clin Exp Immunol 51:299–304

30. Carson DA, Bayer AS, Eisemberg RA, Lawrence S , Theofilopoulos A (1978) IgG rheumatoid factors in subacute bacterial endocarditis: Relationship to IgM rheumatoid factors and circulating immune complexes. Clin Exp Immunol 31:100–113

31. Casali P, Perrin LH, Lambert PH (1979) Immune complexes and tissue injury. *In* Dick G (ed) Immunological Aspects of Infectious Diseases. University Park Press, Baltimore, pp 295–342

32. Dziarski R (1982) Preferential induction of autoantibody secretion in polyclonal activation by peptidoglycan and lypopolysaccharide. II. In vivo studies. J Immunol 128:1026–1030

CHAPTER 17

Transcriptional Regulation by Papillomavirus E2 Gene Products

ALISON A. MCBRIDE, PAUL F. LAMBERT, BARBARA A.
SPALHOLZ, AND PETER M. HOWLEY

Papillomaviruses have been isolated from a variety of higher vertebrates and have a highly species-specific tropism. They induce benign, proliferative, squamous epithelial, and fibroepithelial lesions in their natural hosts. Viral replication is restricted to differentiated squamous epithelial cells, and this characteristic has hampered the propagation of virus in a tissue culture system. In recent years papillomaviruses have gained considerable attention because of their association with specific human cancers.

These viruses have double-stranded circular DNA genomes, and the organization of those that have been analyzed is very similar. Bovine papillomavirus 1 (BPV-1) has served as the prototype, due to its ability to transform certain rodent cell lines (for review, see ref. 1). The viral genome is maintained extrachromosomally within these cells, and this latent, non-productive infection is thought to reflect the virus–host interaction in the basal epithelial cells and dermal fibroblasts of a bovine fibropapilloma. Figure 17.1 shows the genetic organization of the BPV-1 genome; the major open reading frames (ORFs) are located on one strand, and only the region corresponding to the early ORFs (designated E1 to E8) is transcribed in transformed cells. Upstream from the early ORFs is the long control region (LCR), a region of about 1000 bp that contains regulatory elements involved in the control of transcription and DNA replication [2–4]. In transformed cells a low level of transcription can be detected from several promoters located in the LCR and from promoters that generate RNAs with 5' ends in the vicinity of nucleotides 890, 2443, and 3080 (M. Botchan, personal communication) [5]. Many different mRNA species can be found in transformed cells resulting from the use of multiple promoters and complex patterns of splicing [6,7].

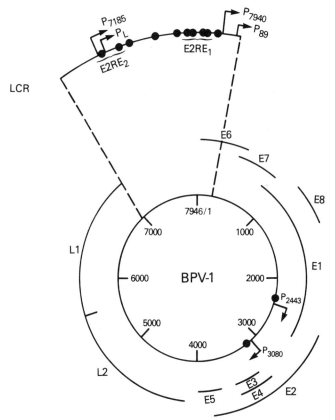

Figure 17.1. Genomic organization of BPV-1 DNA. The nucleotide numbers are noted within the circular map, and the major ORFs (designated E1 to E8, L1 and L2) are shown. Promoters are indicated by P followed by the approximate position of the RNA start site, except for the late promoter P_L. The $ACCN_6GGT$ motifs are represented by circles.

Functions and Polypeptides Encoded by the E2 ORF

Specific functions have now been assigned to many of the ORFs of BPV-1. Mutations in the E2 ORF are pleiotrophic, affecting both transformation and replication functions [8–12]. These pleiotrophic effects may not be direct but may rather be due to the transcriptional regulatory functions encoded by this ORF. Genetic analysis of BPV-1 identified a transactivator encoded by the full-length E2 ORF that activates specific transcriptional enhancers within the LCR [3]. Further analysis of the LCR located two independent E2 responsive elements, designated $E2RE_1$ and $E2RE_2$ [14]. The interaction of E2 with $E2RE_1$ is required for activation of transcription from the LCR promoters P_{89} and P_{7940} [14,15], and recent evidence indicates that the P_{2443}

promoter (from which the E2 transactivator is expressed) is also regulated by E2 through the LCR [16]. In addition to the transactivator, the BPV-1 E2 ORF encodes a second transcriptional regulatory function, which can inhibit viral-mediated transformation and repress E2 *trans*-activation [17]. This function also depends on the E2 responsive elements in the LCR. Genetic and functional studies indicate that two different repressor molecules are encoded by BPV-1. One repressor, E2-TR, is expressed from the P_{3080} promoter and utilizes an internal methionine within the E2 ORF at nucleotide 3091. A second repressor, E8/E2, is translated from a spliced mRNA linking an upstream ORF (E8) to the 3' half of the E2 ORF [17,18]. The latter message is most likely to encode 11 amino acids from the E8 ORF spliced to the E2 ORF via the major splice acceptor at nucleotide 3225. Three E2-encoded proteins, with apparent molecular weights of 48, 31, and 28 kd, have been identified in transformed cells [19]. These proteins share common C-terminal sequences, and their observed molecular weights correlate well with those predicted for the full-length transactivator and each repressor. Mutational analysis has confirmed that the largest protein is the *trans*-activator, the smallest the E8/E2 repressor, and the 31K species is the repressor expressed from the P_{3080} promoter (P. Lambert, N. Hubbert, P. Howley, and J. Schiller, J. Virol. (1989) in press). The structure of the E2 gene products are shown in Figure 17.2. The 31-kd repressor expressed from P_{3080} is in excess over the other E2 gene products [19]. Given the E2 dependence of several viral promoters, this ratio of repressor to *trans*-activator factors could account for the low level of viral transcription observed in BPV-1–transformed cells.

E2 DNA Binding Site

The E2 *trans*-activator is a DNA-binding protein with specificity for a 12-bp sequence, $ACCN_6GGT$, which is repeated several times in the BPV-1 genome [20–22]. There are 12 copies of the E2 binding site, and 10 of these are located within the LCR; additional sites are adjacent to the 2443 and 3080 promoters (see Figure 17.1). The major E2 responsive element E2RE$_1$ has been finely mapped and shown to consist of a 196-bp fragment with two E2 DNA–binding motifs located at either end [14]. Two $ACCN_6GGT$ motifs are sufficient for E2-dependent *trans*-activation [23–25], but the motifs at either end of E2RE$_1$ cooperate to create a highly responsive element [26]. Therefore, the E2-dependent enhancer is similar to other enhancers in that it contains multiple sequence elements, but it is one of the simplest so far described. The precise region of DNA recognized and protected by the E2 protein has been analyzed by several methods, and the contact points of the protein on the $ACCN_6GGT$ motiff have been determined. The protein contacts and protects the guanosine residues on either strand of the motif, which are predicted to be present on the same face of the DNA helix [27].

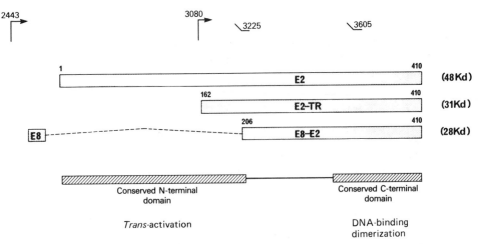

Figure 17.2. Structure of the BPV-1 E2 gene products. The E2 transactivator, consisting of 410 amino acids, is expressed from the P_{2443} promoter; the E2-TR is expressed from the P_{3080} promoter and utilizes a methionine at amino acid 162 of the ORF. The second repressor E8/E2 is encoded by a spliced message that most likely contains 11 amino acids from E8 spliced to the C-terminal half of E2 via the splice acceptor at nucleotide 3225. About 200 amino acids at the N terminus and 100 amino acids at the C terminus are relatively well conserved among papillomaviruses and correspond to functional domains. The *trans*-activation function has been mapped to the N-terminal domain, and the DNA-binding and dimerization properties to the C-terminal domain.

We have recently shown that the E2 protein binds this site as a dimer, which is consistent with this symmetrical binding. Additionally, E2 binding has been demonstrated to bend the DNA in a manner similar to SV40 T antigen and the *Drosophila* heat shock transcription factor [28].

Mechanisms of Repression

The E2 *trans*-activator most likely functions by directly binding to the E2-enhancer elements. The DNA-binding domain is located in the C-terminal 101 amino acids of the E2 *trans*-activator and is therefore present in both repressor proteins [22]. We have recently shown that the E2 proteins bind as a dimer to a single motif and that dimerization is mediated through the DNA-binding domain (A. McBride, J. Byrne, and P. Howley, DNAS 86:510–514 (1989)). There are several mechanisms by which the E2 transcriptional repressors might inhibit *trans*-activation: They could titrate out cellular factors required for *trans*-activation, or they could, by subunit mixing, form inactive complexes with the full-length E2 protein. Perhaps the simplest mechanism by which the repressors block E2 *trans*-activation is by binding competitively to the $ACCN_6GGT$ motif. Since all three E2 regulatory factors contain the C-terminal domain required for DNA binding and dimerization, the latter two models are likely mechanisms for E2 repression. Expression of the C-terminal, DNA-binding domain is sufficient to repress *trans*-activation, and so the role of the additional N-terminal sequences present in both repressors is not clear (P. Lambert, unpublished observations). Each repressor protein is expressed from a different promoter, and although the proteins may have distinct roles in viral gene regulation, this provides a mechanism by which the repressors themselves could be differentially regulated.

Structural and Functional Domains of the E2 *Trans*-Activator

A comparison of the predicted amino-acid sequence of the E2 ORFs of various papillomaviruses indicates that the E2 polypeptide consists of three regions [29,30] that likely reflect functional domains (see Figure 17.2). About 220 amino acids at the N terminus and 100 amino acids at the C terminus are relatively well conserved among papillomaviruses; the internal region is variable in length and in amino-acid composition. The DNA-binding domain is contained in the 101 C-terminal amino acids of the BPV-1 E2 ORF [22] and is not predicted to contain helix–turn–helix structures nor "zinc fingers," which are features characteristic of many other DNA-binding proteins. This domain has some limited homology with the cellular oncogene *mos* [29], but the functional significance of this similarity has yet to be established. The N-terminal conserved region, which is present only in the full-length E2 *trans*-activator encodes the transcriptional activation domain (A. McBride, J. Byrne, and P. Howley, DNAS 86:510–514 (1989); I. Giri and M. Yaniv, EMBO J. 7:2823:2829 (1988)). It is becoming apparent that in many eukaryotic *trans*-activators the sequences important for activation include short stretches of acidic residues. Moreover, the structure of these regions may be more important than their overall charge. One such structure may be an amphipathic helix, one face of which is hydrophobic and another of which bears the acidic residues [31]. The N-terminal domain of the BPV-1 E2 *trans*-activator is large in comparison to these structures, but it does contain two regions within the first 85 residues that would be predicted to form amphipathic helices and to play an important role in transcriptional activation. No function has yet been assigned to the variable internal region that can be deleted with little or no effect on *trans*-activation, although its function may be to provide flexibility between the activation and DNA-binding domains (A. McBride, J. Byrne, and P. Howley, DNAS 86:510–514 (1989)). Thus, it appears that the full-length E2 gene product consists of two independent functional domains linked by a "hinge" region, a structure found in other DNA-binding transcriptional regulatory proteins, such as the yeast *trans*-activator GCN4 and the steroid hormone receptors [32,33].

E2 Trans-Regulation in Other Papillomaviruses

The E2 binding motif $ACCN_6GGT$ is found in the genome of each of the papillomaviruses that have been sequenced [34], and the E2 products of all the papillomaviruses so far examined have been shown to have transcriptional *trans*-activation properties [25,35–39]. Moreover, it seems that factors analogous to BPV-1 transcriptional repressor may also be expressed by other viruses [40]. In those cases studied, the E2 gene product of one papillomavirus is able to *trans*-activate the E2-dependent enhancer of another, indicating that both the binding site and the E2 proteins are functionally conserved among papillomaviruses [35,36,39]. There is, however, a different arrangement of the $ACCN_6GGT$ motifs in the genomes of different classes of papillomaviruses. For example, the distribution of motifs is similar among the animal papillomaviruses that cause fibropapillomas, and another pattern is found in those papillomaviruses associated with anogenital tract lesions. This may indicate that the regulatory pathways involving the E2 factors varies among the different viruses and may play a role in the pathogenesis of viral infection. It also appears that in some circumstances the full-length E2 *trans*-activators may act as repressors. For example, binding of BPV-1 E2 to a site adjacent to the P_{7185} decreases transcription from this promoter [41] and similarly, transcription in HPV18 can be inhibited in the presence of E2 *trans*-activators [42]. In these cases it is most likely that the $ACCN_6GGT$ motifs overlap the binding sites for cellular transcription factors. Other E2-independent transcriptional regulatory elements, which are constitutive, cell-type specific, or glucocorticoid responsive, have been mapped in the papillomavirus LCRs [25,37,43–46]. Thus, the LCRs of the papillomaviruses appear to contain complex regulatory elements. In human cervical carcinomas and their derived cell lines, the HPV genomes are often found integrated, usually in the E1 or E2 ORFs, and the sequences of these ORFs are sometimes deleted. This disruption is thought to play a role in malignant progression, as viral gene expression is no longer under the control of the E2 regulatory factors.

Conclusion

Many questions remain to be answered about E2-dependent *trans*-activation and about eukaryotic gene regulation in general. For example, by what mechanism does the conserved amino-terminal domain activate transcription? Does it interact with other viral or cellular transcription factors? Does each E2 repressor protein play a different role in viral gene regulation? What is the function of the many different E2-binding motifs, and what is the significance of their position and spacing in the viral genome? The papillomavirus system provides an ideal model for the study of both eukaryotic and viral gene regulation.

References

1. Howley PM, Schlegel R (1987) Papillomavirus transformation. *In* Salzman NP, Howley PM (eds) The Papovaviridae 2: The Papillomaviruses. Plenum Press, New York, pp 141–166
2. Sarver N, Rabson MS, Yang YC, Byrne JC, Howley PM (1984) Localization and analysis of bovine papillomavirus type 1 transforming functions. J Virol 52:377–388
3. Waldeck S, Rosl F, Zentgraf H (1984) Origin of replication in episomal bovine papilloma virus type 1 DNA isolated from transformed cells. EMBO J 3:2173–2178
4. Lusky M, Botchan MR (1984) Characterization of the bovine papilloma virus plasmid maintenance sequences. Cell 36:391–401
5. Baker CC, Howley PM (1987) Differential promoter utilization by the bovine papillomavirus in transformed cells and productively infected wart tissues. EMBO J 6:1027–1035
6. Yang Y-C, Okayama H, Howley PM (1985) Bovine papillomavirus contains multiple transforming genes. Proc Natl Acad Sci USA 82:1030–1034
7. Stenlund A, Zabielski J, Ahola H, Moreno-Lopez J, Pettersson U (1985) Messenger RNAs from the transforming region of bovine papilloma virus type 1. J Mol Biol 182:541–554
8. Rabson MS, Yee C, Yang Y-C, Howley PM (1986) Bovine papillomavirus type 1 3' early region transformation and plasmid maintenance functions. J Virol 60:626–634
9. Groff DE, Lancaster WD (1986) Genetic analysis of the 3' early region transformation and replication functions of bovine papillomavirus type 1. Virology 150:221–230
10. DiMaio D (1986) Nonsense mutation in open reading frame E2 of bovine papillomavirus DNA. J Virol 57:475–480
11. Hermonat PL, Howley PM (1987) Mutational analysis of the 3' open reading frames and the splice junction at nucleotide 3225 of bovine papillomavirus type 1. J Virol 61:3889–3895
12. DiMaio D, Settleman J (1988) Bovine papillomavirus mutant temperature sensitive for transformation, replication and transactivation. EMBO J 7:1197–1204
13. Spalholz BA, Yang Y-C, Howley PM (1985) Transactivation of a bovine papillomavirus transcriptional regulatory element by the E2 gene product. Cell 42:183–191
14. Spalholz BA, Lambert PF, Yee CL, Howley PM (1987) Bovine papillomavirus transcriptional regulation: Localization of the E2-responsive elements of the long control region. J Virol 61:2128–2137
15. Haugen TH, Cripe TP, Ginder GD, Karin M, Turek LP (1987) Trans-activation of an upstream early gene promoter of bovine papilloma virus 1 by a product of the viral E2 gene. EMBO J 6:145–152

16. Hermonat PL, Spalholz BA, Howley PM (1988) The bovine papillomavirus P_{2443} promoter is E2 trans-responsive: Evidence for E2 autoregulation. EMBO J 7:2815–2822

17. Lambert PF, Spalholz BA, Howley PM (1987) A transcriptional repressor encoded by BPV-1 shares a common carboxy terminal domain with the E2 transactivator. Cell 50:69–87

18. Lambert PF, Spalholz BA, Howley PM (1987) Evidence that bovine papillomavirus type 1 may encode a negative transcriptional regulatory factor. In Steinberg BM, Brandsma JL, Taichman LB (eds) Papillomaviruses. Cold Spring Harbor Laboratory, Cold Spring Harbor, New York, pp 15–22

19. Hubbert NL, Schiller JT, Lowy DR, Androphy EJ (1988) Bovine papillomovirus transformed cells contain multiple E2 proteins. Proc Natl Acad Sci USA 85:5864–5868

20. Androphy EJ, Lowy DR, Schiller JT (1987) Bovine papillomavirus E2 trans-acting gene product binds to specific sites in papillomavirus DNA. Nature 325:70–73

21. Moskaluk C, Bastia D (1987) The E2 "gene" of bovine papillomavirus encodes an enhancer-binding protein. Proc Natl Acad Sci USA 84:1215–1218

22. McBride AA, Schlegel R, Howley PM (1988) The carboxy-terminal domain shared by the bovine papillomavirus E2 transactivator and repressor proteins contains a specific DNA binding activity. EMBO J 7:533–539

23. Hawley-Nelson P, Androphy EJ, Lowy DR, Schiller JT (1988) The specific DNA recognition sequence of the bovine papillomavirus E2 protein in an E2-dependent enhancer. EMBO J 7:525–531

24. Harrison SM, Gearing KL, Kim SY, Kingsman AJ, Kingsman SM (1987) Multiple cis-active elements in the long control region of bovine papillomavirus type-1 (BPV-1). Nucleic Acids Res 15:10267–10284

25. Hirochika H, Hirochika R, Broker TR, Chow LT (1988) Functional mapping of the human papillomavirus type 11 transcriptional enhancer and its interaction with the trans-acting E2 protein. Genes Develop 2:54–67

26. Spalholz BA, Byrne JC, Howley PM (1988) Evidence for co-operativity between E2 binding sites in the E2 trans-regulation of bovine papillomavirus type 1. J Virol 62:3143–3150

27. Moskaluk C, Bastia D (1988) Interaction of the BPV-1 E2 transcriptional control protein with the viral enhancer: Purification of the DNA binding domain and analysis of its contact points with DNA. J Virol 62:1925–1931

28. Moskaluk C, Bastia D (1988) DNA bending is induced in an enhancer by the DNA-binding domain of the bovine papillomavirus E2 protein. Proc Natl Acad Sci USA 85:1826–1830

29. Giri I, Danos O, Yaniv M (1985) Genomic structure of the cottontail rabbit (Shope) papillomavirus. Proc Natl Acad Sci USA 82:1580–1584

30. Baker CC (1987) Sequence analysis of papillomavirus genomes. In Salzman NP, Howley PM (eds) The Papovaviridae. 2. The Papillomaviruses. Plenum Press, New York, pp 321–385

31. Giniger E, Ptashne M (1987) Transcription in yeast activated by a putative amphipathic alpha helix linked to a DNA binding unit. Nature 330:670–672

32. Hope IA, Mahadevan S, Struhl K (1988) Structural and functional characterization of the short acidic transcriptional activation region of yeast GCN4 protein. Nature 333:635–640

33. Evans, RM (1988) The steroid and thyroid hormone receptor superfamily. Science 240:889–895
34. Dartmann K, Schwarz E, Gissmann L, zur Hausen H (1986) The nucleotide sequence and genome organization of human papilloma virus type 11. Virology 151:124–130
35. Phelps WC, Howley PM (1987) Transcriptional transactivation by the human papillomavirus type 16 E2 gene product. J Virol 61:1630–1638
36. Hirochika H, Broker TR, Chow LT (1987) Enhancers and trans-acting E2 transcriptional factors of papillomaviruses. J Virol 61:2599–2606
37. Cripe TP, Haugen TH, Turk JP, Tabatabai F, Schmid III PG, Durst M, Gissmann L, Roman A, Turek LP (1987) Transcriptional regulation of the human papillomavirus-16 E6-E7 promoter by a keratinocyte-dependent enhancer, and by viral E2 trans-activator and repressor gene products: Implications for cervical carcinogenesis. EMBO J:3745–3753
38. Gius D, Grossman S, Bedell MA, Laimins LA (1988) Inducible and constitutive enhancer domains in the noncoding region of human papillomavirus type 18. J Virol 62:665–672
39. Giri I, Yaniv M (1988) Study of the E2 gene product of the cottontail rabbit papillomavirus reveals a common mechanism of transactivation among papillomaviruses. J Virol 62:1573–1581
40. Chin MT, Hirochika R, Hirochika H, Broker TR, Chow LT (1988) Regulation of the human papillomavirus type 11 enhancer and E6 promoter by activating and repressing proteins from the E2 open reading frame: Functional and biochemical studies. J Virol 62:2994–3002
41. Stenlund A, Bream GL, Botchan MR (1987) A promoter with an internal regulatory domain is part of the origin of replication in BPV-1. Science 236:1666–1671
42. Thierry F, Yaniv M (1987) The BPV1-E2 trans-acting protein can be either an activator or a repressor of the HPV 18 regulatory region. EMBO J 6:3391–3397
43. Rando RF, Lancaster WD, Han P, Lopez C (1986) The noncoding region of HPV-6VC contains two distinct transcriptional enhancing elements. Virology 155:545–556
44. Gloss B, Bernard HU, Seedorf K, Klock G (1987) The upstream regulatory region of the human papilloma virus-16 contains an E2 protein-independent enhancer which is specific for cervical carcinoma cells and regulated by glucocorticoid hormones. EMBO J 6:3735–3743
45. Swift FV, Bhat K, Younghusband HB, Hamada H (1987) Characterization of a cell type-specific enhancer found in the human papilloma virus type 18 genome. EMBO J 6:1339–1344
46. Thierry F, Heard JM, Yaniv M (1987) Characterization of a transcriptional promoter of human papillomavirus 18 and modulation of its expression by simian virus 40 and adenovirus early antigens. J Virol 61:134–142

Transgenic Mice: Expression of Viral Genes

CHAPTER 18

Expression of SV40 Early Region Genes in Transgenic Mice

ARNOLD J. LEVINE

In attempting to understand how and why viral agents cause pathological changes in their host, virologists have commonly found that both the route of introduction of the virus and the nature of the host or the host cell play a critical role. This is well illustrated by a series of observations made over the past 28 years with simian virus 40 (SV40). SV40 was originally isolated from rhesus monkey kidney cells where it is carried in high concentration without any deleterious effects on the cells in culture or the monkeys that harbor this virus [1]. Newborn hamsters injected with SV40, however, develop a wide variety of tumors (carcinomas, sarcomas, lymphomas, or leukemias), depending on the site of injection [2]. This virus was a contaminant of the early poliovirus vaccines produced in rhesus monkey cells, and it was introduced into millions of children in schools and adults in the armed forces with no clear cases of tumors or disease in the human population resulting from this exposure, with one possible exception [3]. The injection of SV40 into newborn or adult mice, which are nonpermissive for virus replication, has not resulted in any reported examples of tumors or disease. The introduction of these same viral genes into mice via the germ line (as a transgene), however, produces a wide variety of tumors depending on the *cis*-acting regulatory signals controlling transcription of the SV40 oncogenes in transgenic mice [4–12].

Enhancer–Promoter Control of the SV40 T Antigen Genes in Transgenic Mice

Transgenic mice that harbor the SV40 early region genes for the large T and small t antigens directed by the SV40 enhancer and promoter region reproducibly develop pappillomas of the choroid plexus [4,11]. Deletion of the SV40 enhancer reduces the incidence of tumors and those tumors that are observed are less frequently of choroid plexus origin [9]. The SV40 large T antigen gene, but not the small t antigen gene, is required for tumorigenesis [9]. When different enhancer–promoter signals are used, the SV40 T antigen oncogene can produce tumors of a variety of tissues: (1) The insulin promoter directs tumors to the beta cells of the pancreas [5]; (2) the elastase promoter directs tumors to the acinar cells of the pancreas [6]; (3) the alpha lens crystalline gene promoter directs tumors to the lens of the eye [7]; (4) the albumin enhancer–promoter with the T antigen oncogene produced hepatomas [8], and the prolactin promoter directed tumors to the pituitary cells [8]. Hybrid enhancer–promoter constructions often give confusing results. The immunoglobulin, heavy-chain gene enhancer with the SV40 T antigen promoter directs the SV40 oncogene in transgenic mice to produce choroid plexus papillomas, histocytic tumors, and B cell lymphomas (along with some T cell lymphomas) [12]. An SV40 T antigen promoter (no enhancer) adjacent to a metallothionein (MT)-human growth hormone gene, directed tumors to the liver and the beta cells of the pancreas, as well as inducing a peripheral neuropathy due to Schwann cell degeneration [10]. The SV40 enhancer–promoter directing T antigen, present in the same plasmid with the human growth hormone genes, induced a thymic hyperplasia not observed when the growth hormone gene was absent [4]. Completely switching the enhancer–promoter signals and using the murine cytomegalovirus immediate early gene 1 enhancer–promoter to direct the SV40 T oncogenes resulted in five independent transgenic mice with papillomas of the choroid plexus (J. Marks and A.J. Levine, unpublished observations).

Clearly SV40 T antigen can initiate tumors in many different tissues. In several cases it is clear that the enhancer–promoter signals are directing this tissue pathology. The repeated examples of choroid plexus tumors [4,11,12] might suggest that the brain is a particularly susceptible target. The appearance of a tumor early in life usually results in death. This precludes further investigations of tumor incidence with long latency times in these same mice.

SV40 Enhancer–Promoter Control of Viral Oncogene Expression

In the first experiments in which the SV40 T antigen genes, under the regulation of the viral enhancer–promoter, were introduced into animals, 25 different transgenic animals were produced, with 16 dying of choroid plexus tumors; and the remaining 9 founder mice had no pathology [4]. One of these 9 mice (number 419) has been examined in more detail. The progeny of this mouse, the 419 family, failed to express SV40, mRNA, or protein in any tissue [13]. The DNA of this transgene was heavily methylated at cytosine residues, and this pattern was inherited [13]. A rare mouse in this lineage did develop a papilloma of the choroid plexus that was expressing T antigen, and the DNA was undermethylated at the cytosine residues [13]. Cells derived from a variety of tissues in these 419 mice expressed the SV40 T antigen when they were maintained in culture (in vitro). It seemed likely then that the 419 family of mice carry a wild-type T antigen oncogene that is not expressed in vivo and that the site or position of integration played a role in regulating expression of this transgene. Families derived from the other founder animals that did die of a papilloma of the choroid plexus; all expressed T antigens in the tumor tissue and in some cases in the kidney and thymus. In these mice viral mRNA or protein was not expressed in other tissues [4]. When T antigen was expressed in the kidney, polycystic kidney disease was observed, and in some cases this developed into carcinomas of the kidney [4]. When T antigen was expressed in the thymus a hyperplasia of the thymic epithelia was observed [4].

The SV11 Mouse Family: Life History

The SV11 founder mouse (SV40 enhancer, promoter, and a small t antigen deletion [9]) produced a family that was studied in some detail [14]. Eight to ten copies of the SV40 transgene were integrated in tandem at a single site in the mouse genome. The SV40 T antigen gene is expressed during the first 2 weeks after birth, as detected in a Western blot of brain tissue [14]. During the first month of life, no abnormal pathology was detected, but from 36 to 45 days, almost every mouse began to develop multiple foci of anaplastic cells (5 to 20 cells per focus) in the choroid plexus. These foci commonly contained dividing cells never observed in normal mouse choroid plexus. By 77 to 91 days, these foci began to grow rapidly, and the tumors could fill the ventricle of the brain. Mice in this family died in a fairly narrow time span of 104 ± 12 days (for 119 mice bred over 10 backcross generations with C57Bl/6J mice). The levels of the SV40 T oncogene mRNA and antigen also appeared to increase (on a per cell basis and because tumor mass increases) with time. T antigen genes were apparently not expressed in all choroid plexus cells but rather became activated in a few cells to form the multiple foci. Alternatively, T antigen may be produced in all choroid plexus cells at a very low level (below the level of detection), and a second, independent event resulted in the production of multiple foci. It appears likely that one or more of the foci expand clonally and cells expressing higher levels of T antigen are selected for in the tumor tissue. Of the animals that inherit the transgene, 100% die of papillomas of the choroid plexus.

The Effects of Gene Dosage and Genetic Background on Tumor Formation

When SV11 heterozygotes (T +/−) in a C57B1/6 genetic background were mated, three categories of mice (T +/+) homozygotes, (T +/−) heterozygotes, and (T −/−) negatives are seen in the same litter. The approximate 1 : 2 : 1 ratio shows no penetrance problems and no fetal lethality. However, T +/+ homozygotes developed neurological symptoms and loss of weight 15 to 30 days earlier than their heterozygote (T +/−) littermates. Of the homozygotic mice, 50% died at 90 days; of the heterozygotic mice, 50% died at 108 days. In general, heterozygotic mice contained one tumor of the choroid plexus in the fourth ventricle or either of the two lateral ventricles. Homozygotic (T +/+) mice with multiple tumors in several ventricles were observed. Homozygous (T +/+) mice were runted and had a tufted stripe hair pattern across their ventral surface. Some of these phenotypes may be attributable to recessive mutations caused by the inserted DNA in SV11 mice (J. Marks and A.J. Levine, unpublished observations).

When SV11 mice, which are in a C57B1/6J genetic background, were crossed with mice from several diverse genetic backgrounds, tumor appearance and time of death in the resultant F_1 hybrid mice that carried the SV40, large T antigen transgene varied widely. The times that 50% of the mice died in crosses were as follows:

Cross	Time of death (days)
C57B1/6J	108
C57B1/10J	109
SJL/J	125
AKR	145
NZW/lacJ	180

Clearly, different genetic backgrounds determine when tumor pathogenesis occurs. In a C57B1/6J background, all the mice were dead by 125 days; in C57B1/6J × NZW/lacJ F-1 hybrids, all mice were alive at 125 days with few or no symptoms of the tumor. The mechanisms that confer a delay in turmoigenesis could be immune mediated (NZW/lacJ contributes an immune response gene) or involve some negative control over expression of the SV40, large T antigen oncogene. Further studies will be required to choose between these alternatives.

The Differentiated State of Choroid Plexus Papillomas

Histologically, SV40-induced papillomas retain the multi-villi, epitheliod appearance of well-differentiated choroid plexus cells, and tumors express a number of gene products that are specific for or partially limited to choroid plexus cells.

The mRNA and receptor protein for the 5-HT$_{1C}$ serotonin receptor gene is also expressed in these tumors [15,16]. Indeed, the cDNA cloning for this mRNA species is derived from a library made from the mRNA from this tumor [17]. The mRNA for transthyretin, a protein that aids in the transport of vitamin A and thyroid hormone across the blood-brain barrier, is found in abundance in these tumors [16], as are the p53 oncogene mRNA and protein and the erb-A oncogene (thyroid hormone receptor) [16]. Upon rare occasions, less-differentiated portions of a tumor invade normal brain tissue and become malignant carcinomas [14].

It appears likely that there are multiple steps in the development of this tumor. The SV40 transgene is expressed within the first 2 weeks of life at very low levels in some cells of the choroid plexus. Little or no tissue pathology results. By 36 to 45 days, foci of 5 to 20 anaplastic cells develop at multiple sites in the choroid plexus epithelia. At 90 days, one or more of these foci grow rapidly and fill the ventricle with a well-differentiated tumor. An increased oncogene dosage accelerates these events, and tumors in older mice express more T antigen per cell than the earlier foci [14]. There appears to be a selection for increased levels of T antigen and secondary events that permit or enhance tumor growth. The genetic background of the host animal can have a dramatic effect on the timing of these events. In some cases well-differentiated papillomas develop into poorly differentiated carcinomas, possibly through the selection of further oncogene-activating events.

Acknowledgments

The author thanks J. Marks and M. Moore for their critical reading and advice, A.K. Teresky for her technical assistance, and K. James for help in preparation of this manuscript. The work reported and reviewed here was supported by NIH Grants 1 P01-CA41086 and R01-CA38757.

References

1. Sweet BH, Hellman MR (1960) The vaculating virus, SV40. Proc Soc Expt Biol Med 105:420–425
2. J. Tooze (ed) (1981) The Molecular Biology of Tumor Viruses: The DNA Tumor Viruses, 2nd ed, rev, Cold Spring Harbor Press, New York, p 208
3. Heinonen OP, Shapiro S, Mouson RR, Hartz SC, Rosenberg L, Sloan D (1973) Immunization during pregnancy against poliomyelitis influenza in relation to childhood malignancy. Int J Epidemiol 2:229–231
4. Brinster RL, Chen HY, Messing A, van Dyke T, Levine AJ, Palmiter RL (1984) Transgenic mice harboring SV40 T antigen genes develop characteristic brain tumors. Cell 37:367

5. Hanahan D (1985) Cell specific expression of recombinant insulin/SV40 onco-genes produces beta cell tumors in transgenic mice. Nature 315:115

6. Ornitz DM, Palmiter RD, Messing A, Hammer RE, Pinkert CA, Brinster RL (1985) Elastase I promoter directs expression of human growth hormone and SV40 T antigen genes to pancreatic acinar cells in transgenic mice. Cold Spring Harbor Symp Quant Biol 50:399–409

7. Malion KA, Chepelinsky AB, Khillan JS, Overbeck PA, Piatigorsky J, Westphal H (1987) Oncogenesis of the lens in transgenic mice. Science 325:1622–1628

8. Palmiter RD, Brinster RL (1986) Germ-line transformation of mice. Ann Rev Genet 20:465–499

9. Palmiter RD, Chen HY, Messing A, Brinster RL (1985) SV40 enhancer and large T antigen are instrumental in development of choroid plexus tumors in transgenic mice. Nature 316:457–460

10. Messing A, Chen HY, Palmiter R, Brinster RL (1985) Peripheral neuropathies, hepatocellular carcinomas and islet cell adenomas in transgenic mice. Nature 316:461–463

11. Small JA, Blair DG, Showalter SD, Scangos GA (1985) Analysis of a transgenic mouse containing simian virus 40 and v-myc sequences. Mol Cell Biol 5:642–648

12. Suda Y, Aizawa S, Hirai S, Inoue T, Furuta Y, Suzuki M, Hirohashi S, Ikawa Y (1987) Driven by the same Ig enhancer and SV40 T promoter ras induced lung adenomatious tumors, myc induced pre-B-cell lymphomas and SV40 large T gene a variety of tumors in transgenic mice. EMBO J 6:4055–4065

13. van Dyke T, Finlay C, Levine AJ (1985) A comparison of several lines of transgenic mice containing the SV40 early genes. Cold Spring Harbor Symp Quant Biol 50:671–678

14. van Dyke TA, Finlay C, Miller D, Marks J, Lozano G, Levine AJ (1987) Relationship between SV40 large tumor antigen expression and tumor formation in transgenic mice. J Virol 61:2029–2032

15. Yagaloff KA, Lozano G, van Dyke T, Levine AJ, Hartig PR (1986) Serotonin 5-HT$_{1C}$ receptors are expressed at high density on choroid plexus tumors from transgenic mice. Brain Res 385:389–394

16. Marks J, Lin J, Miller D, Lozano G, Herbert J, Levine AJ (1987) The expression of viral and cellular genes in papillomas of the choroid plexus induced in transgenic mice. In Harris SE (ed) 3rd Int Symp on Cellular Endocrinology. Cell factors in development and differentiation—Embryos, teratocarcinomas and differentiated tissues. Alan R. Liss, New York

17. Lubbert H, Hoffman BJ, Snutch TP, van Dyke T, Levine AJ, Hartig PR, Lester HA, Davidson N (1987) cDNA cloning of a serotonin 5-HT$_{1C}$ receptor using electrophysiological assays of mRNA injected xenopus oocytes. Proc Natl Acad Sc USA 84:4332–4336

CHAPTER 19
Analysis of Hepatitis B Virus Gene Expression in Transgenic Mice

FRANCIS V. CHISARI

The hepatitis B virus (HBV) is a small, enveloped DNA virus (hepadnavirus) that replicates, primarily, in hepatocytes, causes acute and chronic hepatocellular injury, and is associated with the development of primary hepatocellular carcinoma (PHC). Similar viruses have been found in several nonhuman species, including the woodchuck [1], the ground squirrel [2], and the Pekin duck [3]. All hepadnaviruses have the same general genomic organization (for review, see ref. 4) characterized by four open reading frames (ORF) (envelope, nucleocapsid, X, and polymerase), which encode at least seven different polypeptides.

The contribution of these animal models to hepadnavirus biology has been enormous. For example, much of what is known about the viral replication strategy (for review, see ref 5) has derived from studies in the animal models. Hepadnaviruses replicate by transcription of a nuclear pool of covalently closed, circular, double-stranded HBV DNA to produce a greater-than-genome-length RNA pregenome. The RNA pregenome is then reverse-transcribed in the cytoplasm by a virus-specified polymerase to produce a single-stranded DNA copy (minus strand). The pregenomic RNA is degraded, and plus strand DNA synthesis ensues. The resultant, partially double-stranded, relaxed circular DNA is then either packaged into an infectious virus particle and exported from the cell or it is transported to the nucleus where it joins the pool of covalently closed circular DNA that serves as the transcriptional template for the RNA pregenome. This pool of nuclear HBV DNA (approximately 50 copies per cell in persistently infected hepatocytes) is also the presumed origin of the viral sequences that are found randomly integrated into the host genome in chronic HBV infection of the liver and in PHC.

Despite such advances the mechanisms responsible for acute and chronic

Table 19.1. Transgenic mouse model: Viral biology and pathogenesis

Tissue specificity of viral regulatory sequences
Synthesis, processing, transport, and secretion of viral gene products
Minimal requirements for virus particle assembly
Cytopathic and transforming potential of viral gene products
Pathogenetic potential of immune response to viral gene products
Mechanisms of immunological tolerance to cellular and secreted viral gene products (autoantigens)
Model of viral replication to further define viral life cycle and to assess new strategies for antiviral therapy

hepatocellular injury in HBV infection, and the events leading to the development of PHC, are unknown. With the development of mouse embryo microinjection technology, it became apparent that these questions and others might be addressed by creating a model in which integrated copies of the HBV genome or its individual genes are present and expressed and transmitted in genetically defined strains of mice. Such a transgenic mouse model can serve to address an assortment of unsolved problems related to viral biology and pathogenesis including those listed in Table 19.1.

In an effort to examine some of these problems, at least three groups of investigators have independently generated several lineages of transgenic mice containing HBV sequences. Using constructs containing only HBV-derived regulatory sequences, Pourcel [6] and Burk [7] and their colleagues have produced several lineages of transgenic mice that demonstrate preferential synthesis in the liver a dominant 2.1-kb mRNA species corresponding to the transcript that encodes the major and the middle-envelope polypeptides of HBV. These results indicate that HBV contains *cis*-acting regulatory elements that are responsible for the predominant expression of HBV in the liver, and they help to explain the relative tissue specificity of the virus during natural infection. Nonetheless, recent evidence from Pourcel's group [8] suggests that other factors, possibly at the level of viral entry, are also important determinants of the species and tissue specificity of HBV. Following microinjection of a construct containing a head-to-tail dimer of the entire HBV genome, they obtained evidence of HBV gene expression and DNA replicative intermediates in the liver, kidney, and heart of one transgenic mouse [8]. Taken together with mounting evidence of extrahepatic virus expression in each of the hepadnavirus models (for review, see ref. 9), these data strongly suggest that the relative liver specificity of HBV is a reflection of multiple constraints at the levels of viral entry, replication, and gene expression and that none of these constraints is absolutely specific for the human hepatocyte.

One of the characteristics of HBV infection is its relative prevalence among males. It is not clear, however, whether this is due to a greater incidence of high risk behavior among males or to other factors (e.g., hormonal). Farza et al. [10] have shown that HBV envelope expression is much greater in male than in female transgenic mice, and that castration in both sexes causes a decline in expression that is reversible by the adminis-tration of androgens or estrogens. Furthermore, the Pourcel group [10] and our group [11] have shown that in transgenic mice HBV gene expression is also regulated by glucocorticoids as had been previously demonstrated in transfected cell lines [12]. Thus, there is abundant evidence from this system that HBV gene expression is hormonally regulated and one might conclude, from this datum, that the male prevalence of HBV infection in the natural setting is at least partially determined at this level.

Our own studies [11,13,14] have focused primarily on questions related to HBV envelope particle assembly and the pathogenesis of liver cell injury as a direct or indirect consequence of HBV gene expression. Thus far we have concentrated on the HBV envelope region. We have used constructs in which the entire HBV envelope coding region has been placed under the transcriptional control of relatively liver-specific regulatory elements that are either inducible (metallothionein, MT) or highly constitutively active (albumin, Alb) in the mouse. The advantages of this system relate to the ability to study expression of the large envelope polypeptide of HBV that is not expressed by the other transgenic lineages derived thus far.

The large envelope polypeptide is the product of the entire HBV envelope ORF (see review, ref. 4). It contains 389 amino acids (*ayw* subtype) that define three antigenic domains. The NH_2-terminal domain of the *ayw* viral subtype we used consists of 108 amino acids containing the preS(1) antigen. The middle domain consists of 55 amino acids containing the preS(2) antigen. The carboxy-terminal domain consists of 226 amino acids contain-ing the hepatitis B surface antigen (HBsAg). Translation start condons at the beginning of the three domains define the NH_2 termini of the large, middle, and major envelope polypeptides, respectively. An internal HBV promoter just upstream of the middle domain start codon directs expression of the middle and major polypeptides, whereas the exogenous MT or Alb pro-moters direct expression of the large polypeptide in our system [11,14].

It is important to point out that during natural HBV infection all three envelope polypeptides are present on the surface of infectious virions and on noninfectious subviral filamentous forms in the circulation [15,16], whereas the large envelope polypeptide is not present on the subviral spherical forms (which represent over 90% of the circulating, HBV-derived particles) in the absence of viral replication. It thus appears, from polypeptide analysis of viral particles purified from infected serum, that the large envelope polypep-tide influences particle structure and secretion in some poorly defined fashion. Our transgenic mouse studies have helped to elucidate the molecu-lar basis for these events.

We demonstrated that the large envelope polypeptide exerts significant structural constraints on HBsAg particle formation [11]. When produced at a roughly equimolar ratio with respect to the major envelope polypeptide, the large envelope polypeptide prevents the formation of small spherical particles, which are readily secreted, and causes the formation of long, branching filaments (containing all three envelope polypeptides), which become trapped in the endoplasmic reticulum (ER) of the hepatocyte thereby inhibiting Hepatitis B Surface Antigen (HBsAg) secretion from the hepatocyte. Similar observations have been made by other investigators in cell culture systems [17–21]. From these observations, it is conceivable that the structure and the relative rarity of complete viral particles and subviral filamentous forms in the circulation during HBV infection in man may relate, at least partially, to their content of the large envelope polypeptide.

Next we showed that the progressive accumulation of these filaments leads to expansion of the ER of the hepatocyte [11]. The storage of these filaments is reflected in ultrastructural and histological changes characteristic of "ground glass" hepatocytes that are observed in the liver of patients with chronic HBV infection [22]. The similarities between the human and the transgenic mouse ground glass cells are so strong that it is possible that they share a common molecular pathogenesis. If this is true, it implies that ground glass cell formation in natural infection may result from dysregulation of the expression of the HBV envelope proteins. In turn, this raises the possibility that the ground glass cell merely represents one point on a continuous spectrum and that some of the pathological manifestations of HBV infection might be due to aberrant regulation of HBV gene expression of a greater or lesser degree.

Along these lines we have shown that after prolonged storage in high concentrations, these long subviral filaments are directly cytotoxic to hepatocytes, initiating a lesion characterized by chronic hepatocellular necrosis and a secondary inflammatory response [14]. These data demonstrate that HBV has the potential to be directly cytotoxic to the hepatocyte under certain conditions and they raise the possibility that similar mechanisms might contribute to the pathogenesis of liver cell injury in natural infection. If this does occur, however, we suspect that it is relatively limited in scope, since overproduction of the large envelope polypeptide to this degree is not known to occur in natural infection.

The chronic hepatocellular injury observed in this model leads to a regenerative response characterized by multifocal nodular hyperplasia of HBsAg-negative hepatocytes, and these foci appear clonal histologically [14]. These observations suggest that the cells in these foci may have a selective growth advantage because they do not contain cytotoxic concentrations of HBsAg. Importantly, this microscopic lesion progresses to the formation of massive tumor nodules histologically characterized by nodular regenerative hyperplasia, which is a premalignant lesion [23].

We conclude from these observations that high-level expression of this single HBV gene product can initiate a sequence of events that mimics some of the features of HBV infection in man. Obviously, mice that express each of the remaining HBV gene products must also be studied, and these experiments are under way. Furthermore, immunological manipulation of this model by adoptive transfer of antigen-primed lymphoid cells from immunized syngeneic mice must be done to study the possible immuno-pathogenetic mechanisms involved in HBV-induced liver disease. Finally, the transgenic mouse model offers an excellent opportunity to examine mechanisms of immunological tolerance that may be relevant to a better understanding of the chronic HBV carrier state and also to the immune response in general.

It is important to emphasize that one of the strengths of the transgenic mouse model with respect to viral biology relates to the insights it provides into the potential properties of individual viral gene products in the context of the intact animal. As with any system, one must be careful not to overinterpret the observations. This mouse model is an extraordinarily powerful tool that, properly applied and carefully interpreted, can help to unravel many of the mysteries of host–virus interactions and viral patho-genesis.

Acknowledgments

I thank Pierre Filippi, Carl Pinkert, Richard Palmiter, and Ralph Brinster for collaboration in the production and analysis of transgenic mice. I thank Janette Sanders for manuscript preparation. This research was supported by NIH Grants CA40489, CA38635, AI20001, AI20720, HD17321, HD0912, and HD07155 and by North Atlantic Treaty Organization Grant 675/84 for International Collaboration. This is publication no. 5366BCR of the Research Institute of Scripps Clinic, La Jolla, California.

References

1. Summers J, Smolec J, Snyder R (1987) A virus similar to human hepatitis B virus associated with hepatitis and hepatoma in woodchucks. Proc Natl Acad Sci USA 75:4533–4537
2. Marion PL, Oshiro LS, Regnery DC, Scullard GH, Robinson WS (1980) A virus in Beechey ground squirrels which is related to hepatitis B virus of man. Proc Natl Acad Sci USA 77:2941–2945
3. Mason WS, Seal S, Summers J (1980) Virus of Pekin ducks with structural and biological relatedness to human hepatitis B virus. J Virol 36:829–836
4. Tiollais P, Pourcel C, Dejean A (1985) The hepatitis B virus. Nature 317:489–495
5. Robinson WS, Miller RH, Marion PL (1987) Hepadnaviruses and retroviruses share genome homology and features of replication. Hepatology 7(1):64S–73S
6. Babinet C, Farza H, Morello D, Hadchouel M, Pourcel C (1985) Specific expression of hepatitis surface antigen (HBsAg) in transgenic mice. Science 230:1160–1163
7. Burk RD, DeLoia JA, ElAwady MK, Gearhart JD (1988) Tissue preferential expression of the hepatitis B virus (HBV) surface antigen gene in two lines of HBV transgenic mice. J Virol 62:649–654
8. Pourcel C, Hadchouel M, Scotto J, Tiollais P, Babinet C, Farza H (1987)

Replication of hepatitis B virus (HBV) in a transgenic mouse. Abstracts of papers presented at the 1987 meeting on Hepatitis B Viruses, Cold Spring Harbor Laboratory, Cold Spring Harbor, New York, p 92

9. Korba BE, Wells F, Tennant BC, Yoakum GH, Purcell RH, Gerin JL (1986) Hepadnavirus infection of peripheral blood lymphocytes *in vivo:* Woodchuck and chimpanzee models of viral hepatitis. J Virol 58:1–8

10. Farza H, Salmon AM, Hadchouel M, Moreau JL, Babinet C, Tiollais P, Pourcel C (1987) Hepatitis B surface antigen gene expression is regulated by sex steroids and glucocorticoids in transgenic mice. Proc Natl Acad Sci USA 84:1187–1191

11. Chisari FV, Filippi P, McLachlan A, Milich DR, Riggs M, Lee S, Palmiter RD, Pinkert CA, Brinster RL (1986) Expression of hepatitis B virus large envelope polypeptide inhibits hepatitis B surface antigen secretion in transgenic mice. J Virol 60:880–887

12. Tur-Kaspa R, Burk RD, Shaul Y, Shafritz DA (1986) Hepatitis B virus contains a glucocorticoid-responsive element. Proc Natl Acad Sci USA 83:1627–1631

13. Chisari FV, Pinkert CA, Milich DR, Filippi P, McLachlan A, Palmiter RD, Brinster RL (1985) A transgenic mouse model of the chronic hepatitis B surface antigen carrier state. Science 230:1157–1160

14. Chisari FV, Filippi P, Buras J, McLachlan A, Popper H, Pinkert CA, Palmiter RD, Brinster RL (1987) Structural and pathological effects of synthesis of hepatitis B virus large envelope polypeptide in transgenic mice. Proc Natl Acad Sci USA 84:6909–6913

15. Heermann KH, Goldmann U, Schwartz W, Seyffarth T, Baumgarten H, Gerlich WH (1984) Large surface proteins of hepatitis B virus containing the pre-s sequence. J Virol 52:396–402

16. Takahashi K, Kishimoto S, Ohnuma H, Machida A, Takai E, Tsuda F, Miyamoto H, Tanaka T, Matsushita K, Oda K, Miyakawa Y, Mayumi M (1986) Polypeptides coded for by the region pre-s and gene s of hepatitis B virus DNA with the receptor for polymerized human serum albumin: Expression on the hepatitis B particles produced in the HBeAg or anti-HBe phase of hepatitis B virus infection. J Immunol 136:3467–3472

17. Standring DN, Ou J-H, Rutter WJ (1986) Assembly of viral particles in Xenopus oocytes: Pre-surface–antigens regulate secretion of the hepatitis B viral surface envelope particle. Proc Natl Acad Sci USA 83:9338–9342

18. Ou J-H, Rutter WJ (1987) Regulation of secretion of the hepatitis B virus major surface antigen by the preS-1 protein. J Virol 61:782–786

19. Marquardt O, Heermann K-H, Seifer M et al (1987) Cell type specific expression of pre-S1 antigen and secretion of hepatitis B virus surface antigen. Arch Virol 96:249–256

20. Ou J-H, Rutter WJ (1987) Regulation of secretion of the hepatitis B virus major surface antigen by the pre-S1 protein. J Virol 61:782–796

21. Molnar-Kimber KL, Jarocki-Witek V, Dheer SK, Vernon SK, Conley AJ, Davis AR, Hung PP (1988) Distinctive properties of the hepatitis B virus envelope proteins. J Virol 62:407–416

22. Gerber MA, Hadziyannis S, Vissoulis C, Schaffner F, Paronetto F, Popper H (1974) Electron microscopy and immunoelectronmicroscopy of cytoplasmic hepatitis B antigen in hepatocytes. Am J Pathol 75:489–502

23. Edmondson HA (1958) Tumor-like lesions. *In* Tumors of the Liver and Intrahepatic Bile Ducts. Armed Forces Institute of Pathology, Washington, DC, pp 191–192.

CHAPTER 20
Expression of HTLV-1 in Transgenic Mice

Michael I. Nerenberg and Gilbert Jay

Transgenic technology has been used to determine the biological effects of individual viral gene products in an intact organism. This approach may minimize much of the pathogenic complexity seen during natural infection of animals as a result of variable targeting, differential expression of antigens, and the complicating effects of attack by the immune system. Specific transcriptional promoters may be chosen to target gene expression to certain tissues of interest or to broaden the repertoire of tissue-specific expression beyond what is commonly seen in natural infections. This technology may also be used to derive models of human disease and to create useful cellular and biochemical reagents for the investigation of viral pathogenesis.

We have used transgenic methodology to investigate the biological effects of the HTLV-1 *tax* gene and the pathogenetic role of HTLV-1 in human diseases.

Human Diseases Associated with HTLV-1

Human T lymphotropic virus type 1 (HTLV-1) is endemic in southwestern Japan, in the Caribbean basin, in Africa, and on the northern coast of South America (for review, see ref. 1). Although it occurs to a lesser extent in the southeastern United States, it has recently been detected in an alarming percentage of drug addicts and in 0.035% of blood donors [2] in this country. The virus has been associated with an adult form of T cell leukemia with several unusual characteristics. It has a low penetrance rate, with less than 0.1% of infected individuals developing leukemia [3], and a latency of 10 to 40 years [1]. Patients may either develop a highly malignant disease or follow a slow, smoldering course.

The virus has also been associated with a neurological disease known as tropical spastic paraparesis (TSP) in the Caribbean [4–6] and as HTLV-1–associated myelopathy (HAM) in Japan [7]. Affected patients frequently develop chronic disabling symptoms of spinal cord disease, primarily with motor nerve disorders. The presence of central nervous system (CNS) symptoms, HTLV-1 specific antibodies, and virus-infected lymphocytes in the cerebral lspinal fluid (CSF) of these individuals suggests that the virus may infect the nervous system as well as the lymphatic system [8].

Molecular Biology of HTLV-1

The provirus consists of approximately 9 kb of DNA and encompasses the *gag, pol,* and *env* regions, as do other mammalian retroviruses. In addition, there is a 3' open reading frame known as the X or LOR region that has been shown to encode three proteins. The *tax* or *pX* gene product is a 40-kd protein that localizes at least partially to the nucleus of transformed cells. It has also been shown to mediate transcriptional *trans*-activation of the viral long terminal repeat (LTR), as well as several cellular genes involved in the regulation of growth (interleukin-2, interleukin-2 receptor, and granulocyte macrophage colony-stimulating factor [9–14]. Thus, it was likely that the *tax* protein would have important biological effects in vivo and possibly be a transforming protein.

Studies of Tissue Tropism

Adult T cell leukemia (ATL) is characterized by a monoclonal proliferation of CD4$^+$ mature peripheral T cells, resulting from a common site of integration of the provirus. However, in the majority of cases, fresh peripheral leukemic cells express no detectable HTLV-1 antigens. Expression may be seen when these cells are stimulated and fused with other cells in culture. Thus, constant viral antigen expression may not be necessary to maintain the malignant phenotype in vivo.

Virus may enter a variety of cells by direct cell-to-cell fusion. This has been achieved in vitro for fibroblasts, lymphocytes, endothelial cells, and osteocytes [15–17]. However, the virus appears to immortalize only T cells. Since the virus is carried in the host by circulating T cells and cell-type restriction in infection has not been observed, infection of a variety of cells in vivo would be expected. Although CNS symptoms suggested tropism to the nervous system, a cellular target within the CNS has not been identified.

Transgenic Studies

The plasmid HTLV *tax*-1 was used for expression in transgenic mice. In this construct, the HTLV-1 LTR promoter was placed upstream of an intronless reconstruction of the *tax* coding region. This vector expresses a 40-kd protein that is indistinguishable from the protein encoded by infected human lymphocytes [18]. This *tax* protein also transactivates the HTLV-1 LTR when tested in vitro.

Initial transgenic studies were performed in collaboration with K. Reynolds, S. Hinrichs, and G. Khoury of the National Institutes of Health. The chimeric *tax* gene was microinjected into fertilized mouse oocytes, and transgenic mice that stably transmitted the *tax* gene were established. Four lines were created, and three were studied intensively. These mice varied in the site of integration of the *tax* gene, and in the level of expression of *tax* protein by several-fold. They were grossly indistinguishable from their littermates until approximately 3 months of age, at which time they developed grossly deforming, nerve sheath-associated tumors [19]. These mice died between 4 and 5 months of age (Figure 20.1).

Figure 20.1. Transgenic HTLV-*tax* 1 mouse. Multiple nerve sheath-associated tumors of extremities, tail, and ear can be seen, while massive tumors distort the left hind leg.

RNA and protein studies showed expression of the *tax* gene in muscle, peripheral nerve, salivary gland, and skin. This resulted in peripheral nerve tumors, muscle atrophy, and salivary gland tumors.

Peripheral nerve tumors were responsible for the deformities and were associated with nerve sheaths, mostly of the feet, ears, face, and tail. These are common sites of trauma in mice. They were less frequently seen in association with cranial or perispinous nerves, but they always occurred outside the CNS. The majority of tumors had a fibroblastic morphology, a low mitotic index, and appeared benign. Detailed histological and immunocytochemical characterization was performed in collaboration with Dr. C. Wiley (UCSD). In some regions of these tumors, the cells had prominent lysosomal granules and surface projections resembling histiocytes. Western blot and immunocytochemical analysis showed tumor cells to be positive for vimentin and factor XIIIA and negative for laminin. Ultrastructural analysis suggests that the proliferating cell resides in the endoneurium. These studies showed that these neural sheath tumors are composed of endoneurial fibroblasts and not of Schwann cells. Tumors were heavily infiltrated by polymorphonuclear granulocytes in the absence of obvious tumor necrosis. Prominent granulocytic proliferation also occurred in the bone marrow and spleens of animals with large tumor burdens. The overall appearance of the disease is similar to type 1 neurofibromatosis (NF) of humans (von Recklinghausen's disease), with several exceptions. We have not observed pigmentary changes in the skin or nodules in the iris of affected animals similar to those seen in human NF. We have not observed proliferation of Schwann cells, as is seen in approximately two-thirds of human NF cases.

Expression of the *tax* gene in muscle has been localized to oxidative skeletal fibers (i.e., type 1 and type 2a fibers), which lead to profound atrophy of these fibers. The type 2b fibers were spared. The *tax* protein appears to localize to the nuclei of proliferating cells and occurs exclusively in the cytoplasm of atrophic muscle fibers [20].

Implications

Restriction of tissue tropism during viral infection may occur through a number of mechanisms and at virtually every stage of the viral cycle. Important factors include the presence of viral receptors on the target cells and regulation of viral gene expression through transcriptional and posttranscriptional mechanisms. Because of the lack of in vitro evidence for restricted viral entry into cells, we chose to focus on the promoter-determined tropism of the virus. By placing the HTLV-1 *tax* gene under the control of its own transcriptional promoter and placing this transcription unit into every cell of the mouse, we were able to investigate this as well as the biological consequences of this protein in a number of tissues. Restriction of expression to the perineurium, and salivary duct cells, would not have been predicted based on in vitro studies or with known, human-associated disease. In contrast, recent evidence [21,22] suggests an association of HTLV-1 with muscle disease. It is possible that expression of other viral proteins not previously thought to affect transcription may play a role in determining tropism or that species differences between mouse and human may account for the different tropisms. However, in vitro studies suggest qualitatively similar patterns of transcription in mouse and human. Numerous transgenic experiments have suggested that species differences have little effect on tissue-specific expression of cellular genes. It is equally possible that association of HTLV-1 with diseases of organs affected in transgenic mice has not been recognized in humans. In this case transgenic studies could be used to provide clues to HTLV–associated diseases for which association in humans is difficult to recognize. This could occur if the more common syndromes of ATL and TSP obscured diagnosis of these syndromes. For example, muscular and peripheral nerve dysfunction may be thought to be secondary manifestations of leukemia. Alternatively, these syndromes may be caused less frequently by HTLV-1 than ATL and the epidemiological association missed. We are currently testing these hypotheses.

Whether or not an association of the virus is seen in similar human diseases, these mice provide a useful model and biochemical reagents for the study of molecular mechanisms in disease. Neural sheath tumors are a common malady of humans, and their study has suffered for lack of good animal models. Transgenic mice provide a reproducible way of tracing the steps leading to the genesis of these and other tumors. Since the mechanism of transcriptional *trans*-activation appears to be a likely mode of induction of pathogenesis, activation of cellular genes within affected tissues may be traced.

The ability of the *tax* protein to induce both proliferation and atrophy may allow links between these two apparently disparate processes to be elucidated. Are similar genes induced in the two tissues that have radically different effects, or are different sets of genes activated in the two tissues? The mechanism of tissue-specific differences in compartmentalization of this protein, and its relation to pathogenesis, may also be traced. Does tissue-specific compartmentalization lead to different pathological effects, or are the subcellular differences caused by another underlying pathogenic mechanism? What are the cellular signals that initiate these differences in compartmentalization? These are just some of the questions for which this transgenic system will provide invaluable reagents for analysis at the molecular level.

Acknowledgments

This is Publication no. 5290-IMM of the Scripps Clinic and Research Foundation, La Jolla, CA 92037. This research was supported by USPHS Training Grant AG-00080. We would like to thank Mrs. Gay Schilling for help in preparation of this manuscript.

References

1. Gallo RC (1986) The first human retrovirus. Sci Amer 255:88–98
2. Williams AE, Fang CT, Slamon DJ, Poiesz BJ, Sandler SG, Darr WF, Shulman G, McGowan EI, Douglas DK, Bowman RJ, Peetoom F, Kleinman SH, Lenes B, Dodd RY (1988) Seroprevalence and epidemiological correlates of HTLV-1 infection in U.S. blood donors. Science 240:643–646
3. Ito Y (1985) The epidemiology of human T-cell leukemia/lymphoma virus. Curr Top Microbiol Immunol 115:99–112
4. Gessain A, Vernant JC, Maurs L, Barin F, Gout O, Calendar A, DeThe G (1985) Antibodies to human T-lymphotropic virus type I in patients with tropical spastic paraparesis. Lancet ii:407–409
5. Rodgers-Johnson P, St O, Morgan C, Zaninovic V, Sarin P, Graham DS (1985) HTLV-I and HTLV-II antibodies and tropical spastic paraparesis. Lancet ii:1247–1248
6. Bartholomew C, Cleghaorn F, Charles W, Ratan P, Roberts L, Maharaj K, Janke N, Daisley H, Hanchard B, Blattner W (1986) HTLV-I and tropical spastic paraparesis. Lancet ii:99–100
7. Osame M, Usuku K, Izumo S, Ijichi N, Amitani H, Igata A, Matsumoto M, Tara M (1986) HTLV-I associated myelopathy, a new clinical entity. Lancet ii:1032
8. Jacobson S, Raine CS, Mingioli ES, McFarlin DE (1988) Isolation of an HTLV-1-like retrovirus from patients with tropical spastic paraparesis. Nature 331:540–543
9. Sodroski JG, Rosen CA, Haseltine WA (1984) Trans-acting transcriptional transactivation of the long terminal repeat of human T-lymphotropic viruses in infected cells. Science 225:381–385
10. Fujisawa J, Seiki M, Kiyokawa T, Yoshida M (1985) Functional activation of the long terminal repeat of human T-cell leukemia virus type I by a trans-acting factor. Proc Natl Acad Sci USA 82:2277–2281
11. Felber BK, Paskalis H, Kleinman-Ewing C, Wong-Staal F, Pavlakis GN (1985) The pX protein of HTLV-I is a transcriptional activator of its long terminal repeats. Science 229:675–679
12. Cann AJ, Rosenblatt JD, Wachsman W, Shaw NP, Chen ISY (1985) Identification of the gene responsible for human T-cell leukemia virus transcriptional regulation. Nature 318:571–574
13. Greene WC, Leonard WJ, Wano Y, Svetlik PB, Peffer NK, Sodroski JG, Rosen CA, Goh WC, Haseltine WC (1986) Trans-activator gene of HTLV-II induces IL-2 receptor and IL-2 cellular gene expression. Science 232:877–880
14. Maruyama M, Shibuya H, Harada H, Hatakeyama M, Seiki M, Fujita T, Inoue J, Yoshida M, Taniguchi T (1987) Evidence for aberrant activation of the interleukin-2 autocrine loop by HTLV-1 encoded p40 and T3/Ti complex triggering. Cell 48:343–350
15. Nagy K, Clapham P, Cheingsong-Popov R, Weiss R (1983) Human T-cell leukemia virus type I: Induction of syncytia and inhibition by patients sera. Int J Cancer 32:321–328
16. Ho DD, Rota TR, Hirsch MS (1984) Infection of human endothelial cells by human T-lymphotropic virus type I. Proc Natl Acad Sci USA 81:7588–7590
17. Hoxie JA, Matthews DM, Cines DB (1984) Infection of human endothelial cells by human T-cell leukemia virus type I. Proc Natl Acad Sci USA 81:7591–7595

18. Nerenberg MI, Hinrichs SH, Reynolds RK, Khoury G, Jay G (1987) The tat gene of human T lymphotropic virus type 1 induces mesenchymal tumors in transgenic mice. Science 237:1324–1329
19. Hinrichs SH, Nerenberg M, Reynolds RK, Khoury G, Jay G (1987) A transgenic mouse model for human neurofibromatosis. Science 237:1340–1343
20. Nerenberg MI, Wiley CA (1988) The HTLV-1 tat protein induces tissue specific disorders of neoplasia and atrophy in transgenic mice (manuscript submitted)
21. Wiley CA, Nerenberg M, Cros D, Soto-Aguilar MC (1989) HTLV-1 Polymyositis. N Engl J Med (in press)
22. Mora CA, Garruto RM, Brown P, Guiroy D, Morgan OS, Rodgers-Johnson P, Ceroni M, Yanagihara R, Goldfarb LG, Gibbs CF Jr, et al (1988) Seroprevalence of antibodies to HTLV-1 in patients with chronic neurological disorders other than tropical spastic paraparesis. Ann Neurol 23:S192–S195

CHAPTER 21
Targeted Expression of Viral Genes to Pancreatic β Cells in Transgenic Mice

SUSAN ALPERT, DOUGLAS HANAHAN, AND VICTORIA L. BAUTCH

Viral agents have been implicated in destructive disorders of β cells, the insulin-producing cells of the pancreatic islets of Langerhans. The clinical manifestations of insulin-dependent diabetes mellitus (IDDM) result from β cell destruction. Several epidemiological studies have implied a correlation between viral infection and diabetes, although the etiology of the disease is certainly more complex than a simple cause and effect relationship [1]. Mice exposed to certain viral agents (e.g., encephlomyocarditis virus) have been found to exhibit lymphocyte infiltration, necrosis, and degranulation of the pancreatic islets, accompanied by hyperglycemia that persists for varying periods of time [2].

The complexity of virus–host interactions has made it difficult to assess the role(s) of viral components comprehensively in disease processes such as diabetes. Proliferative disorders associated with certain viruses, however, have proved amenable to molecular analyses, in large part through the availability of cell culture systems to study virally induced transformation. These types of experiments have led to the identification of viral and cellular oncogenes that are capable of stimulating cell proliferation [3,4]. The interaction of cells with their environment is more complex in vivo, however, and it is not intrinsically obvious that oncogenes identified in vitro can elicit cell proliferation (and cancer) in vivo.

It is now possible to stably introduce genes (transgenes) into the mouse germ line, and this technology has led to new approaches in the study of a number of processes, including viral pathogenesis. One strategy involves the use of hybrid genes consisting of tissue-specific regulatory sequences linked to heterologous protein-coding regions [5,6]. These hybrid genes are usually expressed only in tissues that normally express the gene from which the regulatory region was derived. Because these transgenes are transmitted vertically via the mouse germ line, it is possible to assess certain aspects of viral gene expression in specific cell types independently of the infectious process. The viral gene products are in some cases antigenic, and thus these transgenic mice can also be used to study perturbations induced by the immune response. The purpose of this short review is to discuss recent studies on the effects of genes targeted to the β cells in transgenic mice and to examine the implications for the etiology of β cell functions and dysfunctions.

Consequences of Viral Oncogene Expression in β Cells

One of the first examples of targeted expression of a transgene in mice was the β cell-specific expression of SV40 T antigen [5]. Transgenic mice harboring hybrid genes composed of the 5' flanking region of the rat insulin II gene linked to the SV40 early region synthesize SV40 Tag in the β cells, and this expression leads to β cell hyperplasia. Although these mice inevitably develop β cell tumors that are subsequently fatal, only a few of the hyperplastic islets in any given animal progress to tumors. This finding suggests that an additional event(s) is necessary for β cell tumor formation in the insulin/SV40 T antigen transgenic mice, and it is consistent with the idea that tumorigenesis is a multistep process [27].

The use of hybrid genes to target gene expression in transgenic mice allows the effects of different oncogenes on β cell tumor formation to be compared. The early region of SV40 encodes both SV40 large T (SV40 LT) and SV40 small T (SV40 ST) antigens. All of the defined transforming functions associated with this virus are provided by SV40 LT antigen, and the role of SV40 ST antigen in transformation is not clear [7]. In contrast, the transforming functions of another papovavirus, polyoma (Py), are divided among at least two of three early region proteins: Py large T (Py LT) antigen is a nuclear immortalizing protein, Py middle T (Py MT) antigen is a membrane-associated transforming protein, and Py small T (Py ST) antigen has no well-defined role in transformation. In tissue culture, Py LT and Py MT antigens can cooperate with each other and with other oncogenes to transform primary fibroblasts [8–10]. Mice carrying different Py T antigen cDNAs linked to the rat insulin promoter have been generated, and they exhibit different phenotypes. Insulin/Py LT mice develop β cell tumors, whereas insulin-promoted Py MT and Py ST mice do not develop tumors (V. Bautch, unpublished results). The inefficacy of Py MT in β cell tumor formation is not universal, since this gene can cause endothelial cell tumors when linked to the Py promoter [11] and both endothelial tumors and adrenocarcinomas [12] when linked to the herpes thymidine kinase promoter in transgenic mice.

The latency of β cell tumor formation in two lineages of insulin/Py LT mice is much longer than the latency of tumorigenesis in any of the four different insulin/SV40 T antigen lineages. The time of onset of transgene expression differs among the insulin/SV40 lineages and probably results from position effects of the different chromosomal integration sites. This parameter is correlated with the latency of tumor formation, and mice that express SV40 T antigen early develop tumors at a relatively young age [13]. The timing of transgene expression, currently under investigation in the Py LT lineages, may be responsible for the latency difference between the SV40 and Py LT lineages. Alternatively, Py LT may be lacking some functions of SV40 T antigens and thus, the longer time required for overt tumor formation in insulin/Py LT mice may reflect a requirement for multiple secondary events. These putative additional events, however, do not appear to be supplied by Py MT, because mice bred to carry both insulin/Py LT and insulin/Py MT do not differ from the parental insulin/Py LT mice in tumor formation or latency (V. Bautch, unpublished results).

The apparent tissue specificty for tumor formation exhibited by the Py viral oncogenes is also shown by two other oncogenes linked to the insulin promoter. Transgenic mice expressing insulin promoted v-fos, a viral oncogene, do not develop tumors, whereas mice expressing insulin pro-moted c-myc, the cellular homolog of the viral oncogene v-myc, develop rare β cell tumors (S. Efrat and D. Hanahan, unpublished results). As with Py MT, the lack of v-fos-induced β cell tumors is probably not due to a general inefficacy of v-fos in inducing tumors, because this oncogene is associated with osteosarcomas in virally infected animals, and its cellular homolog (c-fos) induces bone and thymic abnormalities in transgenic mice [14,15]. The c-myc-induced tumors are analogous to those induced by Py LT in that they have a very long latency, again suggesting that tumorigenesis promoted by nuclear oncogenes in β cells requires one or more secondary events (S. Efrat and D. Hanahan, unpublished results). Thus, different viral oncogenes appear to have different effects on β cell proliferation and tumor progression, with SV40 T antigen promoting β cell tumor formation more effectively than either Py LT or c-myc.

A number of events can be envisioned that might contribute to the tumor progression seen in transgenic mice carrying insulin-promoted oncogenes. For example, the activation of another oncogene, the induction of an autocrine growth factor, or the induction of an angiogenesis factor could augment the formation of tumors. In principle, once a factor is identified, its effects on tumor formation can be assessed by targeting expression of the gene product to β cells in transgenic mice. The interaction of the putative progression factor with the viral gene product could then be tested in genetic complementation experiments, whereby two lines of mice, one carrying a hybrid insulin/progression factor and one carrying a hybrid insulin/viral gene, are mated.

One of the goals of targeting oncogene expression to rare cell types in transgenic mice is to develop cell lines from largely inaccessible cell types for further analysis. This has recently been accomplished for the pancreatic

β cells with the isolation of insulin-producing, glucose-inducible β cell lines from the insulin/SV40 T antigen transgenic mice [16]. These cells remain more "differentiated" than other β cell–derived lines when passaged, presumably because the selection for insulin-promoted oncogene expression requires cells permissive for endogenous insulin expression. These cells can now be used to dissect β cell function and to evaluate their interaction with other cell types, including components of the immune system. IDDM is thought to have an autoimmune component (see below), but the inaccessibility of β cells has retarded direct investigation of their antigenicity. The ability of the cultured β cells to specifically stimulate T cells from diabetic mice would support the autoimmune model, and it would provide a system for selecting T cell clones directed against β cell components.

Interactions of the Immune System With Transgenes Expressed in β Cells

It is well established that the immune system can recognize β cell antigens in patients with IDDM [17]. Transgenic mice provide a new approach to the study of the immune response to viral antigens on specific cell types. It might be expected that this response would be even more pronounced with a foreign antigen expressed in the β cells. In transgenic mice, however, the transgene is genetically "self" because it is transmitted through the germ line. This mode of transmission differs from normal virus entry and could affect presentation of the viral antigen and the nature of the immune response to the protein.

The immunogenicity of a transgene expressed in β cells has been studied in lines of mice harboring the insulin/SV40 T antigen constructs. Two lines of mice that express T antigen during embryonic development are immunologically tolerant to it, whereas nontolerant mice are found in two other lineages that do not express the antigen until adult life (10 to 12 weeks of age). The lines of mice that exhibit a late onset of expression can produce autoantibodies to T antigen, exhibit lymphocyte infiltration in the islets of Langerhans, and develop tumors at an older age than the lines with early onset of expression [13,18]. The autoimmune response is apparent shortly after the antigen appears and well before nascent tumors form. It is likely, therefore, that the response results from a delayed appearance of a "self" antigen rather than a general tumor immunity.

IDDM is a disease characterized by inadequate amounts of circulating insulin and, in general, results from β cell elimination. The pathology appears to have an autoimmune component, given that patients develop islet cell antibodies, exhibit lymphocyte infiltration of the islets, and subsequently show specific elimination of β cells, but not other islet cell types. Like other autoimmune diseases, the onset of IDDM is thought to be affected by both genetic and epigenetic factors. One mechanism whereby viral infection or other environmental agents may potentiate the onset of IDDM is by T cell activation. Activated T cells could induce the production of γ interferon (γIFN), which in turn might stimulate expression of major histocompatibility complex (MHC) class I or class II molecules on the surface of the pancreatic β cells. The β cells, which normally express only low levels of class I molecules and do not express class II molecules, could then present either the viral antigen(s) or other autoantigens to the immune system and thus induce an autoimmune response [19]. Several studies have reported expression of class II molecules or overexpression of class I molecules in β cells of patients with IDDM [20,21]. Recently MHC molecules, viral antigens, and γ interferon have been individually expressed in the β cells of transgenic mice to investigate their putative role in the development of diabetes.

To examine the effects of inappropriate MHC expression, transgenic mice have been created that express class II or class I molecules under the regulation of the insulin promoter [22–24]. Although the MHC molecules are expressed in the pancreatic β cells and the mice develop severe insulin dependence in all cases, no lymphocytic infiltration of the islets of Langerhans can be demonstrated. This finding rules out an obvious role of the T cell–mediated cellular response in development of the disease in these model systems. The lack of a detectable immune response to the transgenic MHC molecules can be explained by several hypotheses. These genes may be expressed in the β cells during development of the immune system, thereby allowing the introduced genes to be viewed immunologically as self. In fact, some mice harboring the insulin/class II genes are tolerant to the gene product [23]. Alternatively, it is possible that a T cell response to β cells cannot be induced by class I or class II molecule expression alone and that some other, possibly viral antigen is required. In this scenario the diabetic phenotype of the transgenic mice would have a nonimmunological etiology. For example, high levels of expression of class I or class II molecules may directly inhibit β cell function and lead to their inability to produce, process, or secrete insulin. Indeed, in one of these studies, the transgenic mice expressing the class I genes on β cells had an impaired ability to secrete insulin [24]. The mice carrying insulin/MHC genes may, therefore, provide a new model to study aberrations in cellular secretory pathways.

β cell death, however, does not inevitably result from the production of either a secreted or a non-MHC cell surface protein. Transgenic mice expressing either insulin-promoted human placental lactogen (M. Walker, S. Alpert, G. Teitelman, W.J. Rutter, and D. Hanahan, unpublished results) or influenza hemagglutinin (HA) (L. Roman, L. Simons, R. Hammer, T. Braciale, V. Braciale, M-J. Gething, and J. Sambrook, unpublished results) do not exhibit a diabetic phenotype or β cell destruction. Moreover, the recent finding that transgenic mice producing low but detectable levels of class II molecules on the surface of the pancreatic β cells do not develop diabetes suggests that the level of ectopic protein production may influence the resulting phenotype (D. Mathis, personal communication). It is not known how the levels of class I and class II molecules produced in the transgenic mice compare to the relative levels found in diabetic patients, nor is it known what level of expression is sufficient to damage β cells.

Lymphocyte infiltration of islets has been documented in two lineages of transgenic mice expressing SV40 T antigen, although the concomitant proliferation of β cells probably prevents the appearance of an overt diabetic phenotype. These lines show delayed onset of β cell–specific T antigen expression, and immunization of these mice with purified T antigen protein reveals that they are not self-tolerant to the antigen [13]. In contrast, the transgenic mice that express another viral antigen, HA, do not exhibit lymphocyte infiltration of the islets (L. Roman, M-J. Gething, and J. Sambrook, personal communication). Although the developmental pattern of HA expression has not been documented, this finding suggests that β cell expression of a viral antigen may not be sufficient to induce a cellular immune response. Therefore, the immune response induced in insulin/SV40 T antigen mice may result from the presentation of T antigen and other islet cell antigens to the immune system during the SV40 T antigen-induced hyperplasia and subsequent cell death [25]. Alternatively, lymphocyte infiltration of the islets suggests that T antigen expression may induce and/or be efficiently presented by MHC components, such as class I or class II molecules, that are involved in the immune response and are expressed in β cell tumors and β tumor lines (S. Baekkeskov, personal communication; E.H. Leiter, personal communication). This model is supported by recent results indicating that a genetic component modulates the degree of immunological responsiveness of the insulin/SV40 T antigen lines, and the pattern of these effects implicates the genes of the major histocompatibility complex (J. Skowronski, personal communication) [18].

Transgenic mice expressing a hybrid gene composed of the rat insulin II promoter linked to the protein-coding region for γIFN develop a generalized inflammatory response against the pancreas, in which lymphocytes can be seen infiltrating the islets as well as the exocrine tissue [22]. γIFN is known to induce expression of class II genes and to augment the expression of class I genes (for review, see ref. 26), and both of these molecules can be recognized by the immune system in conjunction with an antigen. However, γIFN expression in these transgenic mice may induce the ectopic production of other lymphokines that mediate the inflammatory reaction, because cell types that are presumably not expressing γIFN (acinar cells) are also sites of lymphocyte infiltration [22].

These transgenic mouse models are currently in use to study the events that link viral infection and the induction of diabetes. For example, overexpression of the class I and class II molecules can be triggered by a viral infection. Production of the MHC molecules may not trigger a lymphocyte infiltration but may only damage protein secretion in β cells, ultimately resulting in cell death and lysis. The release of the intracellular components of this rare cell type could then lead to the presentation of antigens normally sequestered from the immune system, and an autoimmune response could ultimately occur. The analysis of hybrid transgenic mice carrying both insulin-promoted viral antigens and insulin-promoted MHC genes might elucidate the events occuring in IDDM and allow further analysis of β cell function.

Summary and Conclusions

The construction of hybrid insulin/viral antigen transgenes allows a direct assessment of the effects of the transgene product on β cells in vivo. This strategy has been used to target expression of both viral and cellular oncogenes to β cells, and it appears that these oncogenes have different efficiencies in inducing insulinoma formation. The lines of mice that heritably form β cell tumors are being used to investigate the events involved in neoplastic progression. Moreover, targeting expression of specific cellular genes to the pancreatic β cells can elicit β cell dysfunction, destruction, and/or an immune response against them, and may therefore prove useful in the study of IDDM. It should be noted, however, that perturbations predicted to establish "models" of specific diseases in transgenic mice often provide surprising phenotypes that do not "model" the disease exactly but nevertheless yield useful information. Given this qualification, it can be predicted that the effects of other viral and cellular gene products on β cells can be usefully studied by similar targeting experiments in transgenic mice. This strategy allows the coupling of genetic complementation with classic biochemical and immunological studies to study the role of a particular gene product or combination of gene products in the pathogenesis of specific diseases.

References

1. Barnett AH (1986) Immunogenetics of diabetes. *In* Immunology of Endocrine Diseases. AM McGregor (ed) MTP Press, Ltd, Lancaster, England, pp 103–121
2. Craighead JE (1985) Viral diabetes. *In* The Diabetic Pancreas, 2nd ed, Plenum Publishing, New York, pp 439–466
3. Bishop JM (1985) Viral oncogenes. Cell 42:23–38
4. Weinberg RA (1985) The action of oncogenes in the cytoplasm and nucleus. Science 230:770–776
5. Hanahan D (1985) Heritable formation of pancreatic β cell tumours in transgenic mice expressing recombinant insulin/simian virus 40 oncogenes. Nature 315:115–122
6. Palmiter RD, Brinster RL (1986) Germ-line transformation of mice. Ann Rev Genet 20:465–499

7. Tooze J (1981) Molecular biology of tumor viruses, Part 2, DNA Tumor Viruses. Cold Spring Harbor Laboratory, Cold Spring Harbor, New York

8. Rassoulzadegan M, Crowie A, Carr A, Glaichenhous N, Kamen R, Cuzin F (1982) The roles of individual polyoma virus early proteins in oncogenic transformation. Nature 300:713–718

9. Ruley HE (1983) Adenovirus early region 1A enables viral and cellular transforming genes to transform primary cells in culture. Nature 304:602–607

10. Land H, Parada LF, Weinberg RA (1983) Tumorigenic conversion of primary embryo fibroblasts requires at least two cooperating oncogenes. Nature 304:596–602

11. Bautch VL, Toda S, Hassell JA, Hanahan D (1987) Endothelial cell tumors develop in transgenic mice carrying polyoma virus middle T oncogene. Cell 51:529–538

12. Williams RL, Courtneidge SA, Wagner EF (1988) Embryonic lethalities and endothelial tumors in chimeric mice expressing polyoma virus middle T oncogene. Cell 52:121–131

13. Adams TE, Alpert S, Hanahan D (1987) Non-tolerance and autoantibodies to a transgenic self antigen expressed in pancreatic β cells. Nature 325:223–228

14. Ruther U, Garber C, Komitowski D, Muller R, Wagner EF (1987) Deregulated c-fos expression interferes with normal bone development in transgenic mice. Nature 325:412–416

15. Ruther U, Muller W, Sumida T, Tokuhisa T, Rajewsky K, Wagner E (1988) c-fos expression interferes with thymus development in transgenic mice. Cell 53:847–856

16. Efrat S, Baekkeskov S, Linde S, Kofod H, Spector D, Delannoy M, Grant S, Hanahan D (1988) β-cell lines derived from transgenic mice expressing hybrid insulin-oncogenes. Proc Natl Acad Sci USA 85:9037–9041

17. Rossini AA, Mordes JP, Like AA (1985) Immunology of insulin-dependent diabetes mellitus. Ann Rev Immunol 3:289–320

18. Skowronski J, Alpert S, Hanahan D (1988) The use of transgenic mice to study interactions of a novel β cell antigen with the immune system. In Lessions from Animal Diabetes II. E Safrir and AE Reynolds (eds). John Libbey & Co, London

19. Bottazzo GF, Pujoi-Borrell R, Hanafusa T (1983) Hypothesis: Role of aberrant HLA-DR expression and antigen presentation in induction of endocrine autoimmunity. Lancet ii:1115–1119

20. Bottazzo GF, Dean BM, McNally JM, MacKay EH, Swift PGF, Gamble DR (1985) In situ characterization of autoimmune phenomena and expression of HLA molecules in the pancreas in diabetic insulitis. N Engl J Med 6:353–360

21. Foulis AK, Farquharson MA, Hardman R (1987) Aberrant expression of class II major histocompatibility complex molecules by insulin containing islets in Type I (insulin-dependent) diabetes mellitus. Diabetologia 30:333–343

22. Sarvetnick N, Liggitt D, Pitts SL, Hansen SE, Stewart TA (1988) Insulin-dependent diabetes mellitus induced in transgenic mice by ectopic expression of class II MHC and interferon-gamma. Cell 52:773–782

23. Lo D, Burkly LC, Widera G, Cowing C, Flavell RA, Palmiter RD, Brinster RL (1988) Diabetes and tolerance in transgenic mice expressing class II MHC molecules in pancreatic beta cells. Cell 53:159–168

24. Allison J, Campbell IL, Morahan G, Mandel TE, Harrison LC, Miller JFAP

(1988) Diabetes in transgenic mice resulting from overexpression of class I histocompatibility molecules in pancreatic β cells. Nature 333:529–533
25. Teitelman G, Alpert S, Hanahan D (1988) Proliferation, senescence, and neoplastic progression of β cells in hyperplasic pancreatic islets. Cell 52:97–105
26. Rosa F, Fellous M (1984) The effect of γ-interferon on MHC antigens. Immunol Today 5:261–262
27. Knudson (1986) Genetics of human cancer. Ann Rev Gen 20:231

Retroviruses

CHAPTER 22
Evolution of Retroviruses

HOWARD M. TEMIN

The most relevant question about the evolution of retroviruses in terms of viral pathogenesis is "how important is the high rate of variation of retroviruses in pathogenesis?"

The general retrovirus life cycle has been very successful in evolution, as seen by the variety of different elements that use reverse transcription [1–3]. Different elements that use reverse transcription are oncoretroviruses; human T cell leukemia virus type 1-like retroviruses; lentiretroviruses; spumaretroviruses; hepatitis B-like viruses; caulimoviruses; retrotransposons; retrotransposons without *env* genes; retrotransposons without coding sequences; and retroposons (no long terminal repeats) (Figure 22.1). In addition there are many different strains for each of these different elements. For example, most cloned murine leukemia (MLV) or human immunodeficiency virus (HIV) strains have significant nucleotide sequence differences from the other isolates in that subspecies.

In this article, the different types of variation that occur in retrovirus replication and the rates of each type of variation are discussed first. Then this knowledge is applied to a discussion of the role of this rate of variation in retrovirus pathogenesis. I shall conclude that the high rate of variation in retrovirus replication has resulted in a high rate of evolution; that this evolution can be mutation-driven (i.e., the rate of mutation is high so that a variant virus can accumulate multiple mutations before a change in fitness is achieved); and that these characteristics have allowed retroviruses to exploit many selective niches, which explains both the great success of these viruses and their association with pathologies, some inadvertent and some adaptive.

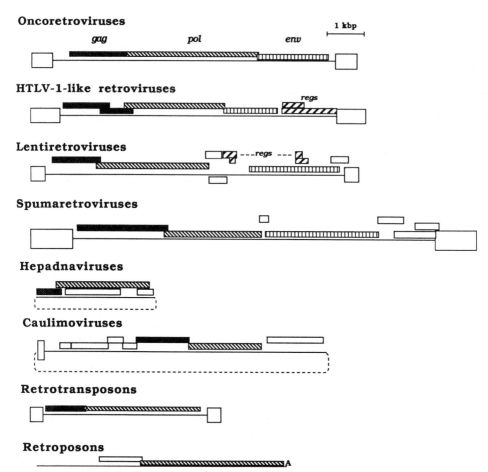

Figure 22.1. Different types of elements using reverse transcription. The open boxes at the ends represent long terminal repeats; the boxes over the lines represent protein coding genes as indicated. The other open boxes represent genes of unknown function or genes specialized for that element. The DNAs of hepadanaviruses and caulimoviruses are circular (dashed lines). The A at the end of the retroposons indicates an A-rich region.

Modes of Variation in Retrovirus Replication

Retroviruses vary by recombination, gene conversion, insertion, base-pair substitution, and deletion [4–6].

Recombination occurs between two retroviruses as well as between a retrovirus and a cell genome. The frequency of virus–virus recombination is very high; perhaps 25% to 50% of the progeny are recombinant when different viruses of the same species infect the same cell. Virus–virus recombination between retroviruses of different species has not been reported, although the structure of some retroviral genomes indicates that it might occur at a low rate [7]. Virus–cell recombination gave rise to certain retroviral genomes, particularly the genomes of highly oncogenic retroviruses, which include virus-derived and cell-derived nucleotide sequences.

The generation of recovered avian sarcoma viruses, transforming avian sarcoma viruses formed after replication of particular transformation-defective avian sarcoma viruses on homologous cells [8], the recovery of replication-competent viruses after passage of replication-defective MLV [9] and the formation of mink cell focus-forming viruses [10] may reflect either virus–cell recombination or gene conversion between cell genes and viral genomes. Gene conversion has been found in an experimental system at a frequency of 10^{-3} after 15 cell generations [11]. Mechanistically retrovirus gene conversion differs from retrovirus recombination in that it takes place at the provirus stage and does not depend on the formation of a heterozygous virus particle.

Insertions of small numbers of base pairs are frequently seen when the nucleotide sequences of related retroviruses are compared. In an experimental study, small insertions occurred at a rate of 10^{-7}/base pair per replication cycle, which amounts to 10^{-3}/genome per replication cycle [11].

Base-pair differences also are very frequently seen in comparisons of nucleotide sequences of different viruses in a single retrovirus species. An experimental study involving replication of a spleen necrosis retrovirus-based vector in dog cells found a rate of base-pair substitution of 10^{-5}/base pair per replication cycle [11]. Study of the rate of appearance of temperature-sensitive mutants of *src* [12] indicates that this rate, 10^{-5}/base pair per replication cycle, probably also applies to other oncoretroviruses. [The *src* mutation experiments involved chemical mutagens (5-fluorouracil and 5-azacytidine), but our unpublished data indicate that the conditions under which these experiments were performed did not significantly increase the already very high rate of spontaneous mutations and, thus, that most of the *src* mutations seen at high frequencies in the mutagen experiments were spontaneous.]

Deletions are commonly seen during the growth of Rous sarcoma virus, as well as with different, genetically engineered retroviruses [13–16]. The rate of deletion per replication cycle is over 10% [16].

Evolution of Retroviruses

These high rates of variation ensure that in every retrovirus population over 20% of the viral genomes are variant. (The 20% is the sum of the rates per genome of the different types of variation.) Since there is good phenotypic mixing between retroviruses infecting the same cell, even replication-defective viruses can infect cells and persist in viral populations until they find a selective niche or undergo further evolution. The 80% of unchanged genomes in each replication cycle ensures the continued survival of the wild-type virus if conditions are unchanged.

Thus, there are some retrovirus variants that are able to survive under different selective conditions, and these variant retroviruses are selected by the different conditions. These circumstances result in a high rate of retrovirus evolution.

Relevance of High Rate of Retrovirus Evolution to Pathogenesis

Pathogenesis influences retrovirus evolution when the disease condition increases or decreases the number and type of surviving viruses or the probability of transmission to an uninfected host or both. Retroviruses can spread vertically or horizontally or both [4]. For example, there is vertical transmission of avian leukosis and MLVs by virus infection of the egg or embryo; HIV can be transmitted perinatally; and human T cell leukemia virus type 1 (HTLV-1) and mouse mammary tumor virus are transmitted in mother's milk. Horizontal infection can occur through saliva for feline leukemia virus or in infected blood for HIV and HTLV-1.

For many disease-causing retroviruses, causing the disease is not a selective advantage for the retrovirus because the disease results in the death of the infected individual without increasing the probability of the virus infecting an uninfected host. Thus, the relationship of retroviruses to disease can be adaptive when the disease results in more virus progeny or increased infection of new hosts or both. Alternatively the relationship of retroviruses to disease can be inadvertent when the disease is an unselected by-product of retrovirus replication.

There are several different ways in which retroviruses cause disease: cancer, as a result of the introduction of a modified cell gene; hyperplasia, as a result of the introduction of a modified *env* gene followed by other changes resulting in neoplasia; hyperplasia, as a result of insertional activation of a cell gene(s) followed by other changes resulting in neoplasia; hyperplasia, as a result of transactivation of a cell gene(s) followed by other changes resulting in neoplasia; immune depression and anemia, as a result of cell killing; hyperplasia, as a result of stimulation of a hyperimmune response resulting in later changes leading to neoplasia; and central nervous system (CNS) degeneration [17–19].

Introduction of a modified cell gene by a highly oncogenic retrovirus results in the rapid appearance of cancer, usually followed by death of the infected animal and consequently death of the highly oncogenic retrovirus. Thus, in the absence of propagation of the virus by scientists, the disease is not adaptive for the virus. The formation of highly oncogenic retroviruses occurs as a by-product of the high rate of retrovirus evolution and is an evolutionary dead end until these viruses are rescued and propagated by scientists.

Introduction of a modified *env* gene by retrovirus infection produces hyperplasia, often followed by further changes that lead to neoplasia. As with highly oncogenic retroviruses, this disease etiology is not adaptive for the virus in the absence of scientific interest, but it is rather a by-product of the high rate of retrovirus evolution resulting in such modified retroviruses.

Insertional activation of a cell gene(s) by weakly oncogenic retroviruses can be the first step leading to leukemia, which appears after a long latent period. Causing leukemia does not significantly increase the replication or transmission of the causal virus, and thus the leukemia is also an inadvertent by-product of the retrovirus replication cycle. When such a weakly oncogenic virus is transmitted to a young animal or an embryo, immune tolerance often results. Thus, the infected animals are viremic. This increased viral replication increases the probability of viral transmission as well as the probability of insertional activation. When the infection is of an older animal, an immune response which can abrogate the infection results. Thus, there has been strong selection for vertical transmission, early infection, and other means to avoid the host immune response. The leukemia, however, is an inadvertent unselected by-product of this selection.

The HTLV-1-like retroviruses remain latent in infected blood cells apparently after *trans*-activation causes hyperplasia of the infected cells. Thus, the virus avoids the host immune response after having replicated as a provirus in the hyperplastic cells. (The presence of regulatory genes in HTLV-1-like retroviruses not found in oncoretroviruses probably reflects selection for this type of capability.) At a low frequency, however, the hyperplastic cells become neoplastic after a very long latent period. This neoplasia increases the number of virus-infected blood cells and thus might be adaptive for the virus. However, most transmission of virus to other hosts probably occurs during the latent period for the disease, and so causing the disease is probably not adaptive for the virus. The low probability of occurrence and the long latent period for the disease support this hypothesis. Thus, disease again appears to be an inadvertent by-product of viral replication.

Causing immune depression is adaptive for a virus; it allows continued viral replication and increases the probability of spread to a new host. Immune depression can be caused by virus infection of cells in the immune system, either exclusively or as one of a number of types of cells infected, or by the effects of a virus protein produced in other cells. Thus, many retroviruses cause immune depression, including MLV, feline leukemia virus, avian reticuloendotheliosis virus, and HIV. For example, MLV produces an immunorepressive peptide [20]. Avian reticuloendotheliosis virus infects and kills cells of the immune system as well as other cells, and HIV specifically infects and kills certain cells of the immune system. The appearance of new tissue tropisms, for example, infecting cells of the immune system, involves both changes in the cellular receptors used by the infecting virus and changes in viral transcriptional controls. The high rate of evolution of retroviruses provides variant viruses with different tissue tropisms, allowing the selection of viruses that cause immune depression.

Equine infectious anemia and Visna viruses seem to use the high rate of retrovirus variation to drive antigenic drift away from any one specific neutralizing antibody [21]. Although HIV is not permanently inhibited by the host immune response and has a very high frequency of amino-acid changes in its envelope glycoproteins, it does not seem that HIV is subject to such antigenic drift. Therefore, HIV probably has an additional mode(s) of escaping the host immune response. (The HIV depression of the immune response resulting in clinical acquired immunodeficiency syndrome (AIDS) is a late effect of virus infection.) Compared to other retroviruses, there are more amino acids in the external HIV glycoprotein, gp120 [22], which might have a role in protecting the HIV receptor-binding site from antibodies. In addition, the gp120-CD4 receptor binding may be stronger than the antibody-gp120 binding preventing antibody neutralization [23]. The HIV probably also has additional mechanisms such as virus latency for avoiding the host immune response.

Another type of retrovirus-induced pathology involves the CNS. Although the mechanism of this pathology is not known (it could be the result of a viral product or of direct killing of certain infected cells in the CNS), it does not appear to be adaptive for the virus. Thus, it probably is an inadvertent by-product of retrovirus variation and evolution.

Summary and Conclusions

Thus, the evolution of retroviruses has resulted in viruses that are very successful in replication. Sometimes disease occurs as a result of this replication, but it usually occurs as an inadvertent by-product of the viral replication.

Finally, the appearance of new retroviruses and retrovirus-induced diseases reflects the characteristics, described above, of a potential for rapid evolution and the interaction of retrovirus replication with both the immune system and the cell genome. Thus, it is likely that other chronic diseases and diseases with a long latent period will be found to have a retrovirus etiology.

Note added in proof. Further studies with retrovirus vectors has shown that the rate of recombination is only 1% per kilobase per viral replication cycle and that the rate of gene conversion is about 10^{-5} per cell division (Hu, Yang, and Temin, unpublished).

Acknowledgment

The work in my laboratory was supported by Public Health Service Research Grants CA-22443 and CA-07175. I am an American Cancer Society Research Professor. I thank Drs. C. Gélinas, W.-S. Hu, M. Hannink, and R. Risser for helpful comments on the manuscript.

References

1. Hull R, Covey SN (1986) Genome organization and expression of reverse transcribing elements: Variations and a theme. J Gen Virol 67:1751–1758
2. Weiner AM, Deininger PL, Efstratiadis A (1986) Nonviral retroposons: Genes, psuedogenes, and transposable elements generated by the reverse flow of genetic information. Ann Rev Biochem 55:631–661
3. Temin HM (1988) Evolution of retroviruses and other retrotranscripts. *In* Bolognesi D (ed) Human Retroviruses, Cancer, and AIDS. Approaches to prevention and therapy. Proceedings of an Abbott-UCLA symposium held at Keystone, Colorado, April 1–6, 1987, Alan R. Liss, New York, pp 1–28
4. Weiss R, Teich N, Varmus H, Coffin J (eds) (1982) RNA Tumor Viruses, Molecular Biology of Tumor Viruses, 2nd ed. Cold Spring Harbor Laboratory, Cold Spring Harbor, NY
5. Weiss R, Teich N, Varmus H, Coffin J (eds) (1985) RNA Tumor Viruses, Molecular Biology of Tumor Viruses, 2nd ed, 2/Supplements and Appendixes. Cold Spring Harbor Laboratory, Cold Spring Harbor, NY
6. Linial M, Blair D (1982) Genetics of retroviruses. *In* Weiss R, Teich N, Varmus H, Coffin J (eds) RNA Tumor Viruses, Molecular Biology of Tumor Viruses, 2nd ed. Cold Spring Harbor Laboratory, Cold Spring Harbor, NY, pp 649–783
7. Callahan R, Chiu I-M, Wong JFH, Tronick SR, Roe BA, Aaronson SA, Schlom J (1985) A new class of endogenous human retroviral genomes. Science 228:1208–1211
8. Wang L-H, Beckson M, Anderson SM, Hanafusa H (1984) Identification of the viral sequence required for the generation of recovered avian sarcoma viruses and characterization of a series of replication-defective recovered avian sarcoma viruses. J Virol 49:881–891
9. Schwartzberg P, Colicelli J, Goff SP (1985) Recombination between a defective retrovirus and homologous sequences in host DNA: Reversion by patch repair. J Virol 53:719–726
10. Holland CA, Hartley JW, Rowe WP, Hopkins N (1985) At least four viral genes contribute to the leukemogenicity of murine retrovirus MCF 247 in AKR mice. J Virol 53:158–165
11. Dougherty JP, Temin HM (1988) Determination of the rate of base-pair substitution and insertion mutations in retrovirus replication. J Virol 62:2817–2822
12. Friis RR (1978) Temperature-sensitive mutants of avian RNA tumor viruses: A review. Curr Top Microbiol Immunol 79:259–293
13. Estis LF, Temin HM (1979) Suppression of multiplication of avian sarcoma virus by rapid spread of transformation-defective virus of the same subgroup. J Virol 31:389–397
14. Emerman M, Temin HM (1984) High frequency deletion in recovered retrovirus vectors containing exogenous DNA with promoters. J Virol 50:42–49
15. Rhode BW, Emerman M, Temin HM (1987) Instability of large direct repeats in retrovirus vectors. J Virol 61:925–927
16. Dougherty JP, Temin HM (1986) High mutation rate of a spleen necrosis virus-based retrovirus vector. Mol Cell Biol 6:4387–4395

17. Teich N, Wyke J, Kaplan P (1985) Pathogenesis of retrovirus-induced disease. *In* Weiss R, Teich N, Varmus H, Coffin J (eds) RNA Tumor Viruses, Molecular biology of Tumor Viruses, 2nd ed, 2/Supplements and appendixes, Cold Spring Harbor Laboratory, Cold Spring Harbor, NY, pp 187–248
18. Temin HM (1988) Evolution of cancer genes as a mutation-driven process. Cancer Res 48:1697–1701
19. Temin HM (1988) Retroviruses and the genetics of cancer. Int J Radiation Oncol Biol Phys 15:543–545
20. Schmidt DM, Sidhu NK, Cianciolo GJ, Snyderman R (1987) Recombinant hydrophilic region of murine retroviral protein p15E inhibits stimulated T-lymphocyte proliferation. Proc Natl Acad Sci USA 87:7290–7294
21. Payne SL, Fang F-D, Liu C-P, Dhruva BR, Rwambo P, Issel CJ, Montelaro RC (1987) Antigenic variation and lentivirus persistence: Variations in envelope gene sequences during EIAV infection resemble changes reported for sequential isolates of HIV. Virology 161:321–331
22. Coffin J (1986) Genetic variation in AIDS viruses. Cell 46:1–4
23. Matthews TJ, Weinhold KJ, Lyerly HK, Langlois AJ, Wigzell H, Bolognesi DP (1987) Interaction between the human T-cell lymphotropic virus type III$_B$ envelope glycoprotein gp120 and the surface antigen CD4: Role of carbohydrate in binding and cell fusion. Proc Natl Acad Sci USA 84:5424–5428

CHAPTER 23
New HIV-Related Viruses

FRANÇOIS CLAVEL AND LUC MONTAGNIER

The human immunodeficiency virus type 1 (HIV-1) is the prototype of a recently characterized group of primate lentiviruses and the agent of the acquired immunodeficiency syndrome (AIDS). Since its discovery, our knowledge of the disease and of its agent has grown in complexity and diversity: initially recognized as a disease of the immune system spreading among North American homosexual males, it is now worldwide, and its clinical spectrum has considerably widened. The virus itself, upon detailed molecular analyses, shows extensive genetic variability and flexibility, as well as a unique molecular organization and regulatory mechanisms that are far from being fully understood. The isolation, 2 years after that of HIV-1, of a closely related simian retrovirus and the subsequent isolation of a second human AIDS virus have further complicated the picture. However, this also raised hopes that detailed comparisons could provide a better understanding of the biology of this group of viruses and help define targets for therapy and vaccination.

The Simian Immunodeficiency Viruses

The first isolates of simian immunodeficiency virus (SIV) were obtained from a few captive macaques in the United States, which presented a disease very close to human AIDS [1]. A virus was obtained from cultures of lymphocytes from these animals; these cultures had biological and morphological characteristics very similar to those of HIV-1. The relatedness of this virus to HIV was confirmed by serological analyses. Sera from HIV-1-infected humans reacted with the core (*gag*) antigens of SIV, and sera from SIV-infected monkeys also recognized the *gag* antigens of HIV-1. There was, however, very little, if any, such cross-reactivity in the envelope glycoproteins. Later, these findings were confirmed by nucleotide sequence analysis of several of these macaque viruses [2,3]: The conservation with HIV-1 is higher (around 60%) in the *gag* and *pol* genes that code for the core antigens and the polymerase, respectively, and lower (around 40%) in the *env* gene that codes for the envelope glycoprotein. Overall, the general organization of the genome is very similar to that of HIV-1. Although macaques, which are Asian primates, seem not to be the natural hosts for these viruses in the wild, they are referred to as SIVmac.

A second group of SIVs, called SIVagm, was subsequently identified in wild-caught African green monkeys, first by serological analysis, later by virus isolation [4]. No disease was found in the infected animals. Nucleotide sequence analysis of one of these isolates [5] determined that SIVagm is clearly distinct from both SIVmac (percent of amino-acid identity of the polymerases = 65%) and HIV-1 (55%).

Other SIVs have also been found in other monkeys, either captive or wild animals, such as sooty mangabeys, mandrills, or captive macaque species [6–8]. Some of these viral isolates appear to be part of the SIVmac subgroup, whereas others are clearly distinct from both the SIVmac and the SIVagm subgroups. It is possible that the known isolates of SIV are members of a whole spectrum of related viruses spreading in different monkey species, in the wild or in captivity. The answer to this question will come from more virus isolations from more species of different geographical origins, and by further molecular analysis of these isolates.

HIV-2, The West African AIDS Virus

Before 1985 many AIDS cases recognized in Central Africa were shown to be associated with HIV-1 infection. Western Africa seemed to be relatively spared from the epidemic. But by the end of 1985, several patients from Guinea-Bissau and other neighboring countries of West Africa were found to be infected with a new virus, HIV-2, related to, but clearly different from HIV-1, and which seemed almost undistinguishable from SIVmac [9].

Serology
As for SIVmac-infected animals, serum from most HIV-2-infected individuals fails to react with the envelope antigens of HIV-1. There is, however, a good cross-reactivity between the *gag* antigens of HIV-1 and -2, mostly concerning the larger *gag* protein. Despite the presence of common immunoreactive domains in HIV-1 and HIV-2 *gag* and *pol* genes products, some sera from HIV-2-infected patients with faint or absent reactivity to these antigens will not be detected by most HIV-1-specific routine diagnostic tests. Such tests should therefore include antigens from both viruses whenever HIV-2 infection is suspected. Conversely, all the SIVmac proteins

can be recognized by HIV-2-specific sera. In fact, sera from HIV-2-infected individuals and from SIV-mac-infected animals are virtually undistinguishable. When the viral proteins of HIV-2 and SIVmac are compared, only subtle differences in molecular sizes can differentiate them. There also exists some variation in the size of the transmembrane envelope protein between, and even within, HIV-2 isolates. This phenomenon could result from a certain flexibility in the presence, absence, or position of a stop codon in the 3' end of the *env* reading frame.

Sequence Comparisons

The overall organization of the HIV-2 genome is very similar to that of HIV-1 [10]: In addition to the classic retroviral genes *gag, pol,* and *env,* a "central region" lies between the *pol* and *env* open reading frame (ORF), which contains several shorter ORFs, such as the *vif* (sor or Q) and the *upr* (R) genes, which also exist in HIV-1. In this region of the genome, a unique ORF is found, *upx* (X), which has no HIV-1 counterpart. The central region also contains the first coding exons of the *tat* and *rev* (*art* or *trs*) genes, whose second coding exons are located further downstream at the level of the transmembrane portion of the envelope gene. Both these genes produce *trans*-acting factors that are necessary for the expression of structural viral proteins, but whose precise mode of action is still unknown. The product of the *tat* gene is a *trans*-activator of the viral "long terminal repeat" (LTR), which contains the viral promoter. Interestingly, the *tat* gene product of HIV-1 acts efficiently on both the HIV-1 and HIV-2 LTRs, whereas its HIV-2 equivalent is not as active on the HIV-1 LTR. This incomplete cross *trans*-activation has been attributed to the fact that, in addition to a set of target sequences common to both LTRs, the HIV-2 LTR contains unique sequences that are only responsive to the HIV-2 *trans*-activator [11]. At the 3' end of the genome, overlapping the *env* gene and the 3' LTR, is found the *nef* ORF (F or 3' *orf*), 30% homologous to its HIV-1 counterpart.

When compared with HIV-1, the most conserved proteins are the *gag*- and *pol*-encoded polypeptides (57% and 59% identity, respectively). The envelope, as expected from the serological data, is one of the less conserved, particularly in its external portion (37% amino-acid identity), in which only short scattered stretches of homology can be found. These stretches are located within regions of the HIV-1 envelope gene that are conserved among different individual HIV-1 isolates, and may correspond to domains of functional importance for virus infectivity. These domains are under intense functional and immunological investigation, since they may represent important targets for protective immunity. As already mentionned, an in-frame stop codon can be found in the envelope gene of HIV-2 and of several SIV isolates, which appears to result in shorter transmembrane proteins. The consequences of this feature on the biology and pathogenicity of HIV-2 and SIV are unknown.

In the context of the low homology between HIV-1 and HIV-2, the high homology between HIV-2 and SIVmac is striking. The *gag–pol* homology is 85%, and the envelope homology is 75%, not very different from the envelope homology between individual isolates of HIV-1 or HIV-2. And SIVmac also has a *vpx* gene, 85% homologous to its HIV-2 equivalent.

Biology
Like HIV-1, HIV-2 only infects cells bearing the CD4 molecule, and despite the fact that its envelope glycoprotein differs dramatically from that of HIV-1, it also uses the CD4 molecule as a receptor [9]. Experiments have shown that the infection of lymphocytes by HIV-2 can be blocked by the same epitope-specific, anti-CD4 monoclonal antibodies that also blocked infection with HIV-1, indicating that both viruses bind to the same structure on the CD4 molecule [12]. Most HIV-2 isolates are cytopathic in vitro, inducing syncytium formation and killing cells. However, the intensity of this cytopathic effect varies with the cell type (HIV-2 can be grown on various human lymphoid tumor cell lines) and with the isolate. Recently, HIV-2 isolates with very limited cytopathicity have been reported [13].

Pathogenicity
The first patients from whom HIV-2 was isolated were typical AIDS patients. Since then, other patients with AIDS have been shown to be infected with HIV-2 [14]. The clinical symptoms of these AIDS patients include opportunistic infections and Kaposi's sarcoma, severe diarrhea, and wasting in a general disease pattern that does not differ from that observed in Central African, HIV-1-infected AIDS patients. Immunologically, a profound decrease in the number of the circulating T4 lymphocytes is found in all HIV-2-related AIDS cases. However, it has been suggested that the length of the asymptomatic phase and the survival times of HIV-2 AIDS patients would be longer than in HIV-1-related disease. This may explain why only a few AIDS cases were initially described in areas in which HIV-2 appeared to be highly prevalent, probably because these areas were reached only recently by the epidemic. More time will be needed before one can affirm that HIV-2 is less or as pathogenic as HIV-1.

Epidemiology
The HIV-2 is a West African virus. The country with the highest prevalence is Guinea-Bissau, a former Portuguese territory to the south of Senegal. Other cases have been found in the Cape Verde Islands, 500 km off the coast of Senegal, in Senegal, and in more eastern countries such as Guinea, Mali, Burkina Faso, the Central African Republic, and the Ivory Coast [15]. Countries of central, equatorial, and eastern Africa, in which the infection by HIV-1 has reached endemic proportions, seem relatively free of HIV-2 so far.

Recently there have been reports of individuals with a serological profile suggesting a double infection with HIV-1 and HIV-2 [16]; the serum of these patients is indeed able to react strongly with the envelope protein of both viruses. Some of these cases were from the Ivory Coast, where 20% of AIDS patients have a HIV-1-specific serum reactivity, 30% are HIV-2-reactive, and 50% have double reactivity. Since the Ivory Coast is between the centers of the two epidemics, the data suggest that those patients are infected with both viruses. The spread of HIV-2 outside West Africa so far seems to be limited. Several cases have been reported in France, Portugal, and West Germany, even in individuals who had never traveled to West Africa, but who belong to the classic AIDS high-risk groups. In the United States, only one case of HIV-2 infection has been reported, in a recently immigrated West African woman with AIDS. Since it is very likely that other HIV-2-infected subjects have traveled outside their countries in recent years, more cases of HIV-2 infection will certainly emerge in the near future in countries outside West Africa.

The Origins of the AIDS Viruses

With the discovery of more SIV isolates and the existence of two different HIVs, we are witnessing the spread of a number of related primate lentiviruses in different monkey species and in humans; these can be divided into three subgroups:

1. *The HIV-1 subgroup*, found so far in humans only. This virus probably originated in central Africa, where cases of AIDS have retrospectively been identified as the result of HIV-1 infection as far back as in 1959. The spread of HIV-1 could have started around this time, and it is clear that the present worldwide epidemic was boosted by the occurrence of the virus in social groups with high transmission rates, such as male homosexuals or intravenous drug users in America and Europe.

2. *The HIV-2 and SIVmac groups*, whose origins are unclear. Although SIVmac is found in macaques, these Asian monkeys do not appear to carry the virus in the wild, and it is probable that the virus was transmitted to macaques by other captive monkeys. Interestingly, macaques are the only monkeys that develope disease upon SIV infection. Given the high homology between HIV-2 and SIVmac, and although there is no indication that SIVmac can infect humans, it is tempting to imagine that HIV-2 was transmitted to humans in West Africa by the same monkey species that infected macaques in captivity.

3. *The SIVagm subgroup*, which is probably quite heterogeneous, may represent members of a wider spectrum of different viruses.

Genetic tree analyses derived from sequence comparisons indicate that viruses from all three subgroups evolved from a common ancestor. Computer projections based on such analyses have lead to the proposal that the HIV-1 and HIV-2–SIVmac subgroups may have diverged as late as 40 years ago. However, these projections may not have taken into account all the parameters of genetic divergence of these viruses, such as those that are related to the cycle of their transmission or replication. Although a recent evolution and diversification of the HIV/SIV group of viruses could be one explanation for the recent occurrence of two apparently independent AIDS epidemics, it is also possible that HIVs were present and transmitted at a slow rate in humans for a long period of time. This could have been the case in remote areas of Africa, where scattered cases of AIDS might have been unnoticed for decades before the disease was recognized. Only recently, when the infection reached some fast-growing urban areas, was the rate of transmission accelerated, leading to the present worldwide epidemic.

References

1. Daniel MD, Letvin NL, King NW, Kannagi M, Sehgal PK, Hunt RD (1985) Isolation of T-cell tropic HTLV-III-like retrovirus from macaques. Science 228:1201–1204
2. Hirsh V, Riedel N, Mullins J (1987) The genome organization of STLV-3 is similar to that of the AIDS virus except for a truncated transmembrane protein. Cell 49:307–319
3. Chakrabarti L, Guyader M, Alizon M, Daniel MD, Desrosiers RC, Tiollais P, Sonigo P (1987) Sequence of simian immunodeficiency virus from macaques and its relationship to other human and simian lentiviruses. Nature 328:543–547
4. Ohta Y, Masuda T, Tsujimoto H, Ishikawa KI, Kodama T, Morikawa S, Nakai M, Honjo S, Hayami M (1988) Isolation of simian immunodeficiency virus from African green monkeys and seroepidemiologic survey of the virus in various nonhuman primates. Int J Cancer 41:115–122
5. Fukasawa M, Miura T, Hasegawa A, Morikawa S, Tsujimoto H, Miki K, Kitamura T, Hayami M (1988) Sequence of simian immunodeficiency virus from African green monkey, a new member of the HIV/SIV group. Nature 333: 457–461
6. Fultz PN, McClure HM, Anderson DC, Swenson RB, Anand R, Srinivasan A (1986) Isolation of a T-lymphotropic retrovirus from naturally infected sooty mangabey monkeys. Proc Natl Acad Sci USA 83:5286–5290
7. Tsujimoto H, Cooper RW, Kodama T, Fukasawa M, Miura T, Ohta Y, Ishikawa KI, Nakai M, Frost E, Roelants GE, Roffi J, Hayami M (1988) Isolation and characterization of simian immunodeficiency virus from Mandrills in Africa and its relationship to other human and simian immunodeficiency viruses. J Virol 62:4044–4050
8. Benveniste RE, Arthur LO, Tsai CC, Sowder R, Copeland T, Henderson L, Oroszlan S (1986) Isolation of a lentivirus from a macaque with lymphoma: Comparison with HTLV-III/LAV and other lentiviruses. J Virol 60:483–490
9. Clavel F, Guetard D, Brun-Vezinet F, Chamaret S, Rey MA, Santos-Ferreira, Laurent AG, Dauguet C, Katlama C, Rouzioux C, Klatzmann D, Champalimaud JL, Montagnier L (1986) Isolation of a new human retrovirus from West African patients with AIDS. Science 233:343–346
10. Guyader M, Emerman M, Sonigo P, Clavel F, Montagnier L, Alizon M (1987)

Genome organization and transactivation of the human immunodeficiency virus type 2. Nature 326:622–669

11. Emerman M, Guyader M, Montagnier L, Baltimore D, Muesing M (1987) The specificity of the human immunodeficiency virus type 2 transactivator is different from that of human immunodeficiency virus type 1. EMBO J 6:3755–3760

12. Sattentau Q, Beverley PCL, Halabi FA, Montagnier L, Gluckman JC, Klatzmann D (1987) A comparison of the interaction of HIV-1 and HIV-2 with the CD4 antigen. Paper presented at the Third International Conference on AIDS, Washington, DC

13. Kong LI, Lee SW, Kappes JC, Parkin JS, Decker D, Hoxie JA, Hahn BA, Shaw GM (1988) West African HIV-2 related retrovirus with attenuated cytopathicity. Science 240:1525–1529

14. Clavel F, Mansinho K, Chamaret S, Guetard D, Favier V, Nina J, Santos-Ferreira MO, Champalimaud JL, Montagnier L (1987) Human immunodeficiency virus type 2 infection associated with AIDS in West Africa. N Engl J Med 316:1180–1185

15. Kanki PJ, M'Boup S, Ricard D, Barin F, Denis F, Boye C, Sangare L, Travers K, Albaum M, Marlink R, Romet-Lemonne JL, Essex M (1987) Human T lymphotropic virus type 4 and the human immunodeficiency virus in West Africa. Science 236:827–831

16. Ouattara A, Groupe Ivoirien de travail sur le SIDA, Rey MA, Brun-Vezinet F, DeThe G (1987) HIV-1 and HIV-2 are present in AIDS patients in Ivory Coast. Paper presented at the Third International Conference on AIDS, Washington DC

CHAPTER 24
The HTLV-I Model and Chronic Demyelinating Neurological Diseases

ALEXANDER KRÄMER AND WILLIAM A. BLATTNER

The discovery that human T cell leukemia virus (HTLV-1) is associated with adult T cell leukemia (ATL) and tropical spastic paraparesis (TSP) documents for the first time that human retroviruses can result in diseases of long latency in man. The epidemiological pattern of this virus and associated diseases provides a model of how a transmissible agent may enter a host through a number of transmission routes and, after many years, cause disease through direct and indirect mechanisms. The lessons learned from this pathogenic model may be useful in understanding other diseases of unknown etiology, particularly chronic degenerative neurological diseases of man.

Human T Cell Leukemia Virus and the Pathogenesis of Lymphoproliferative Malignancies

Adult T cell leukemia is a clinical pathological entity characterized by a usually aggressive, mature, T cell leukemia/lymphoma in adults between the ages of 20 and 60. Its features often include hypercalcemia, peripheral blood involvement, and cutaneous infiltration. Although the spectrum of clinical presentations of the syndrome is broad and sometimes leads to misclassification, the etiological role of HTLV-1 has been strongly supported by molecular and epidemiological evidence. In particular, HTLV-1 is monoclonally integrated into the target mature T cell in virtually all cases of ATL in Japan and in the Caribbean Basin. This integration is random, since there are no common flanking sequences detectable between cases. Within a particular tumor, however, the virus is monoclonally integrated and stably present. Additional laboratory evidence that HTLV-1 is etiological in ATL comes from in vitro observations that HTLV-1 can transform the target T lymphocytes. However, some enigmas remain, particularly the observation that in most cases the tumor tissue itself does not appear to express the HTLV-1 genes when studied using currently available detection techniques. In addition, the proviral DNA integrated into the tumor tissue is methylated, which suggests that the viral genes in these circulating and fixed tissue tumors are not being actively expressed.

Epidemiological data that support the role of HTLV-1 in ATL come from population-based surveys that show a marked concordance between HTLV-1 occurrence and ATL clusters in the same areas. In addition, a recently completed case-control study of ATL in Jamaica documented a 35-fold increased risk for adult T cell leukemia among HTLV-1 positives [1]. In Jamaica and in Japan, however, there are clear examples of HTLV-1 negative ATL, which suggests a possible role for other etiological factors or the presence of undetected virus in the host. The lifetime risk for ATL among HTLV-1 positive patients appears to be low, with estimates ranging from approximately 1% lifetime risk to 3% to 5% lifetime risk in persons affected before the age of 20. A recent model linking HTLV-1 seroprevalence in the population to ATL incidence suggests that early life infection may be particularly important in the subsequent risk for leukemia [2].

A number of unanswered questions concerning the pathogenetic mechanism by which HTLV-1 causes ATL remains to be answered, but it is postulated that the virus acts as an inducer of ATL risk and that other factors, including host genetic factors, possible immunosuppressive effects of the virus, and cytogenetic alterations, may also be required for progression to disease. However, the monoclonal integration of HTLV-1 in most ATL cases suggests a direct pathogenetic role for HTLV-1 in ATL.

Recently, cases of B cell chronic lymphocytic leukemia associated with HTLV-1 have been reported [3]. An indirect role for HTLV-1 in leukemogenesis is suggested by the finding that although HTLV-1 does not directly infect the B cell tumor population, it may contribute to leukemogenesis through polyclonal infection of T cells. Evidence for this concept comes from the finding that the immunoglobulin expressed by the B cells in these cases reacts to HTLV-1-specific antigens. As shown in Figure 24.1, an indirect model for HTLV-1-induced leukemogenesis may involve a number of possible mechanisms. These include chronic antigenic exposure of B cells to the virally infected T lymphocytes; constitutive "help" by infected T lymphocytes, resulting in unregulated stimulation of B lymphocytes; the release of B cell growth factors by HTLV-1-infected T lymphocytes, resulting in the expansion of B cell clones; and/or an effect on T lymphocytes, resulting in the abrogation of suppressor function and leading to unchecked amplification of B lymphocyte populations. The suggestion that HTLV-1 may have an indirect effect through immunological mechanisms on the etiology of a lymphoproliferative malignancy may have particular relevance to arguments that HTLV-1 and related viruses are involved in the pathogenesis of some neurological diseases, as discussed below.

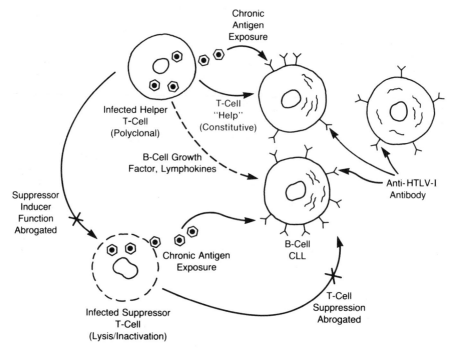

Figure 24.1. HTLV-1 models for indirect leukemogenesis. HTLV-1 may induce B cell leukemogenesis through a variety of mechanisms. For example, expansion and proliferation of the B cell compartment of HTLV-1-committed B cells may result from chronic antigenic exposure to HTLV-1 expressed by infected T cells. In addition, HTLV-1-infected T helper cells may express constitutive help to B cells if normal T regulatory processes are altered by HTLV-1 infection, or proliferation may be stimulated by B cell growth factor and other lymphokines. The effects on suppressor function could result either from abrogation of suppressor induction or through direct abrogation of T suppressor cell function by HTLV-1 infection. The potential targets of HTLV-1 are multiple.

HTLV-1 and Myelopathy

Tropical spastic paraparesis (TSP) is a chronic myelopathy of unknown etiology. Its association with HTLV-1 was first described in 1985 by Guy de The's group [4]. Neurological features, as originally recognized by Cruickshank [5] and Rodgers-Johnson [6], are predominantly those of TSP, with varying degrees of proprioceptive impairment or superficial sensory impairment or both. As is the case for ATL, TSP tends to cluster in areas of HTLV-1 endemicity, overlapping areas in which ATL also occurs. Most cases of TSP have been reported to occur on tropical islands (Jamaica, Trinidad, and Tobago; Martinique; Tumaco, Columbia; and the Seychelles). Clusters have also been observed in nontropical areas, including the more temperate latitudes of Kyushu and Shikoku in southern Japan as well as the tropical low altitude areas of Panama [7]. Cases in Africa and India and cases among migrants residing in the United Kingdom, France, the United States, and Canada have also been reported. In contrast to ATL, which has a relatively low prevalence—for ATL, incidence and prevalence are approximately equal due to short survival of ATL patients—the range of prevalence for TSP ranges from 12 (Jamaica) to 98 per hundred thousand in Tumaco and 128 per hundred thousand in the Seychelles (Table 24.1). The incidence for ATL is in the range of two to seven per hundred thousand. This difference between ATL and TSP undoubtedly reflects the long survival of patients with TSP versus the rather fulminant and acute course for ATL, which often results in the death of the affected individual within months of the onset of symptoms, compared to years to decades for those with TSP. Although formal epidemiological data have not been analyzed to document the link of HTLV-I to TSP, the descriptive data are sufficiently compelling to suggest such an association. In particular, the geographic clustering of TSP in areas of HTLV-I endemic clustering, and the high frequency of HTLV-I among patients with TSP, which ranges as high as 90% to 100% of all cases, support this association.

Table 24.1. Comparison of epidemiological features.

Feature	Adult T cell leukemia	Tropical spastic paraparesis	Multiple sclerosis
Pattern	Endemic	Endemic with occasional hyperendemic foci	Endemic with occasional epidemic foci
Prevalence/ 100,000	2–7	12–128	30–80 (max 300)
Gender (F:M)	1:1	*ca* 3:1	*ca* 2:1
Age (years)	30–60	30–60	20–60
Race	Black, Asian (low in whites)	Black, Asian (low in whites)	White (low in Blacks, Asians)
Modes of transmission			
Mother to child	Likely	Likely	Possible
Sexual intercourse	Possible	Likely	Improbable
Transfusion	Possible	Likely	Improbable
Familial aggregation	Yes	Yes	Yes
Disease risk in migrant populations	Yes	Yes	Yes

Natural History of HTLV-1 Infection

The modes of transmission of HTLV-1 are similar to those for the human immunodeficiency virus (HIV). It is well documented, primarily from studies in Japan, that HTLV-1 is efficiently transmitted from mother to child, with approximately 25% of children of seropositive mothers seroconverting within the first 2 years of life if they are breast fed [8]. Since the virus has not been consistently detectable in cord blood cells, it has been postulated that breast feeding is responsible for most perinatal transmission of the virus. As for HIV, sexual transmission occurring primarily from males to females and, to a lesser degree, from females to males accounts for the adult, age-dependent rise in HTLV-1 seroprevalence, although reactivation of silent HTLV-1 infection in later life has also been suggested. For female-to-male transmission, ulcerative genital lesions are an important cofactor [9]. In Japan among primarily monogamous couples, the transmission of HTLV-1 from husband to wife was estimated to be approximately 6% per year, whereas female-to-male transmission was less than 1% per year [10]. Transfusion transmission of HTLV-1 [11] has also been documented, as has parenteral transmission through the sharing of needles among intravenous drug users [12]. The finding of anecdotal evidence for transfusion-associated TSP has been provocative and raises the possibility that the latent period between HTLV-1 infection and disease occurrence may be shorter than that suspected for ATL. The importance of transfusion in the overall incidence of TSP is probably low, with the majority of cases occurring as the result of perinatal exposure and adult sexual intercourse. There is no evidence to support the concept that HTLV-1 is transmitted by vectors, such as mosquitoes, although this possibility has been raised, based on descriptive epidemiological data not controlled for potentially confounding factors [13] and the observation that biting flies are involved in the transmission of a related bovine leukemia virus [14]. Environmental cofactors may amplify the risk for transmission (e.g., by stimulating viremia through immune activators), and this may explain associations of HTLV-1 with certain venereal and parasitic infestations. In this regard the association of TSP with certain other cofactors, such as treponema and borrelia [15], may be the result of cofactors that influence HTLV-1 transmission or cofactors that amplify the subsequent risk for TSP in concert with HTLV-1 (see model below).

Following infection, the determinants of subsequent risk for ATL and TSP are unclear. However, the incidence of disease among infected individuals appears to be relatively low. The distinguishing epidemiological features between ATL and TSP, as summarized in Table 24.1, point to the fact that females are more frequently afflicted with TSP then are males, while age, race, familial clustering, and genetic predisposition are very similar. In particular, recent studies from Sonoda and colleagues [16] in Southern Japan have documented differences in the immunogenetic background of persons susceptible to TSP and those at risk for ATL, raising the possibility that disease occurrence is influenced by the genetic response of the individual to viral infection. In particular, it has been noted that the immune response of patients with ATL or with the ATL-associated HLA (Human Lymphocyte Antigen) haplotype. These data suggest that immune response to the virus, modulated in part by immunogenetic factors, may influence subsequent risk for ATL or TSP. The finding of familial clusters of ATL and TSP lend support to this concept.

Both ATL and TSP have been observed in migrant populations, but a female predominance for TSP was seen in all populations studied; the incidence of ATL is more or less equal between males and females. This observation, coupled with the reports of transfusion-associated TSP, has suggested that adult-acquired as well as childhood-acquired HTLV-1 infection may confer risk on the individual for TSP. The more equal male-to-female risk, despite an adult predominance of HTLV-1-infected females, suggests that childhood exposure may be most important for ATL [2].

HTLV-1 as a Model for Multiple Sclerosis

The etiology of multiple sclerosis (MS) continues to be an enigma. Data from Koprowski and colleagues, however, have demonstrated important new information suggesting a possible role for HTLV-1 or a closely related retrovirus in the etiology of this disease [17]. The finding of antibodies in the cerebrospinal fluid (CSF) of MS patients reactive with the p24 antigen, a conserved epitope of the HTLV-1 virus, and the recognition that some cases of TSP have clinical features indistinguishable from those of MS suggest the possibility that HTLV-1 or a related virus may be etiologically linked to some cases of MS. This assumption is further supported by the recent detection of HTLV-1 related proviral genome sequences in some MS patients [18,19]. The epidemiological features of ATL, TSP, and MS are compared and summarized in Table 24.1. HTLV-1-associated diseases occur primarily in tropical areas, whereas the incidence of MS is highest in temperate climates. Similarly, Caucasians are more at risk for MS; the risk among blacks and Asians is low. Epidemiological features that suggest a transmissible agent for the putative MS virus come from evidence that the disease clusters in families [20]. In addition, migrant studies in the Faroe Islands by Kurtzke and Hyllested have documented apparent epidemics of MS following the stationing of British troops in this distant outpost during World War II; this suggests the introduction of a new etiological agent into the population [21]. Clinical MS began between 1943 and 1973 in this population and comprised three epidemics, each one significantly later in time and lower in incidence than the preceding ones. The initial very high incidence peak in 1945 was reflected as a point source epidemic of MS brought to the island population by the British troops. Locations of troop encampments strongly correlated with places of residence of the Faroe Island MS patients. The average incubation period between putative acquisition of a transmissible agent and the clinical onset of disease was 6 years.

Cases occurred equally in men and women and persons in different age groups, suggesting that the putative transmissible agent was transmitted in households in a way that was analogous to enteric virus transmission. This pattern contrasts with the suggested pattern of HTLV-1 transmission that involves intimate sexual contact, primarily male to female, or other parenteral routes. Thus, if the MS agent has the characteristics of a household-transmitted agent, its transmission characteristics are very different from those reported for HTLV-1, which is relatively inefficiently transmitted. However, other epidemiological data suggest a number of shared features between HTLV-1 and its related diseases and endemic MS cases, including similarities in disease prevalence, female predominance, and age of occurrence. The recent report that MS occurs excessively among members of families with a variety of primarily B cell lymphoid malignancies [22] provides an interesting comparison to the HTLV-1- TSP-lymphoid malignancy link. Features often shared between MS and HTLV-1 diseases include HLA-linked genetic markers and the occurrence of both diseases in migrant populations. These migrant studies also suggest a long latency for both HTLV-1 disease and MS, a finding that contrasts with the Faroe Island experience. Epidemiological studies of MS have not provided evidence that the putative MS virus is transmitted through sexual intercourse or blood transfusions. However, these studies were conducted some years ago, and a reevaluation of this issue should be undertaken in light of the findings that are emerging from epidemiological studies of HTLV-1. If there is a similarity between the etiology of TSP and MS, one would expect that modes of transmission and disease risk factors would be similar. In the absence of evidence for sexual transmission, maternal-to-infant transmission of a putative MS agent, such as is observed for HTLV-1, with a long latent period between exposure and disease occurrence is an attractive alternative and is consistent with the endemic pattern of MS in northern latitudes, except for the first Faroe Island outbreak. As is likely for TSP and ATL, additional cofactors (e.g., genetic, other infections) that influence the virus–host interaction may then determine the risk for occurrence of disease. Finally, it is probable, as for HTLV-1, that the MS rate is low compared to the prevalence of putative retrovirus infection in the population.

Models for the Role of Retroviruses in Chronic Neurodegenerative Diseases

The pathogenic mechanisms for HTLV-1 and TSP have not been fully elucidated. Several alternative models have been postulated. In the first, HTLV-1 itself may be neuropathic, having direct cytopathic effects on neuronal tissue analogous to that observed for the related human HIV that causes AIDS-associated dementia. The finding of highly positive antibody titers to HTLV-1 in the CSF, isolation of HTLV-1 from CSF [23–25], and identification of HTLV-1-like particles in spinal cord cells [26] are suggestive evidence for this model. The destruction of neuronal tissue would be analogous to the finding of subacute encephalomyelitis in HIV infection resulting in AIDS dementia complex [27]. The neuropathic properties of HIV have already been established. If the neuropathic model is valid, it should be possible to identify HTLV-I from specific sites within the central nervous system (CNS) at a much higher frequency than is seen in other neurological disorders. Up until now this has not been convincingly shown. Such a "direct" model would be analogous to the postulated "direct" role of HTLV-I in ATL. However, as is also the case for ATL, there are antibody-negative cases of TSP for which other mechanisms or cryptic virus infection must be postulated.

Models for HTLV-I-associated demyelinating diseases involving indirect effects of the virus on various immune functions are presented in Figure 24.2. In the cell-mediated models, a T cell targeted to specific myelin proteins could effect demyelination through a number of possible mechanisms. First, as postulated for HTLV-I-associated B cell chronic lymphocytic leukemia, infection of T helper cells could activate the T effector cell to cause neurological damage. Second, abrogation of specific T suppressor cell function by HTLV-I infection could result in uncontrolled and harmful T effector cell function. Additionally, there are a number of humoral-mediated mechanisms for demyelination. Through "molecular mimicry" [28], antibodies to HTLV-I could secondarily become directed against myelin basic protein (MBP). The cycle would continue, since the autoimmune response would lead to tissue injury that, in turn, would release more self-antigens. This chronic autoimmune process could then cause demyelination.

Moreover, there is an additional possibility: Popovic and colleagues have shown that HTLV-I-infected, cytotoxic T lymphocytes lose their cytotoxic function [29]. This may result in persistent infections by a variety of different pathogens. Again, the immune response against any one of these pathogens may theoretically damage myelin through molecular mimicry. A number of viral proteins were shown to have homologous sequences with the encephalitogenic site of MBP, with which antibodies may cross-react [28]. The same mechanism may also be considered for exogenous antigens, such as are seen in treponemal and borrelial infections, to explain their potential cofactor role in TSP in which a high prevalence of these infections has been reported [15].

This concept is supported by the finding that HTLV-I directly affects the host immune system, possibly modulating the immune response to other antigens. Upon antigenic stimulation, a retrovirus may lead to another dimension of virus–host interactions by increasing antibody responses qualitatively and quantitatively.

The pathological findings of lymphocytic infiltration in some areas of demyelination, reports of transfusion-associated cases, the predominance of female cases possibly linked to male-female sexual transmission, and sporadic reports that immunosuppressive corticosteroid therapy results in clinical improvement are consistent with an immunologically mediated mechanism. However, no conclusive data support one or another etiology.

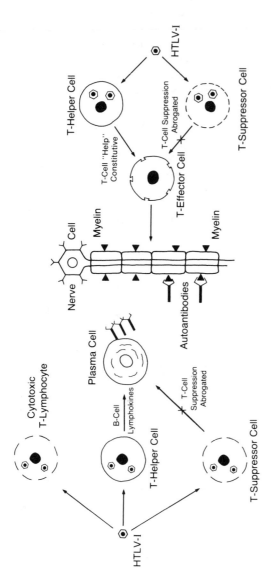

HUMORAL MEDIATED MODELS   CELL MEDIATED MODELS

Figure 24.2. HTLV-1 models in demyelinating diseases, such as tropical spastic paraparesis and multiple sclerosis. HTLV-1 may damage the CNS in these diseases by several humoral-mediated and cell-mediated mechanisms, in which HTLV-1-transformed cytotoxic T lymphocytes, T helper cells, and T suppressor cells have abrogated functions. Antibodies against HTLV-1 or other persistent pathogens (e.g., treponemal and borrelial infections in TSP, other viruses in MS) may cross-react with MBP protein through molecular mimicry. For further explanation, see the text.

Summary

The discovery of the first human retrovirus has provided an important model for considering retroviruses in the etiology of chronic diseases of unknown etiology. Lessons learned from epidemiological and pathobiological studies of the prototype virus provide a framework for testing the hypothesis that a retrovirus is involved in these other conditions. The current data available from comparative serology and epidemiological studies of MS, TSP, and ATL show compelling similarities that should further stimulate the investigation of whether cryptic human retrovirus is involved in the etiology of other chronic degenerative neurological diseases.

References

1. Murphy EL, Hanchard B, Lofters WS, Campbell M, Clark J, Cranston B, Hartge P, Gibbs WN, Blattner WA (1987) Adult T-cell leukemia/lymphoma and human T-cell lymphotropic virus type I in Jamaica. Proc Am Soc Clin Oncol 6:4
2. Murphy EL, Hanchard B, Figueroa JP, Gibbs WN, Goedert JJ, Blattner WA (1989) Modelling the risk of adult T-cell leukemia/lymphoma in persons infected with human T-lymphotropic virus type I. Int J Cancer 43:250–253
3. Mann DL, DeSantis P, Mark G, Pfeifer A, Newman M, Gibbs N, Popovic M, Sarngadharan MG, Gallo RC, Clark J, Blattner W (1987) HTLV-I-associated B-cell CLL: Indirect role for retrovirus in leukemogenesis. Science 236:1103–1106
4. Gessain A, Barin F, Vernant JC, Gout O, Maurs L, Calender A, DeThe (1985) Antibodies to human T-lymphotropic virus type-I in patients with tropical spastic paraparesis. Lancet ii:407–410
5. Cruickshank EK (1956) A neuropathic syndrome of uncertain origin. West Indian Med J 5:147–158
6. Rodgers PEB (1965) The clinical features and aetiology of the neuropathic syndrome in Jamaica. West Indian Med J 14:36–47
7. Roman GC (1988) The neuroepidemiology of tropical spastic paraparesis. Ann Neurol 23(suppl):S113–120
8. Hino S, Yamaguchi K, Katamine S, Sugiyama H, Amagasaki T, Kinoshita K, Yoshida Y, Doi H, Tsuji Y, Miyamoto T (1985) Mother-to-child transmission of human T-cell leukemia virus type-I. Gann 76:474–480
9. Murphy EL, Figueroa JP, Gibbs WN, Brathwaite A, Holding-Cobham M, Waters D, Cranston B, Hanchard B, Blattner WA Sexual transmission of HTLV-I. (submitted)
10. Kajiyama W, Kasniwagi S, Ikematsu H, Hayashi J, Nomura H, Okochi K (1980) Intrafamilial transmission of adult T-cell leukemia virus. J Infect Dis 154:851–856
11. Osame M, Izumo S, Igata A, Matsumoto M, Matsumoto T, Sonoda S, Tara M, Shibata Y (1986) Blood transfusion and HTLV-I associated myelopathy. Lancet ii:104–105
12. Robert-Guroff M, Weiss SH, Giron JA, Jennings AM, Ginzburg HM, Margolis IB, Blattner WA, Gallo RC (1986) Prevalence of antibodies to HTLV-I, -II, and -III in intravenous drug abusers from an AIDS endemic region. JAMA 255:3133–3137
13. Miller GJ, Pegram SM, Kirkwood BR, Beckles GL, Byam NT, Clayden SA,

Kinlen LJ, Chan LC, Carson DC, Greaves MF (1986) Ethnic composition, age and sex, together with location and standard of housing as determinants of HTLV-I infection in an urban Trinidadian community. Int J Cancer 38:801–808

14. Bech-Nielsen S, Piper CE, Ferrer JF (1978) Natural mode of transmission of the bovine leukemia virus: Role of bloodsucking insects. Am J Vet Res 39:1089–1092
15. Rodgers-Johnson P, Morgan OStC, Mora C, Sarin P, Ceroni M, Piccardo P, Garruto RM, Gibbs CJ, Gajdusek DC (1988) The role of HTLV-I in tropical spastic paraparesis in Jamaica. Ann Neurol 23(suppl):S121–S126
16. Usuku K, Sonoda S, Osame M, Yashiki S, Takahashi K, Matsumoto M, Sawada T, Tsuji K, Tara M, Igata A (1988) HLA haplotype-linked high immune responsiveness against HTLV-I in HTLV-I-associated myelopathy: Comparison with adult T-cell leukemia/lymphoma. Ann Neurol 23(suppl):S143–S150
17. Koprowski H, DeFreitas E (1988) HTLV-I and chronic nerve diseases: Present status and a look into the future. Ann Neurol 23(suppl):S166–S170
18. Reddy PE, Magnhild S, Mettus RV, Ray PE, DeFreitas E, Koprowski H (1989) Amplification and molecular cloning of HTLV-I sequences from DNA of Multiple Sclerosis patients. Science 243:529–533
19. Greenberg, SJ, Ehrlich GD, Abbott MA, Hurwitz BJ, Waldmann TA, Poiesz BJ Detection of sequences homologous to human retroviral DNA in Multiple Sclerosis by gene amplification. Proc Natl Acad Sci (in press)
20. Kranz JS, Kurland LT (1982) General overview of the epidemiology of multiple sclerosis with emphasis on the geographic pattern and long-term trends. In Kuroiwa Y, Kurland LT (eds) Multiple sclerosis: East and West. Karger, Basel, New York, pp 3–29
21. Kurtzke JF, Hyllested K (1987) MS epidemiology in Faroe Islands. Riv Neurol 57:77–87
22. Bernard SM, Cartwright RA, Darwin CM, Richards IDG, Roberts B, O'Brien C, Bird CC (1986) A possible epidemiological association between multiple sclerosis and lymphoma/leukaemia. Br J Haematol 65:122–123
23. Hirose S, Uemura Y, Fujishita M, Kitagawa T, Yamashita M, Imamura J, Ohtsuki Y, Taguchi H, Miyoshi I (1986) Isolation of HTLV-I from cerebrospinal fluid of a patient with myelopathy. Lancet ii:397–398
24. Jacobson S, Raine CS, Mingioli ES, McFarlin DE (1988) Isolation of an HTLV-I-like retrovirus from patients with tropical spastic paraparesis. Nature 331:540–543
25. Bhagavati S, Ehrlich G, Kula RW, Kwok S, Sninsky J, Udani V, Poiesz BJ (1980) Detection of human T-cell lymphoma/leukemia virus type I DNA and antigen in spinal fluid and blood of patients with chronic progressive myelopathy. New Engl J Med 318:1141–1147
26. Liberski PP, Rodgers-Johnson P, Char G, Piccardo P, Gibbs CJ, Gajdusek C (1988) HTLV-I-like viral particles in spinal cord cells in Jamaican tropical spastic paraparesis. Ann Neurol 23:S185–S187
27. de la Monte SM, Ho DD, Schooley RT, Hirsch MS, Richardson EP (1987) Subacute encephalomyelitis of AIDS and its relation to HTLV-III infection. Neurology 37:562–569
28. Oldstone MBA (1987) Molecular mimicry and autoimmune disease. Cell 50:819–820
29. Popovic M, Flomenberg N, Volkman DJ, Mann D, Fauci AS, Dupont B, Gallo RC (1984) Alteration of T-cell functions by infection with HTLV-I or HTLV-II. Science 226:459–462

CHAPTER 25
Mechanism of CD4+ Cell Killing by the Human Immunodeficiency Virus

JAMES A. HOXIE AND JAY A. LEVY

Infection by the human immunodeficiency virus (HIV) leads to progressive abnormalities in the host immune system that, in its most advanced stage, is diagnosed clinically as the acquired immunodeficiency syndrome (AIDS). Although HIV has been shown to infect a number of cell types that are essential components of the immune system, the mechanism by which infection leads to disease is unclear. Several studies have indicated that this process likely involves direct effects of the virus on infected cells, as well as indirect effects of the host immune response. This review will discuss the cytopathic effects of HIV infection on CD4+ lymphocytes observed in vitro and potential mechanisms for their loss during infection in vivo.

Cytopathic Effects of HIV Infection In Vitro

The early events of HIV infection of lymphoid cells, including viral binding to and penetration of target cells, are highly dependent on the interaction between the 120-kd viral envelope glycoprotein gp120 and the 56-kd cell surface molecule CD4 [1]. The specificity of this interaction is mediated by a region near the carboxy terminus of gp120, which contributes to the formation of a high-affinity binding site for the amino terminus of CD4 [2–4]. Monoclonal antibodies reactive with distinct epitopes on CD4 [5], as well as soluble forms of the CD4 protein itself [6,7], have been shown to neutralize HIV by inhibiting binding of the virus to cells. Moreover, human B cells and HeLa cells that lack CD4 and are resistant to HIV infection can be rendered susceptible to infection following transfection of the CD4 gene and expression of this protein on the cell surface [8].

This envelope–CD4 interaction also contributes to the cytopathic effects that occur during viral replication. HIV infection of cultured peripheral blood CD4+ lymphocytes and lymphoid cell lines leads to cell fusion and the formation of multinucleated syncytia [9,10]. Fusion clearly contributes to the reduced viability of cultured cells [11] and is mediated specifically by the interaction of the viral envelope with the CD4 molecule [10,12]. This interaction has also been shown to be important in producing cell death, as indicated by the finding that the expression of the HIV envelope gene in CD4+ T cell lines leads to both cell fusion and killing but has no effect when transfected into B cell lines and HeLa cells that lack CD4 [10]. In addition, neither killing nor fusion has been seen during HIV replication in CD8+ T lymphocytes, even with viruses that are highly cytotoxic for CD4+ lymphocytes [13,14]. Some reports have noted a correlation between the extent of cell fusion and killing induced by HIV with the amount of CD4 expressed on the cell surface [15]. Moreover, peripheral blood monocytes, which express low levels of CD4 compared to T lymphocytes, often show minimal or no cytopathic effects during HIV infection in vitro [16,17]. Nevertheless, as discussed below, some CD4+ cell lines may be killed by HIV without undergoing cell fusion [18].

The mechanism by which the HIV envelope–CD4 interaction leads to fusion and cell death has not been fully elucidated. However, mutational analyses of the HIV envelope have identified highly conserved domains on gp120 that are involved in binding to CD4 [2,3] and on the amino terminus of gp41 that induce fusion between the viral membrane and the cellular plasma

Table 25.1. Mechanism of CD4+ cell killing by HIV.

Possible mechanisms for killing of CD4+ cells in vitro
   Syncytia formation
   Cytopathic effects of the viral envelope on single cells
      Activation of envelope fusogenic domains within cells
      Increased permeability of the plasma membrane causing balloon
         degeneration
   Accumulation of unintegrated forms of proviral DNA
   Second-round superinfection by HIV
   Alterations in host cell protein synthesis

Possible mechanisms for depletion of CD4+ cells in vivo
   Direct killing of CD4+ cells by HIV
   Syncytia formation
   Autoimmune response against activated and virus-infected CD4+
      cells
   Antibody-dependent cellular cytotoxicity (ADCC) against
      virus-infected cells and cells to which the viral envelope has
      attached
   Direct infection of lymphoid stem cells or other cells necessary
      for stem cell proliferation and differentiation
   Emergence of more virulent, highly cytopathic viruses in the host

membrane [2,19,20]. As with paramyxoviruses and orthomyxoviruses, the generation of a functional fusion domain on the envelope is absolutely dependent on cleavage of the gp160 envelope precursor molecule by a cellular protease [19]. In addition to the role of these envelope determinants in mediating fusion during viral entry and the formation of syncytia between CD4+ cells, it is also probable that the interaction of the viral envelope with CD4 is cytopathic for single cells. Complexes of CD4 with envelope precursor molecules have been detected in lysates of infected cells, which suggests that the interaction of CD4 with viral envelope occurs intracellularly during processing of the envelope glycoproteins [21]. Such an interaction could be cytopathic to single cells by exposing fusogenic domains on the envelope that could disrupt cell membranes within the Golgi or on the cell surface. Thus, susceptibility of a cell to killing by HIV may reflect a critical balance between the amounts of viral envelope and CD4 protein present on or within cells during infection [10]. Interestingly, in studies of the avian leukosis virus it has been determined that the cytopathic determinants of these viruses also localize to receptor-binding regions on the viral envelope [22].

Although the interaction of the viral envelope with CD4 and the subsequent generation of a functional fusion domain is probably important in mediating HIV-induced cytopathic effects, several studies have indicated that additional cellular factors are important in determining the susceptibility of a cell to infection as well as to the cytopathic effects that follow infection. For example, some results have indicated that although HIV can bind to nonhuman cells expressing a transfected human CD4 gene, productive infection does not occur [2,8]. Moreover, differences in the susceptibility of T cell lines to infection by a single isolate of HIV that did not correlate with the amount of CD4 expressed on the cell surface have been described [18,23,24]. In addition, several studies have questioned the extent to which cell fusion and killing of cells by HIV are necessarily related, including reports describing both noncytopathic [25] and cytopathic [18] infection without fusion or syncytia formation.

The role of cellular factors in addition to CD4 that determine cell susceptibility to infection has also been suggested by a study of the HIV-related simian immunodeficiency virus, SIVmac. This virus was found to induce syncytia and to kill several human CD4+ lymphoid cell lines, but was unable to infect, fuse, or kill the CD4+ cell line, CEM [26]. However, a somatic cell hybrid between CEM and a B lymphoblastoid cell line was highly susceptible to both fusion and infection by SIVmac, indicating that cell surface molecules in addition to CD4 were likely required by this virus for entry. This study also found that SIVmac produced a completely noncytopathic infection of the Sup-T1 T cell line, even though these cells were shown to be exquisitely sensitive to the cytopathic effects of HIV-1 [21]. Because Sup-T1 cells were also found to be resistant to cell fusion when cocultured with SIVmac-infected lines, it was proposed that Sup-T1 cells lacked determinants that, in addition to CD4, contributed to the cytopathic effects during viral replication.

Viral factors have also been implicated in the cytopathic effects of HIV infection. Isolates of HIV-1 can differ in their ability to infect, fuse, and kill various lymphoid cell lines [24,27] and peripheral blood lymphocytes [24]. In addition, biological differences with respect to rate and extent of replication and ability to induce cytopathic effects have been found for viruses isolated from asymptomatic seropositive individuals (replicating slowly and with low levels of reverse transcriptase) as compared to viruses from patients with advanced immunodeficiency (replicating rapidly and with high levels of reverse transcriptase) [28]. Similar observations have been made in comparisons of viral isolates from the same individual before and after the onset of clinical symptoms. These results suggest that the development of clinical disease can be linked to variations in the virulence of the virus itself [29]. Moreover, viral isolates from individuals with predominantly neurological symptoms rather than immunodeficiency have been found to replicate in lymphocytes without inducing cell killing [30]. Finally, recently described isolates of HIV-2 have been found to replicate efficiently in some lymphoid cell lines and peripheral blood lymphocytes without inducing substantial cell fusion or killing [31,32]. Although the viral determinants responsible for this biological heterogeneity have not been precisely identified, it is likely that they will involve the viral envelope, which is known to vary markedly among different isolates [33].

A number of other potential mechanisms for cell killing have been implicated by HIV, as well as for other cytopathic retroviruses, including the accumulation of unintegrated forms of proviral DNA [34] and the more general effects of viral infection on the synthesis of cellular proteins [35]. It has been proposed that, following the initial infection and production of infectious virions, cell killing may result from a high level of viral replication when cells are reinfected by the virions they produce; this process has been termed *second-round superinfection* [34,36]. Viewed in this context, the ability of the HIV gene products to bind to and to down modulate the CD4 receptor on cells that survive infection [21] could be an important mechanism in preventing the cytopathic effects of viral superinfection. Consistent with this hypothesis is the finding that the HIV envelope gene transfected into CD4+ cells was able to protect cells from cytopathic effects when these same cells were subsequently infected with HIV [36]. It was proposed that this protection is the result of partial interference by the transfected envelope gene products of subsequent rounds of viral entry and replication. Contrary to this hypothesis, however, is the observation that a noncytopathic strain of HIV-2 replicates to high titer in CD4+ cells but does not modulate CD4 expression [31]. Finally the possible effect of HIV gene products on cell membrane permeability and ion exchange should be considered as a cause for the ballooning and degeneration of single cells observed in vitro [37].

Cytopathic Effects of HIV Infection on CD4+ Lymphocytes In Vivo

One of the earliest immunological abnormalities recognized in patients with HIV infection was a progressive reduction in the number of CD4+ lymphocytes. The cytopathic effects of HIV infection observed on these cells in vitro initially suggested that CD4 cells were depleted in vivo as a direct consequence of viral infection. Several observations, however, indicate that other mechanisms are likely involved. The small number of infected CD4+ lymphocytes detected in the peripheral blood or lymph nodes (roughly 0.01%) does not explain the continual and progressive loss of these cells over time in infected individuals [38]. In addition, although it has been proposed that viral-induced fusion and syncytia formation could be an important mechanism for CD4 cell depletion, histological studies of a variety of lymphoid tissues from HIV-infected individuals have not shown a substantial number of multinucleated cells [39]. Nevertheless, progression to disease has been linked to a relatively acute drop in the number of CD4+ cells over a short period of time [38]. This observation may be related to the recognition that more cytopathic variants of HIV emerge with time in infected hosts [29]. These latter viruses replicate rapidly and kill infected cells quickly. It seems clear, however, that other mechanisms for CD4+ cell depletion are also involved. For instance, some HIV-infected individuals have immunological disorders and reduced numbers of CD4+ cells in the presence of relatively noncytopathic HIV isolates [30,38].

Autoimmune Responses to CD4+ Cells

Several investigators have reported that antilymphocyte antibodies can be detected in a large number of individuals with AIDS [40–42]. In general, these immunoglobulins react with T helper cells, but antibodies to T suppressor and B lymphocytes have also been described. Stricker and colleagues [41] examined the autoantibodies to T helper cells and detected their reactivity with an 18,000 Mr (p18) protein expressed on the surface of virus-infected as well as activated CD4+ lymphocytes and macrophages. The IgG antibodies involved are cytotoxic in the presence of complement. This observation suggests that autoantibody response can be a contributing factor in the decrease in CD4+ cells during HIV infection.

The reason for the induction of autoantibodies is unclear. Polyclonal activation of B lymphocytes occurs early in HIV infection, perhaps secondary to mitogenic effects of the envelope protein, cytokine production, or a disruption in normal T cell control [38]. The hypergammaglobulinemia could cause the observed pathology. However, the selective production of anti-lymphocyte antibodies in certain individuals and its association with disease suggests an alternate mechanism. Several years ago Lindenmann and colleagues showed that virus infection of tumor cells enhanced the recognition of normal cellular proteins on the surface of malignant cells [43]. Mice immunized with virus-infected tumor cells or purified virus from the tumors recognized and eliminated uninfected tumor cells that had previously killed the host cells. This observation suggests that a carrier-hapten mechanism may be involved. Similarly, HIV infection of CD4+ cells may induce the production of autoantibodies against a normal cellular protein [42]. Although this mechanism should be studied further, an approach to reduce the autoantibody production may benefit infected individuals by protecting them from CD4+ cell depletion.

Depletion of Uninfected CD4+ Lymphocytes by
Cellular Immune Responses
Another potential mechanism for the loss of T helper lymphocytes is antibody-dependent cellular cytotoxicity (ADCC). First described using effector cells from normal donors and antibodies from HIV-infected individuals, this mechanism is mediated principally by antibodies to the HIV envelope molecules and could be helpful early after infection when relatively few virus-infected cells are present. However, the viral envelope produced by relatively few infected cells can also bind to the CD4 molecule on uninfected lymphocytes, and lead to the killing of these cells in the absence of infection [44]. Finally some studies have suggested that CD4+ cells, by binding gp120 to their surface, can process this antigen so it is recognized by antiviral cytotoxic cells [45,46]. This phenomenon, if present in vivo, could also contribute to the depletion of CD4+ cells in AIDS. In addition, Ziegler and Stites [47] have proposed that autoreactive T suppressor cells that recognize major histocompatibility (MHC) antigens on activated T cells or virus-infected T cells may show a cytotoxic response involving these normal cellular proteins. In this regard, autoantibodies to regions of the viral gp41 that resemble the HLA class II molecule have recently been shown to suppress CD4+ cell function [48].

Infection of Stem Cells
It has been suggested that HIV infection could indirectly lead to a depletion of CD4+ lymphocytes by infecting a stem cell critical for the regeneration of the lymphoid population. In this regard, CD4− progenitors of T cells in the bone marrow have been found to be susceptible to HIV infection [49,50]. Alternatively, HIV infection could impair the function of a cell that has an essential role in lymphocyte maturation. CD4+ lymphocytes depend on the production of IL-2 and other cytokines for their growth and differentiation, and infection by HIV and the resulting loss of cytokine production could further reduce the number of CD4+ cells [38]. The responsiveness of myeloid progenitor cells to specific cytokines has been shown to be impaired, particularly in the presence of antibodies to HIV envelope glycoproteins [51].

These in vitro and in vivo studies can direct approaches to antiviral therapy. Recognition of the "virulent" genes of the virus may permit selective therapeutic modalities, and suppression of anti-lymphocyte antibodies could greatly enhance CD4+ cell survival and perhaps lead to a long asymptomatic period. Moreover, a strong immune response against HIV might prevent it from undergoing further replication and spread and thus suppress the emergence of more cytopathic variants. In this regard, the antiviral effect of CD8+ cells merits attention [52].

Acknowledgments

The research by the authors cited in this paper was supported by Grants A12360 and AI-24499 from the National Institutes of Health.

References

1. McDougal JS, Kennedy, MS, Sligh, JN, Cort, SP, Kennedy, SM, Mawle, AC (1986) Binding of HTLV-III/LAV to T4+ cells by a complex of 110K viral protein and the T4 molecule. Science 231:382–385
2. Lasky LA, Nakamura G, Smith JH, Fennie C, Shimosaki L, Patzer E, Berman P, Gregory T, Capon DJ (1987) Delineation of a region of the human immunodeficiency virus type 1 (HIV-1) gp 120 glycoprotein critical for interaction with the CD4 receptor. Cell 50:975–985
3. Kowalski M, Potz J, Basiripour L, Dorfman T, Goh WC, Terwilliger E, Dayton A, Rosen C, Haseltine W, Sodroski J (1987) Functional regions of the human immunodeficiency virus envelope glycoprotein. Science 237:1351–1355
4. Jameson BA, Rao PE, Kong LI, Hahn BH, Shaw GM, Hood LE, Kent SBH (1988) Locaton and chemical synthesis of a binding site for HIV-1 on the CD4 protein. Science 240:1355–1339
5. Sattentau QJ, Dalgleish AJ, Weiss RA, Beverley PCL (1986) Epitopes of CD4 antigen and HIV infection. Science 234:1120–1122
6. Smith DH, Byrn RA, Marsters SA, Gregory T, Groopman JE, Capon DJ (1987) Blocking of HIV-1 infectivity by a soluble, secreted form of the CD4 antigen. Science 238:1704–1707
7. Weiss RA (1988) Receptor molecule blocks HIV. Nature 331:15
8. Maddon PJ, Dalgleish AG, McDougal JS, Clapham PR, Weiss RA, Axel R (1986) The T4 gene encodes the AIDS virus receptor and is expressed in the immune system and the brain. Cell 47:333–348
9. Lifson JD, Reyes GR, McGrath MS, Stein BS, Engleman EG (1986) AIDS retrovirus-induced cytopathology: Giant cell formation and involvement of the CD4 antigen. Science 232:1123–1127
10. Sodroski J, Goh WC, Rosen C, Campbell K, Haseltine WA (1986) Role of the HTLV-III/LAV envelope in syncytium formation and cytopathicity. Nature 322:470–474
11. Yoffe B, Lewis DE, Petrie BL, Noonan CA, Melnick JL, Hollinger FB (1987) Fusion as a mediator of cytolysis in mixtures of uninfected CD4+ lymphocytes and cells infected by human immunodeficiency virus. Proc Natl Acad Sci USA 84:1429–1433
12. Lifson JD, Feinberg MB, Reyes GR, Rabin L, Banapour S, Chakrabarti S, Moss

B, Wong-Staal F, Steimer KS, Engelman EG (1986) Induction of CD4-dependent cell fusion by the HTLV-III/LAV envelope glycoprotein. Nature (London) 323:725–728

13. DeRossi A, Franchini G, Aldovini A, DelMistro A, Chieco-Bianchi L, Gallo RC, Wong-Staal F (1986) Differential response to the cytopathic effects of human T-cell lymphotropic virus type III (HTLV-III) superinfection in T4+ (helper) and T8+ (suppressor) T-cell clones transformed by HTLV-1. Proc Natl Acad Sci USA 83:4297–4301

14. Cheynier R, Soulha M, Loure F, Vol JC, Reveil B, Gallo RC, Sarin PC, Zagury D. (1988) HIV-1 expression by T8 lymphocytes after transfection. AIDS Res Human Retroviruses 4:43–50

15. Åsjö B, Ivhed I, Gidlund M, Fuerstenberg S, Fenyo EM, Nilsson E, Wigzell H (1987) Susceptibility to infection by the human immunodeficiency virus (HIV) correlates with T4 expression in a parental monocytoid cell line and its subclones. Virology 157:359–365

16. Gartner S, Markovitz P, Markovitz DM, Kaplan MH, Gallo RC, Popovic M (1986) The role of mononuclear phagocytes in HTLV-III/LAV infection. Science 23:215–219

17. Ho DD, Rota TR, Hirsch MS (1986) Infection of monocyte/macrophages by human T lymphotropic virus type III. J Clin Invest 77:1712–1715

18. Somasundaran M, Robinson HL (1987) A major mechanism of human immuno-deficiency virus-induced cell killing does not involve cell fusion. J Virol 61:3114–3119

19. McCune JM, Rabin LB, Feinberg MB, Lieberman M, Kosek JC, Reyes GR, Weissman IL (1988) Endoproteolytic cleavage of gp160 is required for the activation of human immunodeficiency virus. Cell 53:55–67

20. Stein BS, Gowda SD, Lifson JK, Penhallow RC, Bensch KG, Engleman EG (1987) pH-independent HIV entry into CD4-positive T cells via virus envelope fusion to the plasma membrane. Cell 49:659–668

21. Hoxie JA, Alpers JD, Rackowski JLJ, Huebner K, Haggarty BS, Cedarbaum AJ, Reed JC (1986) Alterations in T4 (CD4) protein and mRNA synthesis in cells infected with HIV. Science 234:1123–1127

22. Dorner AJ, Coffin JM (1986) Determinants for receptor interaction and cell killing on the avian retrovirus glycoprotein gp85. Cell 45:365–374

23. Kikukawa R, Koyanagi Y, Harada S, Kobayashi N, Hatanaka M, Yamamoto N (1986) Differential susceptibility to the acquired immunodeficiency syndrome retrovirus in cloned cells of human leukemic T-cell line Molt-4. J Virol 57:1159–1162

24. Evans LA, McHugh TM, Stites DP, Levy JA (1987) Differential ability of human immunodeficiency virus isolates to productively infect human cells. J Immunol 138:3415–3418

25. Casareale D, Stevenson M, Sakai K, Volsky DJ (1987) A human T-cell line resistant to cytopathic effects of the human immunodeficiency virus (HIV). Virology 156:40–49

26. Hoxie JA, Haggarty BS, Bonser SE, Rackowski JL, Shan H, Kanki PJ (1988) Biological characterization of a simian immunodeficiency virus-like retrovirus (HTLV IV): Evidence for CD4-associated molecules required for infection. J Virol 62:2557–2568

27. Dahl K, Martin K, Miller G (1987) Differences among human immunodeficiency

virus strains in their capacities to induce cytolysis or persistent infection of a lymphoblastoid cell line immortalized by Epstein-Barr virus. J Virol 61:1602–1608

28. Åsjö B, Morfeldt-Manson L, Alberg J, Biberfeld G, Karlsson A, Lidman K, Fenyo EM (1986) Replicative capacity of human immunodeficiency virus from patients with varying severity of HIV infection. Lancet ii:660–662

29. Cheng-Mayer C, Seto D, Tateno M, Levy JA (1988) Biological features of HIV-1 that correlate with virulence in the host. Science 240:80–82

30. Anand R, Siegal F, Reed C, Cheung T, Forlenza S, Moore J (1987) Non-cytocidal natural variants of human immunodeficiency virus isolated from AIDS patients with neurological disorders. Lancet ii:234–238

31. Evans LA, Moreau J, Odehouri K, Legg H, Barboza A, Cheng-Mayer C, Levy JA (1988) Characterization of a noncytopathic HIV-2 strain with unusual effects on CD4 expression. Science 240:1522–1525

32. Kong LI, Lee S-W, Kappes JC, Parkin JS, Decker D, Hoxie JA, Hahn BH, Shaw GM (1988) West African HIV-2-related human retrovirus with attenuated cytopathicity. Science 240:1525–1528

33. Wong-Staal F, Shaw GM, Hahn BH (1985) Genomic diversity of human T-lymphotropic virus type III (HTLV-III). Science 229:759–762

34. Temin HM (1988) Mechanisms of cell killing/cytopathic effects by nonhuman retroviruses. Rev Infect Dis 10:399–405

35. Stevenson M, Zhang X, Volsky DJ (1987) Downregulation of cell surface molecules during noncytopathic infection of T cells with human immunodeficiency virus. J Virol 61:3741–3748

36. Stevenson M, Meier C, Mann AM, Chapman N, Wasiak A (1988) Envelope glycoprotein of HIV induces interference and cytolysis resistance in CD4+ cells: Mechanism for persistence in AIDS. Cell 53:483–496

37. Garry RF, Gottlieb AA, Zuckerman KP, Pace JR, Frank TW, Bostick DA (1988) HIV-induced alterations in intracellular monovalent cations. IV International Conference on AIDS: Abst no. 1039

38. Levy JA (1988) The mysteries of HIV: Challenges for therapy and prevention. Nature 333:519–522

39. Cohen MB, Beckstead JH (1988) Pathology of AIDS. In Levy JA (ed) AIDS: Pathogenesis and Treatment. Marcel Dekker, New York (in press)

40. Kiprov DD, Anderson RE, Morand PR, Simpson DM, Chermann JC, Levy JA, Moss AR (1985) Antilymphocyte antibodies and seropositivity for retroviruses in groups at high risk for AIDS. N Engl J Med 312:1517

41. Stricker RB, McHugh TM, Moody DJ, Morrow WJW, Stites DP, Shuman MA, Levy JA (1987) An AIDS-related cytotoxic autoantibody reacts with a specific antigen on stimulated CD4+ T cells. Nature 327:710–713

42. Levy JA (1989). The human immunodeficiency viruses: Detection and pathogenesis. In Levy JA (ed) AIDS: Pathogenesis and Treatment. Marcel Dekker, New York

43. Lindenmann J (1974) Viruses as immunological adjuvants in cancer. Biochim Biophys Acta 49:355–375

44. Weinhold KJ, Lyerly HK, Matthews TJ, Tyler DS, Ahearne PM, Stine KC, Langlois AJ, Durack DT, Bolognesi DP (1988) Cellular anti-gp120 cytolytic reactivities in HIV-1 seropositive individuals. Lancet i:902–904

45. Lanzavecchia A, Roosnek E, Gregory T, Berman P, Abrignani S (1988) T cells

can present antigens such as HIV gp120 targeted to their own surface molecules. Nature 334:530–532

46. Siliciano RF, Lawton T, Knall C, Karr RW, Berman P, Gregory T, Reinherz EL (1988) Analysis of host–virus interactions in AIDS with anti-gp120 T cell clones: Effect of HIV sequence variation and a mechanism for CD4+ cell depletion. Cell 54:561–575

47. Ziegler JL, Stites DP (1986) Hypothesis: AIDS is an autoimmune disease directed at the immune system and triggered by a lymphotropic retrovirus. Clin Immunol Immunopathol 41:305–313

48. Golding H, Shearer GM, Hillman K, Lucas P, Manischewitz J, Zajac RA, Clerici M, Gress RE, Boswell RN, Golding B (1989) Common epitope in human immunodeficiency virus (HIV) 1-GP41 and HLA class II elicits immunosuppressive autoantibodies capable of contributing to immune dysfunction of HIV 1–infected individuals. J Clin Invest 83:1430–1435

49. Folks TM, Kessler SW, Orenstein JM, Justement JS, Jaffe ES, Fauci AS (1988) Infection and replication of HIV-1 in purified progenitor cells of normal human bone marrow. Science 242:919–922

50. Lunardi-Iskandar Y, Nugeyre MT, Georgoulias V, Barre-Sinoussi F, Jasmin C, Chermann JC (1989) Replication of the human immunodeficiency virus 1 and impaired differentiation of T cells after *in vitro* infection of bone marrow immature T cells. J Clin Invest 83:610–615

51. Donahue RE, Johnson MM, Zon LI, Clark SC, Groopman JE (1987) Suppression of in vitro haematopoiesis following human immunodeficiency virus infection. Nature 326:200–203

52. Walker CM, Moody DJ, Stites DP, Levy JA (1986) CD8+ lymphocytes can control HIV infection in vitro by suppressing virus replication. Science 234:1563–1566

CHAPTER 26

Human and Simian Immunodeficiency Viruses in Nonhuman Primates

PRESTON A. MARX

Three Nonhuman Primate Lentivirus Models of AIDS

Human immunodeficiency viruses (HIV-1 and HIV-2) and the simian immunodeficiency viruses (SIVs) are members of the lentivirus subfamily of retroviruses. HIV-1 and HIV-2 were isolated from humans with AIDS, and SIVs, a related group, were isolated from various species of African and Asian monkeys (Table 26.1). These three lentiviruses have been used for three different animals models of AIDS. Each model has different applications; their use and current status are the subject of this review. The three animal models are (1) infection of chimpanzees (*Pan troglodytes*) with HIV-1 [1], (2) infection of rhesus macaques (mac) (*Macaca mulatta*) with HIV-2 [2], and (3) infection of various macaque species (monkeys belonging to the genus *Macaca*) with one of the various SIVs (Table 26.1) found in Asian macaques, West African sooty mangabeys (sm), and East African green monkeys (agm) [3–9].

HIV in Nonhuman Primates

Although numerous animal species, including nonhuman primates [10], have been inoculated with HIV-1, only chimpanzees have become persistently viremic [1]. Thus, the only in vivo model for HIV-1 infection is the chimpanzee. None of these chimpanzees, however, has developed AIDS. Inoculation of the vaginal mucosa with HIV-1, but not the oral mucosa, has induced viremia [11]. Also in this study, HIV did not spread from HIV-infected chimpanzees to uninfected animals housed in the same cage [11]. Infected chimpanzees had an HIV-antibody response similar to that

Table 26.1. Simian immunodeficiency viruses of asian and african monkeys.

Old world monkeys	Current designation[a]	Infected animal location(s)	Natural range of infected monkeys
Asian species			
Rhesus Macaque (*Macaca mulatta*)	SIVmac[b]	NIH primate colony, USA	India and China
Cynomolgous macaque (*M. fascicularis*)	SIVcyn	NIH primate colony, USA	Southeast Asia
Pig-Tailed macaque (*M. nemestrina*)	SIVMne	NIH primate colony, USA	Indonesia and Southeast Asia
Stump-tailed macaque (*M. arctoides*)	SIVStm	NIH primate colony, USA	Indonesia and Southeast Asia
African species			
Sooty mangabey (*Cercocebus atys*)	SIVsm	Primate colonies and zoo, USA	West Africa
African green monkey (*Cercopithecus aethiops*) vervet subspecies	SIVagm	Primate colony, Japan, and primate colony, Kenya	East Africa[c]

[a] A parenthetical identification is sometimes used to distinguish individual isolates, for example, SIVagm(tyo) and SIVagm(ken) distinguishes the Tokyo and Kenyan isolates of SIVagm.
[b] Although rhesus, cynomolgous, pig-tailed, and stump-tailed monkeys are all macaques (members of the monkey genus *Macaca*), by convention, the term SIVmac has been reserved for SIVs isolated from the rhesus macaque. SIVs from macaques other than the rhesus are identified by either a common or scientific name subscript (e.g., SIVStm for SIVStump-tailed macaque and SIVMne for SIVM. nemestrina).
[c] AGMs are found throughout Africa, but thus far, confirmed SIV isolates are from AGMs found in East Africa only [9].

seen in infected humans [1]. In regard to in vitro inoculations, the peripheral blood mononuclear cells (PBMCs) of macaques could be infected [7].

The HIV-2 was inoculated into rhesus macaques, but persistent viremia was established only in a minority of inoculated animals [2]. The HIV-2 infected rhesus macaques have remained healthy.

A major use of the HIV-1/chimpanzee model will be vaccine testing. Thus far, chimpanzees have been immunized with HIV-1 virion envelope preparations, recombinant expressed envelope, and vaccinia vectors expressing the

HIV envelope. In one study, envelope gp120 expressed as a herpesvirus fusion protein in Chinese hamster ovary cells induced neutralizing antibody and cell-mediated immunity, but did not protect chimpanzees from an intravenous HIV-1 challenge [12].

SIV in Nonhuman Primates

The initial isolate of SIV, SIVmac, came from rhesus macaques with either lymphoma or immunodeficiency [3]. Inoculation of SIVmac [13], SIVsm (SIVsooty mangabey) [6–8], or SIVMne (SIV*macaca nemestrina*) [4] into rhesus or pig-tailed macaques has consistently induced AIDS. In contrast, the pathogenesis of the SIV isolate from agms (*Cercopithecus aethiops*) [9] in macaques is not known. The pathogenesis of SIV in their original African host is also unclear, but it appears that most infected East African agms and sms are inapparent carriers [6–9]. Seroepidemiological surveys suggest that other species of African monkeys such as talapoins (*Miopithecus talapoin*), De Brazza's monkeys (*Cercopithecus neglectus*), and mandrills (*Mandrillus sphinx*) may have inapparent infections with SIV-related retroviruses [8]. Other isolates of SIV-like viruses such as simian T cell lymphotropic virus III agm (STLV-IIIagm) and human T cell lymphotropic virus 4 (HTLV-4) are indistinguishable from SIVmac clone 251 [14] and were probably the result of contamination in the laboratory.

Interest in SIV is high because they are the closest known relatives to HIV [9,15–17] and because SIV from macaques and sooty mangabeys induces the full spectrum of clinical signs of AIDS when inoculated into Asian macaques [3–8,13]. However, time of disease appearance and time of death vary greatly. Up to one-half of the animals die 3 to 4 months postinoculation, with three or more of the following clinical signs: generalized lymphadenopathy; persistent maculopapular rash; adenovirus-associated, ulcerative colitis; retroviral encephalitis, with giant cells; oral candidiasis; anemia; depressed T_4/T_8 ratio; wasting; cryptosporidiosis; trichomoniasis; and disseminated cytomegalovirus infection [3–8,13,18–21]. A similar immunodeficiency syndrome develops in animals at 1 year or more postinfection, except that lymphomas may also occur [18]. Although the SIV group is apparently derived solely from African monkey species [8,9], spontaneous SIV infections with disease have occurred in Asian macaques at North American primate centers. These SIV infections were probably the result of accidental or intentional exposure of macaques to SIV-containing African monkey tissue (see ref. 6 for the only documented instance). In the spontaneously induced disease of macaques, *Myocabacterium avium–intracellulare* infection and rarely *Pneumocystis carinii* pneumonia have been observed [5]. There are no known markers to predict the emergence of a specific opportunist in SIV-infected macaques; however, the prognosis is poor for animals that fail to develop a strong antibody response by 4 to 6 weeks postexposure to SIV [18,19,21]. Immunohistochemistry techniques show

SIV antigens in multinucleated giant cells and macrophages of inoculated macaques [22].

SIV Infections in the Central Nervous System

Even though several retroviruses from different subfamilies have been found to infect the central nervous system (CNS) of nonhuman primates, only SIV has been found to induce a primary encephalitis. The CNS disease is similar to that in AIDS patients [13,23,24]. At the New England Primate Research Center, 10 of 18 rhesus macaques inoculated with SIVmac had meningoencephalitis with multinucleate giant cells throughout the brain parenchyma and leptomeninges [23]. Lentiviral particles were present within membrane-bound, cytoplasmic vacuoles in macrophages and giant cells. Budding of viral particles into vacuoles was observed. A similar encephalitis has been associated with inoculation of the SIVsm [24]. In this study five of five animals inoculated with SIVsm(B670) died of an AIDS-like disease, with multinucleate giant cells in the leptomeninges and brain parenchyma [24]. Using an SIV-specific nucleic acid probe, polyclonal antisera, and electron microscopy, the SIV genome, antigens, and lentiviral particles were detected in the brain [24]. Clinically evident neurological abnormalities were not described in either of these studies. None of the animals with encephalopathy had grossly visible abnormalities of the brain. The encephalitis in these animals was assumed to be a primary retroviral disease because of the presence of viral particles and/or virus-specific nucleic acid in the infiltrating macrophages and giant cells [24].

In the mid-1970s, a similar giant cell encephalitis was recognized in stump-tailed macaques (*Macaca arctoides*) at the California Primate Research Center as one manifestation of an epizootic of immunosuppression, opportunistic infection, and lymphoma [5]. The giant cell encephalitis was originally thought to be due to *Mycobacterium avium*, but no organisms could be demonstrated with acid-fast stains. A retrospective seroepidemiological study of these animals showed a high seroprevalence (approximately 70%) of antibodies to SIV [5]. The virus was isolated from the frozen tissue of a stump-tailed macaque [5] infected in the outbreak.

SIV Genome and Gene Products

The RNA genome of SIV is transcribed into integrated DNA that contains numerous open reading frames (ORFs), but 11 are conserved between SIVmac and HIV-2 [9,15–17] (Figure 26.1, rectangles, currently recommended names are underlined [25], and old names are in brackets). Six single ORFs (*GAG, POL, ENV, NEF, VIF, VPX*) code for known products, whereas *TAT* and *REV* gene products are in two ORFs each. No product for

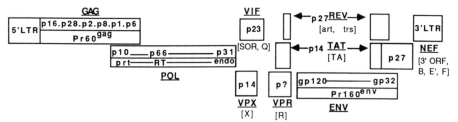

Figure 26.1. Genomic organization of SIVmac. The genomic organization of SIVagm[TYO] is the same except that *VPR* is not yet identified in SIVagm.

the eleventh ORF (*VPR*) is known. The *VPR* gene is not found in SIVagm [9]. Using the 5' prime end of the RNA genome as a starting point, *GAG* codes a precursor polypeptide (Pr60^{gag}), which is cleaved into the proteins p16, p28, p2, p8, p1, and p6 [26]; p16, p28, p2, and p8 occur in the virion core in about equimolar amounts, whereas p1 and p6 occur in lesser amounts [26]. Polysomal frame shifting expresses the *POL* gene as a *gag–pol* precursor, which may exclude C-terminal p1^{gag} and p6^{gag}; this would account for their lower molar ratios in the virion [26]. By analogy to HIV, the *gag–pol* precursor (p66$^{gag-pol}$) is cleaved into three *POL* polypeptides, an N-terminal protease (p10?), the reverse transcriptase (RT), and endonuclease (p31?) [27]. Reverse transcriptase activity is found in both the p66$^{gag-pol}$ and p55$^{gag-pol}$ [27], which suggests that a 10-kd protein is processed away from the RT precursor. This 10-kd protein may be the N-terminal protease. The central region contains five genes, the p23 virion infectivity factor (*VIF*), the p14 virion protein X (*VPX*) that is unique to SIV and HIV-2 [26,28], the putative viral protein R (VPR), and the positive regulatory proteins *TAT* (p14) and *REV* (p19). *ENV* encodes a 160-kd precursor polypeptide (Pr160^{env}), which is cleaved into the external glycopolypeptide (*gp*) gp120 and the subsurface transmembrane gp32 [15]. By analogy with HIV, the p27 negative factor (NEF) product presumably down-regulates viral synthesis and is the 3' prime of other genes [25].

Cell Biology of SIV

The in vitro host range of SIV in PBMCs includes macaques, humans, and gibbons (*Hylobates lar*) [29]. The PBMCs from new world monkeys, baboons (*Papio papio*), and surprisingly, chimpanzees were resistant to infection by SIV [29]. Conversely, macaque PBMCs are in vitro hosts to HIV-1 [7]. There is no apparent correlation between the in vitro and in vivo host range of primate retroviruses and their ability to induce disease. The receptor for SIV was confirmed as CD4 by experiments in which anti-T$_4$, but not anti-T$_8$ monoclonal antibodies blocked SIV infection of human and macaque PBMCs [29]. When cell cultures were depleted of CD8-bearing cells, SIV was isolated more readily [29]. Certain isolates of SIV have a more restricted host cell range than HIV-1. The HTLV-4 isolate, which is indistinguishable from SIVmac (Clone 251) [13], only replicates in cells expressing CD4 and human leukocyte class II antigens [30]. The HTLV-4 was also noncytolytic in certain human T cell lines and replicated without affecting the cell surface expression of CD4 [30].

Summary

Nonhuman primate lentivirus infections are the animal model of choice for AIDS research. The replication scheme of SIV makes it ideal for in vivo testing of potential therapeutic agents, and its pathogenesis makes it suitable for testing vaccine strategies and understanding basic disease mechanisms. Future research is needed on the mode of transmission of simian retroviruses in nature and the development of animal models for the sexual transmission of AIDS.

References

1. Alter HJ, Eichberg JW, Masur H, Saxinger WC, Gallo RC, Macher AM, Lane HC, Fauci AS (1984) Transmission of HTLV-III infection from human plasma to chimpanzees: An animal model for AIDS. Science 226:549–552
2. Fultz PN, Switzer W, McClure HM, Anderson D, Montagnier L (1988) Simian models for AIDS: SIV/SMM and HIV-2 infection of macaques. In Ginsburg H, Brown F, Lerner RA, Chanock RM (eds) Vaccines 88. CSHL Publisher, CSH New York, pp 167–170
3. Daniel MD, Letvin NL, King NW, Kannagi M, Sehgal PK, Hunt RD, Kanki PJ, Essex M, Desrosiers RC (1985) Isolation of T-cell tropic HTLV-III-like retrovirus from macaques. Science 228:1201
4. Benveniste RE, Morton WR, Clark EA, Tsai C-C, Ochs HD, Ward JM, Kuller L, Knott WB, Hill RW, Gale MJ, Thouless ME (1988) Inoculation of baboons and macaques with simian immunodeficiency virus/Mne, a primate lentivirus closely related to human immunodeficiency virus type 2. J Virol 62:2091–2101
5. Lowenstine L, Lerche N, Jennings M, Marx P, Gardner M, Pedersen N (1987) An epizootic of simian AIDS caused by SIV in captive macaques in the 1970's. In Girard M, deThe G, Vallette L (eds) Retroviruses of Human AIDS and Related Animal Viruses Pasteur Vaccine, Lyon. pp 174–176
6. Murphey-Corb M, Martin LN, Rangan SRS, Baskin GB, Gormus BJ, Wolf RH, Andes WA, West M, Montelaro RC (1986) Isolation of an HTLV-III related retrovirus from macaques with simian AIDS and its possible origin in asymptomatic mangabeys. Nature 321:435–437
7. Fultz PN, McClure HM, Anderson DC, Swenson RB, Anand R, Srinivasan A (1986) Isolation of a T-lymphotropic retrovirus from naturally infected sooty mangabey monkeys (Cercocebus atys). Proc Natl Acad Sci USA 83:5286–5290
8. Lowenstine LJ, Pedersen NC, Higgins J, Pallis KC, Uyeda A, Marx P, Lerche

NW, Munn RJ, Gardner MB (1986) Seroepidemiologic survey of captive Old World primates for antibodies to human and simian retroviruses, and isolation of a lentivirus from sooty mangabeys (*cercocebus atys*). Int J Cancer 38:563–575

9. Fukasawa M, Miura T, Hasegawa A, Morikawa S, Tsujimoto H, Miki K, Kitamura T, Hayami M (1988) Sequence of simian immunodeficiency virus from African green monkey, a new member of the HIV–SIV group. Nature 333:457–461

10. Morrow WJW, Wharton M, Lau D, Levy JA (1987) Small animals species are not susceptible to HIV infection. J Gen Virol 68:2253–2257

11. Fultz P, McClure HM, Daugharty H, Brodie A, McGrath CR, Swenson B, Francis DP (1986) Vaginal transmission of human immunodeficiency virus (HIV) to a chimpanzee. J Infect Dis 5:896–900

12. Berman PW, Groopman JE, Gregory T, Clapham PR, Weiss RA, Ferriani R, Riddle L, Shimasaki C, Lucas C, Lasky LA, Eichberg JW (1988) Human immunodeficiency virus type 1 challenge of chimpanzees immunized with recombinant envelope glycoprotein gp120. Proc Natl Acad Sci USA 85:5200–5204

13. Letvin NL, Daniel MD, Sehgal PK, Desrosiers RC, Hunt RD, Waldron LM, MacKoy JJ, Schmidt DK, Chalifoux LV, King NW (1985) Induction of AIDS-like disease in macaque monkeys with T-cell tropic retrovirus STLV-III. Science 230:71–73

14. Kestler III HW, Yen L, Naidu YM, Butler CV, Ochs MF, Jaenel G, King NW, Daniel MD, Desrosiers RC (1988) Comparison of simian immunodeficiency virus isolates. Nature 331:619–620

15. Hirsch V, Riedel N, Mullins JI (1987) The genome organization of STLV-3 is similar to that of the AIDS virus except for a truncated transmembrane protein. Cell 49:307–319

16. Chakrabarti L, Guyader M, Alizon M, Daniel MD, Desrosiers RC, Tiollais P, Sonigo P (1987) Sequence of simian immunodeficiency virus from macaque and its relationship to other human and simian retroviruses. Nature 328:543–547

17. Franchini G, Gurgo C, Guo HG, Gallo RC, Collalti E, Fargnoli KA, Hall LF, Wong-Staal F, Reitz Jr MS (1987) Sequence of simian immunodeficiency virus and its relationship to the human immunodeficiency viruses. Nature 328:539–543

18. Daniel MD, Letvin NL, Sehgal PK, Hunsmann G, Schmidt DK, King NW, Desrosiers RC (1987) Long-term persistent infection of macaque monkeys with the simian immunodeficiency virus. J Gen Virol 68:3183–3189

19. Chalifoux LV, Ringler DJ, King NW, Sehgal PK, Desrosiers RC, Daniel MD, Letvin NL (1987) Lymphadenopathy in macaques experimentally infected with the simian immunodeficiency virus (SIV). Am J Pathol 128:104–109

20. Baskin GB (1987) Disseminated cytomegalovirus infection in immunodeficient rhesus monkeys. Am J Pathol 129:345–352

21. Kannagi M, Kiyotaki M, Desrosiers RC, Reimann KA, King NW, Waldron LM, Letvin NL (1986) Humoral immune responses to T-cell tropic retrovirus simian T-lymphotropic virus type III in monkeys with experimentally induced acquired immune deficiency-like syndrome. J Clin Invest 78:1229–1236

22. Ward JM, O'Leary TJ, Baskin GB, Benveniste R, Harris CA, Nara PL, Rhods RH (1987) Immunohistochemical localization of human and simian immunodeficiency viral antigens in fixed tissue sections. Am J Pathol 127:199–205

23. Ringler DJ, Hunt RD, Desrosiers RC, Daniel MD, Chalifoux LV, King NW

(1988) Simian immunodeficiency virus-induced meningoencephalitis: Natural history and retrospective study. Ann Neurol 23(suppl):S101–S107

24. Sharer LR, Baskin GB, Cho E-S, Murphey-Corb M, Blumberg BM, Epstein LG (1988) Comparison of simian immunodeficiency virus and human immunodeficiency virus encephalitides in the immature host. Ann Neurol 23 (suppl):S108–S112

25. Gallo R, Wong-Staal F, Montagnier L, Haseltine WA, Yoshida M (1988) HIV/HTLV gene nomenclature. Nature 333:504

26. Henderson LE, Benveniste RE, Sowder R, Copeland TD, Schultz AM, Oroszlan S (1988) Molecular characterization of *gag* proteins from simian immunodeficiency virus (SIV$_{Mne}$). J Virol 62:2587–2595

26. di Marzo Veronese F, Copeland TD, DeVico AL, Rahman R, Oroszlan S, Gallo RC, Sarngadharan MG (1986) Characterization of highly immunogenic p66/p51 as the reverse transcriptase of HTLV-III/LAV. Science 231:1289–1291

28. Kappes JC, Morrow CD, Lee S-W, Jameson BA, Kent SBH, Hood LE, Shaw GM, Hahn BH (1988) Identification of a novel retroviral gene unique to human immunodeficiency virus type 2 and simian immunodeficiency virus SIV$_{MAC}$. J Virol 62:3501–3505

29. Kannagi M, Yetz JM, Letvin NL (1985) *In vitro* growth characteristics of simian T-lymphotropic virus type III. Proc Natl Acad Sci USA 82:7053–7057

30. Hoxie JA, Haggarty BS, Bonser SE, Rackowski JL, Shan H, Kanki PJ (1988) Biological characterization of a simian immunodeficiency virus-like retrovirus (HTLV-IV): Evidence for CD4-associated molecules required for infection. J Virol 62:2557–2568

CHAPTER 27
Mechanisms of Persistent Replication of Lentiviruses in Immunocompetent Hosts

PAULINE JOLLY, DAVID HUSO, AND OPENDRA NARAYAN

Lentiviruses are nononcogenic, replication-competent retroviruses. They are genus specific in animal host range and are transmitted exogenously in nature by blood, milk, or inflammatory exudates. The viruses have been known for several years as specific pathogens of domestic animals, causing infectious anemia in horses, visna-maedi in sheep, arthritis–encephalitis in caprine species, and more recently, acquired immunodeficiency syndrome (AIDS) and AIDS-related complex (ARC) in humans [1–5]. Newer members of the lentivirus family include feline and bovine immunodeficiency viruses and lentiviruses of macaques (SIVmac) and African green monkeys (SIVagm). The viruses persist indefinitely in infected hosts and cause disease after variable and often prolonged incubation periods of months to years. Disease is gradual in onset and becomes progressively worse with time. It involves many organ systems in inflammatory and degenerative processes and leads eventually to cachexia and death [6–10].

The genomes of lentiviruses contain the *gag, pol,* and *env* genes common to all retroviruses. The *env* encodes a single, highly glycosylated glycoprotein structure that forms the envelope of the virus. The envelope protects the virus during the extracellular phase of replication and contains the determinants that bind to cellular receptors, that cause cell fusion [11] and that induce neutralizing antibodies [12]. In addition to the structural genes, the lentiviral genomes contain small open reading frames (ORFs) that encode proteins that bind to sequences in the long terminal repeat in the proviral DNA and regulate the rate of virus replication [13]. In addition, factors that regulate immunological activation of T helper lymphocytes and maturation/differentiation of monocytes/macrophages, the two key viral-host cells in the body, also regulate the degree of expression of viral genomes and determine whether the virus life cycle will be productive. Thus, depending on the physiological state of the cell, the viral life cycle may be arrested at different stages preceding viral maturation, or it may be complete but subject to a regulated rate of production of particles [14,15].

The propensity of lentiviruses to cause disease months to years after initial infection is predicated not only on their ability to persist within a cell population (perhaps via integration of proviral DNA) but also on their ability to replicate continuously in immunocompetent hosts. Various factors, including the cells in which the agents replicate, the retroviral mechanism of replication, and the intrinsic anatomical features of the infectious lentivirus particles, combine to endow these agents with remarkable powers for escape from host defenses. The mechanisms of these factors are summarized below.

Host Cells

The tropism of viruses for T helper lymphocytes and monocyte/macrophages compromises the function of these cells. Visna virus infection in T helper lymphocytes of sheep is accompanied by an inability of these cells to respond immunologically to visna viral antigens in lymphoproliferation assays (Kennedy-Stoskopf, unpublished observations). This loss of T helper cell function is confined to responses to the lentivirus and, in effect, amounts to immunosuppression of the host specifically to this agent. The animals remain competent to respond immunologically to other pathogens. This is unlike the global loss of T helper cell function seen in humans with AIDS. The infection in macrophages precludes the ability of these cells to clear virus, since such processes result only in further infection in these cells (see below). Infected macrophages produce virus particles. They also process some of the viral polypeptides they synthesize, and present these within context of their class II major histocompatibility complex (MHC) antigens, to T lymphocytes [16]. This results in the production of cytokines, including interferon. Supernatant fluids containing this interferon have antiproliferative properties (inhibition of maturation of infected monocytes, which retards virus production), antiviral properties (inhibition of completion of the virus life cycle in infected macrophages), and macrophage-activating effects (induction of the expression of MHC class II antigens) [17]. The net effect of the infection in macrophages *in vivo* is therefore to perpetuate infection and simultaneously to reduce the rate of replication.

The Genomic Organization of the Lentiviruses

In addition to the typical *gag, pol,* and *env* structural genes of retroviruses, the lentiviruses have several small ORFs that encode the proteins that regulate the rate of virus replication [13]. Some of these genes are also thought to facilitate replication of the virus in nondividing cells and to activate the cells immunologically [18]to enhance the virus RNA transcription rate.

The Retroviral Scheme of Replication

Reverse transcription of lentiviral RNA to proviral DNA is inherently susceptible to error because of lack of editing by the reverse transcriptase enzyme. This results in mutations in the viral genome. Mutations in the *env* gene are frequently retained without compromising the vital functions of the envelope protein, and this gives rise to genetic variants of the virus. In some cases, these mutations result in changes in antigenicity of the virus; these mutants are recognized as neutralization–escape mutants or antigenic variant viruses [12,19,20]. This provides another mechanism for lentivirus to escape from immune control.

Anatomical Features of the Lentivirus

The infectious lentivirus particle is intrinsically poorly antigenic because its anatomical constitution is such that antigenic proteins are sequestered by poorly antigenic carbohydrate molecules on the surface of the particle. The *env* gene of the virus that encodes the glycoprotein that forms the envelope of the virus has numerous glycosylation sites, and oligosaccharides are attached to these sites by O and N linkages [21] (S. Wain-Hobson, Pasteur Institute, Paris, personal communication). The penultimate carbohydrate moieties are saturated with sialic acids. In some viruses, for example, visna virus, the sialic acids partially obscure neutralization determinants of the virus. The result of this is that animals infected with this virus produce neutralizing antibodies, but these neutralize only the parental visna virus, not other strains. These antibodies select for emerging antigenic variant viruses that arise after mutations in the *env* gene described above. These viruses are neutralization–escape mutants.

Caprine arthritis–encephalitis viruses (CAEVs) are genetically and antigenically closely related to visna viruses but usually do not induce neutralizing antibodies during natural infection. Immunization of goats with ultraviolet inactivated native virus particles also fails to induce neutralizing antibodies except under rigorous conditions of immunization [22]. Examination of such sera in neutralization assays showed that the efficiency of neutralization increased markedly when the virus was pretreated with neuraminidase (Figure 27.1). This enzyme liberated sialic acids from the surface of the particles but did not directly affect infectivity. Sialic acids reduce the affinity of antibodies for neutralization determinants on the virions presumably because of the steric interference or unfavorable electro-

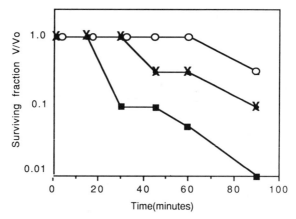

Figure 27.1. Caprine arthritis–encephalitis virus (CAEV) was treated with *Vibrio cholerae* neuraminidase to remove sialic acids or mock-treated with buffer; then each sample was mixed with control lamb serum or CAEV-neutralizing serum. At specific times, an aliquot was removed and diluted in cold MEM, and the surviving virus was titrated. The sample neutralization kinetics were plotted. *Key:* O, virus + neuraminidase + normal lamb serum; **X,** virus + neutralizing serum; ■, virus + neuraminidase + neutralizing serum. Typically slow neutralization kinetics were seen when control (CAEV) was incubated with neutralizing serum (**X**). Although neuraminidase had no significant direct effect on viral infectivity, the kinetics and the extent of neutralization were enhanced when viral sialic acids were first removed by preincubation with neuraminidase (■). Figure adapted from J Virol 62:1974–1980, 1988, with permission.

static interactions between the two reactants. Their removal with neuraminidase helped to correct this problem.

Armed with this new information, we reexamined virus and sera from naturally arthritic goats persistently infected with CAEV. Virus was obtained from cultures of their macrophages and examined for reacivity with homologous sera. The antibodies bound to viral polypeptides after virions had been disrupted with detergents, but as noted previously, the antibodies did not neutralize infectious particles. However, the sera neutralized infectivity after the virions had been treated with neuraminidase (D.L. Huso, unpublished results). Thus, the animals had produced antibodies to neutralization determinants of the virus, but presumably, during maturation of the infectious particles, these determinants had become obscured by the sialic acids. Infection could therefore occur in the presence of "neutralizing antibodies." These results help to explain the phenomenon that the presence of antiviral antibodies in sera of lentivirus-infected animals or humans indicates infection rather than of protective immunity [22,23].

Role of the Macrophages

Initial tests of efficacy of neutralizing antibodies to visna virus and CAEVs in protecting cell cultures against infection with homologous viruses were performed in fibroblastic cultures [22,23]. However, the known tropism of the viruses for monocytes/macrophages *in vivo* required an evaluation of the different antibodies, using the macrophage as target cells in neutralization tests. When fibroblasts were used as indicator cells, neutralizing antibodies prevented binding of virus to the cells, thereby protecting them from infection. When macrophages were used as target cells, however, virus–antibody complexes were internalized faster than control virus suspensions. In these experiments, uncoating of the antibody-treated virions also occurred faster than the control virus suspensions. However, no viral RNA transcripts were produced in these macrophages. The neutralization process thus occurred within the cytoplasm of macrophages in contrast to the extracellular nature of the event in fibroblasts [23].

Examination of sera that failed to neutralize CAEV in fibroblast cultures showed that they enhanced infection in macrophages (P. Jolly, submitted for publication). Antibodies in these sera simulated the effects of neutralizing antibodies in enhancing binding (Table 27.1) and uncoating of virus particles. However, unlike neutralizing antibodies that prevented transcription of viral RNA, the nonneutralizing antibodies (by definition) allowed completion of the virus life cycle. In effect, these antibodies mediated infection in macrophages with a higher multiplicity of inoculation than macrophages treated with control virus suspensions. Subsequent studies showed that the effect of neutralizing and nonneutralizing antibodies in enhancing binding and internalization of virus in macrophages was mediated by the binding of the Fc fragments of the immunoglobulin molecules to Fc receptors on the cells. (F(ab')$_2$ fragments of the immunoglobulins were produced and examined for their effect on the kinetics of binding and uncoating of virus particles in fibroblasts and macrophages. The fragments reduced the binding of virus to fibroblasts and neutralized infectivity in these cells, simulating the activity of whole immunoglobulins. However, these fragments did not enhance binding or uncoating of virus in macrophages. Surprisingly, F(ab')$_2$ fragments had minimal effects in neutralization of virus in macrophage cultures. These data provided new information on the effects of different kinds of antibodies in regulating virus replication in macrophages. First, the Fc portion of the Ig molecules determined whether the infection was aborted or enhanced. Second, F(ab')$_2$ fragments were not nearly as effective in protecting macrophages as they were in protecting fibroblasts against infection. Presumably phagocytosis of these complexes resulted in infection in the macrophages. It is possible that similar events could occur in macrophage precursor cells (*in vivo*) that lack Fc receptors and strengthens the hypothesis that antibodies that protect one type of cell, for example,

Table 27.1. Binding and uncoating of ^{35}S-visna virus in fibroblasts
and macrophages.

	Fibroblast		Macrophage	
Postinfectious serum	Binding (CPM)	Uncoating (ASC)	Binding (CPM)	Uncoating (ASC)
Neutralizing	2,038	268	6,187	3,263
Nonneutralizing 1	2,673	493	4,506	2,845
Nonneutralizing 2	1,773	443	4,722	2,763
Nonneutralizing 3	2,356	583	4,277	2,600
Newborn lamb serum	3,010	605	3,823	1,628

Visna virus-infected cells were isotopically labeled with ^{35}S-methionine; radiolabeled visna virus was purified and concentrated from infectious supernatant fluid, and then mixed with either control newborn lamb serum, neutralizing serum, or one of the three nonneutralizing sera and incubated at 37° C for 1 hour. The ^{35}S-visna virus–antibody samples were inoculated into macrophage or fibroblast cultures at an MOI of 7, incubated at 4° C for 14 hours, and washed thoroughly with cold medium. To determine total binding in each cell type, the cells were immediately lysed in SDS, and counts per minute (cpm) in the cell lysates were determined by scintillation spectroscopy. To determine uncoating of virus following binding at 4° C, fresh media was added to the washed cells, and the cultures were incubated at 37° C for 2 hours. Acid-soluble counts (ASC), representing uncoated viral particles, were determined as follows: ASC = (total cpm in supernatant) − (perchloric acid precipitable cpm in supernatant). Neutralizing serum dramatically decreased binding and uncoating of virus in fibroblast cultures, but in macrophages, binding and uncoating of virus were enhanced (even though the macrophages were still protected from infection). Nonneutralizing sera had variable effects on viral binding and uncoating in fibroblasts, but in macrophages, all three nonneutralizing sera enhanced viral binding and uncoating. Since these sera did not protect cells from infection, these virus-specific Ig may have enhanced infection in the macrophages.

fibroblasts or lymphocytes, from infection may fail to protect another cell, for example, the macrophage, and indeed may enhance the infection in these cells.

Summary

The slowly progressive diseases caused by lentiviruses result from persistent infection and continuous replication of the viruses in the host. Although the viruses cause immunosuppression, they must nevertheless be capable of establishing the means for persistent replication in the immunocompetent host. The agents accomplish this by the use of a combination of unique strategies. First, the viruses infect cells of the immune system, which nullifies potential adverse immunological responses by the host. Second, the genomic organization of lentiviruses allow the agents to infect nondividing, terminally differentiated cells and to establish latency via its proviral DNA. Third, because of the lack of editing functions of reverse transcriptase enzymes, the proviral DNA transcribed from viral RNA often acquire and accumulate mutations, which can lead to the emergence of neutralization–escape mutants. Fourth, the envelope of the lentivirus is comprised of a highly glycosylated protein structure and, in the morphologically intact virus, sialic acid molecules on the surface of the particles obscure protein antigens. This reduces the affinity between neutralizing antibodies and the virus. The delay in binding between antibodies and virus allows the agent to infect cells. In some cases, the loose binding between antibodies and virus facilitates enhanced entry into macrophages via Fc receptors, and this results in infection of the cells. The combined effects of these properties of the lentiviruses thus guarantee infection, provided that the virus had been introduced to the mononuclear phagocytic system of the host parenterally, either as cell-free virus or within infected cells.

Acknowledgments

These studies were supported by Grants NS12127, AI25774, NS21916, and NS0700 from the National Institutes of Health.

References

1. Sigurdsson BH, Thormar H, Palsson PA (1960) Cultivation of visna virus in tissue culture. Arch Ges Virusforsch 10:368–381
2. Crawford TB, Adams DS, Cheevers WP, Cork LC (1980) Chronic arthritis in goats caused by a retrovirus. Science 207:997–999
3. Barre-Sinoussi FC, Cherman JC, Rey F, Nugeryre MT, Chavaret S, Gruest J, Dauget C, Axler-Blin C Vezenet-Brun F, Rouzioux C, Roenbaum W, Montagnier L (1983) Isolation of a T lymphotropic retrovirus from a patient at risk of acquired immunodeficiency syndrome (AIDS). Science 220:868–870
4. Gallo RC, Salahuddin SQ, Popovic M, Shearer GM, Kaplan M, Haynes BF, Palker TJ, Redfield R, Oleske J, Safai B, White G, Foster P, Markham PD (1984) Frequent detection and isolation of cytopathic retroviruses (HTVL-III) from patients with AIDS and at risk for AIDS. Science 224:500–503
5. Gonda MA, Wong-Staal F, Gallo RC, Clements JE, Narayan O, Gilden RV (1985) Sequence homology and morphologic similarity of HTLV III and visna virus, a pathogenic lentivirus. Science 227:173–177
6. Cheevers WP, McGuire TC (1985) Equine infectious anemia virus: Immunopathogenesis and persistence. Rev Infect Dis 7:83–88
7. Narayan O, Cork LC (1985) Lentiviral diseases of sheep and goats: Chronic pneumonia, leukoencephalomyelitis and arthritis. Rev Infect Dis 7:89–98
8. Haase AT (1986) Pathogenesis of lentivirus infections. Nature 322:130–136
9. Price RW, Brew B, Sidtis J, Rosenblum M, Scheck AC, Cleary P (1988) The brain in AIDS: Central nervous system HIV-1 infection and AIDS dementia complex. Science 239:586–592
10. Fauci AS (1988) The human immunodeficiency virus: Infectivity and mechanisms of pathogenesis. Science 239:617–622
11. Crane SE, Clements JE, Narayan O (1988) Separate epitopes in the envelope of

visna virus are responsible for fusion and neutralization: Biological implications for anti-fusion antibodies in limiting virus replication. J Virol 62:2680–2685

12. Scott JV, Stowring L, Haase AT, Narayan O, Vigne R (1979) Antigenic variation in visna virus. Cell 18:321–327

13. Hess JL, Clements JE, Narayan O (1985) Cis- and trans-acting transcriptional regulation of visna virus. Science 229:482–485

14. Narayan O, Kennedy-Stoskopf S, Sheffer, D, Griffin DE, Clements JE (1983) Activation of carpine arthritis–encephalitis virus expression during maturation of monocytes to macrophages. Infect Immun 41:67–73

15. Gendelman HE, Narayan O, Kennedy-Stoskopf S, Kennedy PGE, Ghotbi Z, Clements JE, Stanley J, Pezeshkpour G (1986) Tropism of lentiviruses for monocytes: susceptibility to infection and virus gene expression increase during maturation of monocytes to macrophages. J Virol 58:167–74

16. Kennedy PGE, Narayan O, Ghotbi Z, Hopkins J, Gendelman HE, Clements JE (1985) Persistent expression of Ia antigen and viral genome in visna-maedi virus-induced inflammatory cells. J Exp Med 162:1970–1982

17. Narayan O, Sheffer D, Clements JE, Tennekoon G (1985) Restricted replication of lentiviruses: Visna viruses induce a unique interferon during interaction between lymphocytes and infected macrophages. J Exp Med 162:1954–1969

18. Kannagi M, Kiyotaki M, King NW, Lord CI, Letvin NL (1987) Simian immunodeficiency virus induces expression of class II major histocompatibility complex structures on infected target cells in vitro. J Virol 61:1421–1426

19. Clements JE, Pederson FS, Narayan O, Halsetine WS (1980) Genomic changes associated with antigenic variation of visna virus during persistent infection. Proc Natl Acad Sci USA 77:4454–4458

20. Stanley J, Bhaduri LM, Narayan O, Clements JE (1987) Topographical rearrangements of visna virus envelope glycoprotein during antigenic drift. J Virol 61:1019–1028

21. Huso DL, Narayan O, Hart GW (1988) Sialic acids on the surface of caprine arthritis–encephalitis virus define the biological properties of the virus. J Virol 62:1974–1980

22. Narayan O, Sheffer D, Griffin DE, Clements J, Hess J (1984) Lack of neutralizing antibodies to caprine arthritis–encephalitis lentivirus can be overcome by immunization with inactivated *Mycobacterium tuberculosis*. J Virol 49:349–355

23. Kennedy-Stoskopf S, Narayan O (1986) Neutralizing antibodies to visna lentivirus: Mechanism of action and possible role in virus persistence. J Virol 59:37–44

CHAPTER 28
Generation of Pathogenic Feline Leukemia Viruses

Oswald Jarrett

Of the diseases that occur frequently in domestic cats chronically infected with feline leukemia virus (FeLV), the most intensively studied are the malignant tumors lymphosarcoma and fibrosarcoma and the nonmalignant conditions erythroid aplasia and immunodeficiency. In each case, pathogenesis is associated with the generation of novel viruses from the parental FeLV with which the cat was originally infected. These new viruses arise de novo in individual cats by mutation or recombination and are not transmitted further in nature, although they rapidly produce characteristic disease when inoculated into experimental hosts. The cycle of FeLV infection is maintained in cat communities by the transmission of the parental viruses.

Natural Infection With FeLV

FeLV is spread from cats with chronic productive infections, most frequently through saliva or across the placenta to the fetus [1,2]. Subsequently, the outcome depends to a large extent on the age at which the cat was exposed: Prenatal infection or infection during the first eight weeks of life results in a high proportion of kittens developing a permanent infection, whereas older cats tend to become immune and resist subsequent challenge [3]. Persistently infected cats may remain healthy for long periods, but the prognosis is ultimately poor, since 85% die within 3.5 years of infection [4].

The FeLV-related diseases occur in cats with persistent infections. Since the virus grows in hematopoietic and lymphoid cells, as well as in other tissues with dividing cells (mainly epithelial surfaces), very large amounts of virus are produced over a considerable period during which time there is ample opportunity for mutation of the viral genome or recombination between FeLV and cellular genes. It is likely that most of these events are insignificant, but on rare occasions a variant that is able to exert profound effects is produced.

FeLV-Associated Malignant Disease

Transduction of cellular oncogenes by FeLV is involved in the pathogenesis of leukemia and fibrosarcomas. As a result of this recombination, part of the FeLV genome is replaced by the *onc* gene, which renders the virus replication defective, so that it must rely for growth on complementation by parental, "helper" FeLV replicating in the same cell.

Lymphosarcoma

Lymphosarcoma (LSA) is the most common neoplasm in the cat. Both T and B cell tumors are found, but T cell tumors, particularly thymic lymphosarcoma (T-LSA), have been studied more extensively, since the vast majority of these are FeLV-induced. Recombinants of FeLV and the c-*myc* oncogene have been found in tumor cells from naturally occurring cases of T-LSA [5–7]. Up to 30% of naturally occurring T-LSA cases have this type of virus. A common feature of four independent isolates that have been molecularly cloned is that each has acquired the coding regions of exons 2 and 3 of the cellular c-*myc* gene. However, the junction of the FeLV and v-*myc* sequences is not constant, indicating that each virus represents a unique recombinational event. Thus, in the LC clone, v-*myc* replaces part of the viral *gag* and *pol* genes, whereas in the CT4 and FTT clones part of the *env* gene is replaced.

The *myc*-containing genome is expressed in tumor cells from which viruses with the recombinant genome may be harvested. Although the recombinants have no demonstrable transforming effect in vitro, they do have a dramatic effect in cats, inducing T-LSA in the remarkably short latent period of 10 weeks [8]. The resulting tumors are phenotypically identical to those from which the virus was isolated and are interleukin-2 (IL-2) independent. The cells are monoclonal, or at least oligoclonal, suggesting that a selection process other than transformation of target cells by v-*myc* is involved in leukemogenesis. A possible correlate of the secondary leukemogenic event is release of the T cells from IL-2 dependence. Although there is a good correlation between the presence of the recombinant *myc* virus and IL-2 independence of the tumor cells, this phenotype has not been reproduced by in vitro infection of T cells with the *myc* recombinant viruses. It is not known if *myc* expression activates either IL-2 production or IL-2 receptor expression.

Recently a T cell tumor was found, which, in addition to a recombinant *myc*-containing genome, possessed a genome in which the FeLV had recombined with the chain of the T cell receptor (TCR) [9]. It was proposed that expression of this virus might simulate natural activation and signal transduction by the TCR that drives the T cell into division. Alternatively, the viral *tcr* product may be a high-affinity receptor for some self-antigen, resulting in proliferation by an autoreactive mechanism.

Fibrosarcomas

Many viruses have been isolated from young FeLV viremic cats with multicentric fibrosarcomas and shown to cause a similar disease when inoculated into susceptible kittens and to transform fibroblasts in culture [10]. These feline sarcoma viruses (FeSV) are recombinants between FeLV and cellular oncogenes of which no less than seven have been transduced (*abl, fes, fgr, fms, kit,* K-*ras,* and *sis*). As for the FeLV-*myc* viruses, it is likely that the appearance and expression of these recombinants directly causes the tumors.

FeLV Subgroups

The parental, or common, form of FeLV is subgroup A (FeLV-A), which is present in every isolate and grows very efficiently in cats [11]. Subgroup is determined by a domain on the surface glycoprotein of the virus encoded by the *env* gene, by which the virus binds to a receptor on the cell surface. Consequently, host and tissue tropism and, to some extent, antigenicity are related to subgroup.

Two other subgroups, B and C, have been defined. The FeLV isolates that contain either FeLV-B or FeLV-C also contain FeLV-A [11,12]. The FeLV subgroups B and C depend on FeLV-A not because the former are replication defective, as is the case with the recombinant viruses described above, but probably because they have a restricted tissue tropism in vivo and require assistance from FeLV-A [12]. This is probably achieved by forming phenotypically mixed viral populations with FeLV-A, which permits FeLV-B and FeLV-C to infect cells that are usually permissive only for FeLV-A.

Recombination Generates New FeLV Subgroups

All cat cells have eight to ten copies of an endogenous, FeLV-related genome, but this element has never been recovered as infectious virus. Recombination between the *env* genes of FeLV and the endogenous viral genome, however, occurs in FeLV viremic cats. The recombinants are recognized structurally by changes in nucleotide sequence and functionally by the new host range characteristics by virtue of which they are assigned to subgroup B [13]. The significance of FeLV-B in pathogenesis is unclear. Biologically cloned FeLV-B isolates grow poorly in cats and require help from FeLV-A for extensive replication and for transmission between animals. Nevertheless, FeLV-B is oncogenic: although only a small proportion of kittens inoculated with FeLV-B actually became persistently viremic, of these, two-thirds developed T-LSA [12]. Further evidence of the involvement of FeLV-B in leukemogenesis comes from typing of field isolates of virus, which indicated that lymphosarcoma was twice as common in cats viremic with both FeLV-A and FeLV-B than in cats with FeLV-A alone [14].

Immunodeficiency

Immunodeficiency is common in FeLV viremic cats, and in at least some cases, it is caused by variant viruses. These viruses are probably not recombinants. Instead, they have point mutations, deletions, or insertions of direct repeats. One strain of FeLV with immunosuppressive properties has been intensively studied [15]. The disease in cats infected with this virus may be either acute or chronic. The acute immunodeficiency syndrome occurs within 12 weeks. During this time, a variant virus (variant A) appears in the bone marrow and persists, mainly as unintegrated DNA. The chronic form of the disease develops after one year and is also associated with the appearance of variant genomes. Many cases of the chronic form are found to have LSA.

Both common form and variant genomes have been cloned. In contrast to the common form, which is replication competent, variants are replication defective despite being full length [16]. They can, however, be rescued from cells by coinfection with the common form, and when inoculated into susceptible kittens, they can cause immunodeficiency. In culture the variant viruses are cytopathogenic for T cells, whereas the common form is not.

A notable feature of the replication of the variants is the large quantity of unintegrated viral DNA that accumulates in infected cells. Although a similar situation has been recorded in some cells infected with human immunodeficiency virus (HIV), it remains to be established if this feature is associated with the pathogenicity of these very different immunosuppressive viruses.

Erythroid Aplasia

Anemia is another frequently diagnosed condition of FeLV-infected cats. A form of nonregenerative anemia, analogous to pure red cell aplasia in man, has been described that is specifically associated with FeLV of subgroup C [17]. Four independent isolates of this subgroup have produced an identical disease characterized by a block in the differentiation of early erythroid cells at the level of the BFU-E [18]. The determinants of erythroid specificity of these viruses is not yet known but are likely to reside within the *env* gene that controls cell tropism. One possible pathogenic mechanism is that the receptor for FeLV-C is also the receptor for a factor necessary for erythroid cell differentiation. Binding of the viral glycoprotein may inactivate this vital receptor.

Sequencing of the *env* gene of FeLV-C reveals only small differences between FeLV-A and FeLV-C [19]. It is likely that FeLV-C strains arise as variants in cats viremic with FeLV-A. Although FeLV-C does not normally grow in adult cats, the experimental introduction of FeLV-C into healthy cats that have been viremic with FeLV-A for several months results in the appearance of FeLV-C in the plasma and the simultaneous occurrence of anemia [20]. This finding indicates that if a variant with the characteristics of FeLV-C were to arise, it would produce remarkable pathogenic effects within a short time. These and other experiments suggest that the appearance of FeLV-C is the crucial event in the pathogenesis of erythroid aplasia.

References

1. Hardy WD Jr, Old LJ, Hess PW, Essex M, Cotter SM (1973) Horizontal transmission of feline leukaemia virus. Nature 244:266–269
2. Jarrett W, Jarrett O, Mackey L, Laird H, Hardy WD Jr, Essex M (1973) Horizontal transmission of leukemia virus and leukemia in the cat. J Natl Cancer Inst 51:833–841
3. Hoover EA, Olsen RC, Hardy WD Jr, Schaller JD, Mathes LF (1976) Feline leukemia virus infecion: Age-related variation in response of cats to experimental infection. J Natl Cancer Inst 57:365–369
4. McClelland AJ, Hardy WD Jr, Zuckerman EE (1980) Prognosis of healthy feline leukemia virus infected cats. In Hardy WD Jr, Essex M, McClelland AJ (eds) Feline Leukemia Virus. New York, Elsevier—North Holland, pp 121–131
5. Neil JC, Hughes D, McFarland R, Wilkie NM, Onions DE, Lees G, Jarrett O (1984) Transduction and rearrangement of the myc gene by feline leukaemia virus in naturally occurring T-cell leukaemias. Nature 308:814–820
6. Mullins JI, Brody DS, Binari RC Jr, Cotter SM (1984) Viral transduction of c-myc gene in naturally occurring feline leukaemias. Nature 308:856–858
7. Levy LS, Gardner MB, Casey JW (1984) Isolation of a feline leukaemia provirus containing the oncogene myc from a feline lymphosarcoma. Nature 308:853–856
8. Onions D, Lees G, Forrest D, Neil J (1987) Recombinant feline viruses containing the myc gene rapidly produce clonal tumours expressing T cell antigen receptor gene transcripts. Int J Cancer 40:40–45
9. Fulton R, Forrest D, McFarlane R, Onions D, Neil JC (1987) Retroviral transduction of T cell antigen receptor β-chain and myc genes. Nature 326:190–194
10. Besmer P (1983) Acute transforming feline retroviruses. Curr Top Microbiol Immunol 107:1–27
11. Sarma PS, Log T (1973) Subgroup classification of feline leukemia and sarcoma viruses by viral interference and neutralization tests. Virology 54:160–169

12. Jarrett O (1980) Natural occurrence of subgroups of feline leukemia virus. *In* Essex M, Todaro G, Zur Hausen H (eds) Viruses in Naturally Occurring Cancers. Cold Spring Harbor Laboratory, Cold Spring Harbor, New York, pp. 603–611

13. Stewart MA, Warnock M, Wheeler A, Wilkie N, Mullins JI, Onions DE, Neil JC (1986) Nucleotide sequences of a feline leukemia virus subgroup A envelope gene and long terminal repeat and evidence for the recombinational origin of subgroup B viruses. J Virol 58:825–834

14. Jarrett O, Hardy WD Jr, Golder MC, Hay D (1978) The frequency of occurrence of feline leukaemia virus subgroups in cats. Int J Cancer 21:334–337

15. Hoover, EA, Mullins, JT, Quackenbush SL, Gasper PW (1987) Experimental transmission and pathogenesis of feline AIDS. Blood 70:1880–1892

16. Overbaugh J, Donahue PR, Quackenbush SL, Hoover EA, Mullins JI (1988) Molecular cloning of a feline leukemia virus that induces fatal immunodeficiency disease in cats. Science 239:906–910

17. Mackey L, Jarrett W, Jarrett O, Laird H (1975) Anemia associated with feline leukemia infection in cats. J Natl Cancer Inst 54:209–217

18. Onions D, Jarrett O, Testa N, Frassoni F, Toth S (1982) Selective effect of feline leukaemia virus on early erythroid precursors. Nature 296:156–158

19. Riedel N, Hoover EA, Gasper PW, Nicolson MO, Mullins JI (1986) Molecular analysis and pathogenesis of the feline aplastic anemia retrovirus, FeLV-C-Sarma. J Virol 60:242–250

20. Jarrett O, Golder MC, Toth S, Onions DE, Stewart MF (1984) Interaction between feline leukaemia virus subgroups in the pathogenesis of erythroid hypoplasia. Int J Cancer 34:283–288

CHAPTER 29
Neurovirulent Retrovirus of Wild Mice

JOHN L. PORTIS

A murine retrovirus (WM-E) isolated originally from a population of wild mice in Southern California [1] induces a neurological disease that has gained increasing attention with the advent of the encephalomyelopathies associated with human retrovirus infection. The virus resembles, in genomic structure, typical type C murine leukemia viruses. It exists in the wild as an exogenous infectious agent [2] transmitted from mother to offspring in the milk [3] and establishes a persistent productive infection throughout the life of the mouse. The incubation period of the neurological disease ranges from a few weeks to as long as a year, depending on the virus isolate and the concentration of the inoculum. The disease is characterized clinically by initial tremor and abnormal abduction reflex of the hindlimbs, progressing to frank hindlimb paralysis with neurogenic atrophy of the skeletal muscles. The pathology consists of a progressive loss of motor neurons in the ventral gray matter of the spinal cord, as well as the nuclei of the brainstem and cerebellum, associated with microglial proliferation and astrocytosis, but without inflammatory cell infiltrates [4]. Neurons as well as glial elements undergo a vacuolar degeneration reminiscent of the spongiform encephalopathies caused by the unconventional agents. However, no evidence for accumulation of "prion protein" (syn. SAF protein) has been detected in the murine retroviral disease [5].

Host Determinants of Disease

Inoculation of mice with WM-E during the neonatal period (less than 6 days of age) results in a state of specific immunological unresponsiveness to the viral proteins. The virus initially replicates in the spleen and bone marrow, and within 1 to 2 days after inoculation, a viremia ensues that persists throughout the life of the mouse [6]. The central nervous system (CNS) becomes infected within 2 weeks of inoculation, but the mouse remains asymptomatic for a variable incubation period of 1 to 6 months. The epithelial cells in organs of external secretion, such as epididymis, uterus, vagina, oviducts, and salivary glands, are also infected [7]. These persistently viremic mice excrete large amounts of virus in the milk, semen, and saliva, and adult males can transmit the virus to females with high efficiency by sexual contact [7]. In females so infected, however, virus spread is well controlled by the immune system, and a state of persistent low level (less than 20 infectious centers/10^6 cells) productive infection, primarily in the spleen, is observed without spread to the CNS. These mice are not viremic, they express high levels of antiviral antibody in the serum, they do not develop neurological disease, and they do not excrete virus. Thus, the immune response is a strong resistance factor in this disease.

It is of interest that viremic adult females (i.e., inoculated as neonates) in which the oviductal epithelium as well as the endometrium are productively infected [7] fail to transmit WM-E to the embryo [3]. Thus, although both the pre- and postimplantation embryo is exposed to high concentrations of virus within the reproductive tract, a strong protective barrier to infection exists. This barrier is likely responsible for the absence of WM-E viral sequences in the germ line of wild mice [2].

Some mouse strains (AKR and NZB) exhibit a genetically semidominant, nonimmunological resistance to CNS disease [8]. These mice fail to develop neurological disease after neonatal inoculation despite the replication of virus in both spleen and CNS at levels comparable to levels seen in susceptible mouse strains [9]. This resistance likely involves restriction of virus infection at the level of specific CNS target cells and thus may offer an opportunity to define those target cells that are relevant to disease pathogenesis. Genetic studies [8] suggest a multigenic effect, which may not be amenable to classic genetic analysis.

Viral Determinants of Neurovirulence

Two regions of the retroviral genome have the potential to influence disease specificity, the *env* gene and the viral long terminal repeat (LTR). The *env* gene encodes two envelope proteins that mediate receptor binding and membrane fusion, both functions being required for viral entry into the cell. The U3 region of the viral LTR contains enhancer elements that control the level of transcription of viral structural proteins in a tissue-specific fashion. In the case of the nondefective leukemogenic murine retroviruses, the viral enhancer sequences determine the cell type involved in the proliferative disease [10]. In contrast, the primary determinant of neurovirulence of WM-E resides not in the LTR, but rather in the viral *env* gene [11,12]. The LTR has been shown to influence the severity and quality of the neurological disease [13] but functions in this capacity only in the context of the WM-E *env* gene. This is significant because it suggests the existance of unique receptors for this virus in the CNS. Specific receptors for murine and avian retroviruses have been studied primarily in fibroblastic cell cultures and generally define species specificity. Whether tissue- or cell type-specific receptors exist within the organism has yet to be determined.

Mechanism of Neurovirulence

Unlike CNS diseases caused by the lentiviruses, the neurovirulence of WM-E appears not to have an immunopathological component. Thus, in addition to the lack of significant inflammation in sites of CNS degeneration, both active and passive immunity to the virus prevent disease, whereas immunosuppression enhances neurovirulence [14]. Although the precise mechanism of neurovirulence of WM-E is not understood, it is generally agreed that the degenerative changes are a direct consequence of virus replication or the accumulation of a viral product in the CNS.

Electron microscopic studies of paralyzed mice reveal the existence of neurons [1,4,11,15] and oligodendroglial cells [15] in which C type viral particles are seen budding aberrantly into intracytoplasmic vesicles, probably representing distended endoplasmic reticulum. This striking observation was initially thought to represent a cytopathic effect of the virus. Its significance, however, was brought into question when it was noted that intraneuronal virus budding is primarily seen late in disease [16]. Furthermore, cells exhibiting degenerative changes early in disease generally do not contain these particles [4,16]. These observations have suggested that the intracellular C type particles might represent activation of an endogenous provius and are unrelated to disease pathogenesis. This would resolve a nagging question in this system. How does a retrovirus replicate in a neuron, which is a postmitotic quiescent cell, when the early functions in the replication cycle (i.e., viral DNA integration) appear to depend on cellular DNA synthesis? This apparent roadblock to neuronal infection could be bypassed if viral DNA persists in the unintegrated form, as is common in lentivirus infections.

In an attempt to address the question of virus localization in the CNS, WM-E-specific monoclonal antibodies, which are nonreactive with endogenous viruses, were prepared [9]. Preliminary immunofluorescence studies indicate that WM-E gp70 is located primarily in the hindbrain and spinal cord (thus corresponding to the sites of neuropathology) and can be seen in association with both vascular elements and process-bearing cells resembling neuroglia and perhaps microglia. Occasional large, process-bearing cells resembling neurons are also visualized in the gray matter. The identity of these cells is being studied with neuron-specific markers and in-situ hybridization.

Resolving the question of neuronal infection is critical to understanding the pathogenesis of the disease. Clearly WM-E infection of the CNS ultimately causes the death of neurons, but this effect may not be the result of a lytic neuronal infection. By electron microscopy WM-E has been shown to replicate in vascular endothelial cells of the CNS [17], sometimes forming impressive arrays of virus primarily in a subendothelial location. This observation appears to correlate temporally with the onset of clinical disease and is proposed as playing an essential part in disease pathogenesis [16,17]. Since the endothelial cell is an important component of the blood–brain barrier, it is possible that virus infection of these cells could induce a physiological imbalance within the CNS that leads indirectly to neuronal degeneration. Since human immunodeficiency virus (HIV) also replicates in CNS endothelial cells [18], a similar mechanism may be responsible for the encephalopathy associated with acquired immunodeficiency syndrome (AIDS). Although the ultrastructural studies indicate that WM-E productively infects CNS endothelial cells, the detection of WM-E *env* protein in cells of the CNS parenchyma indicates that extravascular spread of the virus must occur. The lack of ultrastructural evidence of virion assembly in these extravascular elements early in disease [16] suggests the possibility of defective virus replication. In another murine retroviral neurological disease caused by a *ts* mutant of Moloney murine leukemia virus [19], posttranslational processing of the viral envelope protein in vitro is defective at the nonpermissive temperature. Interestingly, both the clinical picture and the pathological picture are very similar to the WM-E-induced disease, and ultrastructurally, there is no evidence of virion assembly within the lesions.

Finer mapping of the *env* sequences that determine neurovirulence of WM-E should provide insight into the nature and location of specific viral receptors in the CNS. These studies promise not only to explain the unique tropism of this virus for the CNS, but also may reveal the nature of its cytopathic effects. Cytopathology induced in fibroblastic cells in vitro by group B avian leukosis viruses have been mapped specifically to a sequence in the *env* gene that determines receptor specificity [20]. These studies suggest that viral envelope protein itself may represent the cytolytic agent, acting by directly affecting the stability of the cell membrane or by interrupting critical cell functions mediated by the viral receptor. In this regard, it is of interest that an acute, noninflammatory spongiform encephalomyelopathy seen in mice inoculated with a *ts* mutant of vesicular stomatitis virus (the replication of which is restricted in the CNS) is associated with the accumulation of viral glycoprotein within the vacuolar lesions [21]. The concept that neuropathology might be induced by the accumulation of a viral product (protein or perhaps nucleic acid) in the absence of overt productive virus replication is appealing, especially in view of the emerging awareness that some slowly evolving neurodegenerative diseases of humans may be associated with cryptic, nonproductive viral infections [22]. The availability of infectious molecular clones of WM-E [12], in addition to specific reagents for the detection of WM-E proteins and nucleic acids in the tissues (exclusive of endogenous retroviruses), is making it possible to address these issues systematically.

References

1. Gardner MB, Henderson BE, Officer JE, Rongey RW, Parker JC, Oliver C, Estes JD, Huebner RJ (1973) A spontaneous lower motor neuron disease apparently caused by indigenous type-C RNA virus in wild mice. J Natl Cancer Inst 51:1243–1254
2. Barbacid M, Robbins KC, Aaronson SA (1979) Wild mouse RNA tumor viruses. A nongenetically transmitted virus group closely related to exogenous leukemia viruses of laboratory mouse strains. J Exp Med 149:254–266
3. Gardner MB, Chivi A, Dougherty MF, Casagrande J, Estes JD (1979) Congenital transmission of murine leukemia virus from wild mice prone to the development of lymphoma and paralysis. J Natl Cancer Inst 62:63–69
4. Andrews JM, Gardner MB (1974) Lower motor neuron degeneration associated with type C RNA virus infection in mice: Neuropathological features. J Neuropathol Exp Neurol 33:285–307
5. Merz PA, Rohwer RG, Kascsak R, Wisniewski HM, Somerville RA, Gibbs CJ, Gajdusek DC (1984) Infection-specific particle from unconventional slow virus diseases. Science 225:437–440
6. Brooks BR, Swarz JR, Johnson RT (1980) Spongiform polioencephalomyelopathy caused by a murine retrovirus. Lab Invest 43:480–486
7. Portis JL, McAtee FJ, Hayes SF (1987) Horizontal transmission of murine retroviruses. J Virol 61:1037–1044
8. Hoffman PM, Morse HC (1985) Host genetic determinants of neurological disease induced by Cas-Br-M murine leukemia virus. J Virol 53:40–43
9. McAtee FJ, Portis JL (1985) Monoclonal antibodies specific for wild mouse neurotropic retroviruses: Defection of comparable levels of virus replication in mouse strains susceptible and resistant to paralytic disease. J Virol 56:1018–1022
10. Li Y, Golemis E, Hartley JW, Hopkins N (1987) Disease specificity of nondefective Friend and Moloney murine leukemia viruses is controlled by a small number of nucleotides. J Virol 61:693–700
11. Oldstone MBA, Jensen F, Dixon FJ, Lampert PW (1980) Pathogenesis of the slow disease of the central nervous system associated with wild mouse virus. II. Role of virus and host gene products. Virology 107:180–193
12. DesGroseillers L, Barrette M, Jolicoeur P (1984) Physical mapping of the paralysis-inducing determinant of a wild mouse ecotropic neurotropic virus. J Virol 52:356–363
13. DesGroseillers L, Rassart E, Robitaille Y, Jolicoeur P (1985) Retrovirus-induced spongiform encephalopathy: The 3′-end long terminal repeat–containing viral sequences influence the incidence of disease and the specificity of the neurological syndrome. Proc Natl Acad Sci USA 82:8818–8822
14. Hoffman PM, Robbins DS, Morse HC (1984) Role of immunity in age-related resistance to paralysis after murine leukemia virus infection. J Virol 52:734–738
15. Oldstone MBA, Lampert PW, Lee S, Dixon FJ (1977) Pathogenesis of the slow

disease of the central nervous system associated with WM 1504 E virus. Am J Pathol 88:193–212

16. Swarz JR, Brooks BR, Johnson RT (1981) Spongiform polioencephalomyelopathy caused by a murine retrovirus. II. Ultrastructural localization of virus replication and spongiform changes in the central nervous system. Neuropathol Appl Neurobiol 7:365–380

17. Pitts OM, Powers JM, Bilello JA, Hoffman PM (1987) Ultrastructural changes associated with retroviral replication in central nervous system capillary endothelial cells. Lab Invest 56:401–409

18. Wiley CA, Schrier RD, Nelson JA, Lampert PW, Oldstone MBA (1986) Cellular localization of human immunodeficiency virus infection within the brains of acquired immune deficiency syndrome patients. Proc Natl Acad Sci USA 83:7089–7093

19. Zachary JF, Knupp CJ, Wong PKY (1986) Noninflammatory spongiform polioencephalomyelopathy caused by a neurotropic temperature-sensitive mutant of Moloney murine leukemia virus TB. Am J Pathol 124:457–468

20. Dorner AJ, Coffin JM (1988) Determinants for receptor interaction and cell killing on the avian retrovirus glycoprotein gp85. Cell 45:365–374

21. Robain O, Chany-Fournier F, Cerutti I, Mazlo M, Chany C (1986) Role of VSV G antigen in the development of experimental spongiform encephalopathy in mice. Acta Neuropathol (Berl) 70:220–226

22. Haase AT, Gantz D, Eble B, Walker D, Stowring L, Ventura P, Blum H, Wiegrefe S, Zupancic M, Tourtellotte W, Gibbs CJ Jr, Norrby E, Rozenblatt S (1985) Natural history of restricted synthesis and expression of measles virus genes in subacute sclerosing panencephalitis. Proc Natl Acad Sci USA 82:3020–3024

Evolving Concepts in Viral Diseases

CHAPTER 30
Mechanisms of Hepatic Oncogenesis During Persistent Hepadna Virus Infection

CHARLES E. ROGLER, OKIO HINO, AND DAVID A. SHAFRITZ

Pathway of Viral DNA Replication in Hepatocytes

One of the earliest events following penetration and uncoating of hepadna viruses is the conversion of the viral DNA into a covalently closed circular (CCC) molecule [1], (Figure 30.1). The CCC DNA accumulates in the nucleus and is believed to serve as the template for synthesis of viral RNAs, including pregenome RNA and mRNA encoding viral proteins. The pregenome RNAs are packaged into particles with viral core protein in the cytoplasm and reverse transcription occurs, leading to the production of a full-length viral DNA minus strand [2]. Plus strand DNA synthesis begins approximately 250 bp from the 5' end of the minus strand and is primed by an RNA oligonucleotide from the 5' end of the RNA pregenome [3]. After plus strand DNA synthesis reaches the 5' end of the minus DNA strand, it "jumps the gap" to the 3' end of the minus DNA strand, resulting in circularization of the minus strand DNA. For reasons that are unclear, virions are formed and secreted before plus strand DNA synthesis is completed, and therefore, DNA molecules in virions are open circular and contain a single strand region. Completion of the viral DNA plus strand is an early event after the virus infects a permissive host cell (Figure 30.1). Transcription and translation of viral genes proceeds in an undirectional fashion, utilizing all three translational reading frames as the result of the overlap of viral genes [4] (Figure 30.2**A**).

It should be emphasized that in contrast to RNA-containing retroviruses, reverse transcription occurs in hepadna viruses during viral assembly rather than viral disassembly. The work of Tuttleman, Pourcel, and Summers [5] also supports a model in which DNA molecules in core particles are recycled into the nucleus where they are converted to the CCC form. This recycling

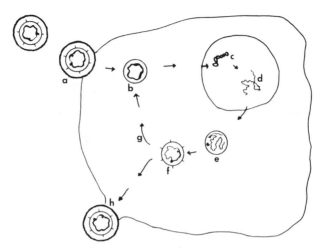

Figure 30.1. Schematic diagram of the pathway of hepadna viral nucleic acids involved in hepadna virus replication. (Viral mRNAs are not included.) **A** Viral attachment. **B** Penetration of hepatocyte, completion of plus strand DNA. **C** Conversion of viral DNA to the CCC form. **D** Transcription of pregenomic RNA. **E** Packaging of pregenome RNA in viral core particles. **F** Reverse transcription in core particles, generation of minus strand viral DNA, and open circular molecules after plus strand synthesis. **G** Recycling of some viral core particles to the nucleus where they provide a source of CCC DNA. **H** Viral maturation and excretion. (Diagram kindly provided by J. Tuttleman.)

mechanism is believed to provide a continuous source of viral DNA for replication of the virus. Unlike retroviruses, integration of viral DNA is not required for the hepadna virus replication cycle. However, hepadna virus integration does occur during persistent infection [6]. Recent studies also suggest that the integration process may be augmented during a state of persistent infection in which viral DNA replication continues but the virus assembly/secretion mechanism is blocked [7].

Viral DNA Integration and Hepatocellular Carcinoma

The association of hepatocellular carcinoma (HCC) with long-term, persistent hepatitis B virus (HBV) infection in humans was recognized soon after discovery of the virus [8]. Prospective studies have shown that HBV carriers have a 200-fold greater risk of developing HCC then uninfected individuals from the same population. Males are more prone than females to develop persistent infection, and male HBV carriers have a greater than 40% lifetime risk of HCC [8]. Persistent infection with each of the animal hepadna viruses, woodchuck hepatitis virus (WHV), ground squirrel hepatitis virus (GSHV), and duck hepatitis B virus (DHBV) also leads to HCC in the respective hosts. The association of persistent infection with HCC is

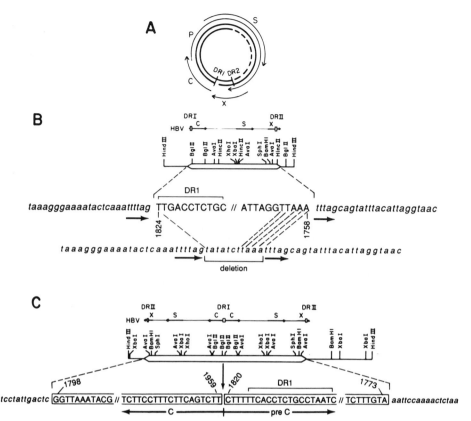

Figure 30.2. Examples of a linear HBV integration with one viral-cell junction at the DR1 sequence and the second viral-cell junction at a site of 5 bp homology with cellular DNA and a second integration containing inverted duplication of integrated HBV sequences at a chromosome translocation. **A** Genetic map of HBV including the virion open circular DNA (solid circular line) with a single-stranded region (dashed line). Arrangement of genes for the viral core protein C, surface antigen S, polymerase P, and X gene are illustrated with arrows to show the direction of transcription. The position of the eleven base directly repeated sequences (DR1 and DR2) and plus strands, respectively, are shown on the viral map at the termini of the viral minus and plus strands, respectively. **B** Example of a prototype linear HBV integration [23]. **Top two lines:** Genetic map of the integrated HBV sequences, and restriction endonuclease map of the integrated HBV sequences. The DNA sequence of the left and right viral-cell junctions is shown below the restriction map. The DNA sequence of normal cellular DNA at the HBV integration site along with the 5-11 Bp deletion is shown on the bottom line. **HBV DNA sequences are in capitals and cellular DNA in lowercase.** The region of homology between HBV DNA and cellular DNA at the right viral-cell junction is noted by dashed lines. The solid arrows under the sequence indicate a 5bp, directly repeated cellular DNA sequence at the integration site. **C** Inverted HBV integration at the site of a chromosome translo-cation t(17,18) (q21,q11). **Top two lines:** Genetic map of inverted HBV sequences and restriction endonuclease map of the integrated HBV. **Bottom line:** DNA sequence at the left and right viral-cell junctions and the HBV DNA inversion point. Nucleotide 1820 of HBV at this inversion point is the 3' end of the HBV DNA minus strand. Nucleotide numbers refer to the unique EcoR I site in the HBV genome.

strongest in woodchucks, which show nearly a 100% lifetime risk. Although primary HCC is the predominant tumor that arises in woodchucks, rare cholangiocarcinomas have also been reported [9,10].

The initial discovery that HCCs arising in WHV and HBV carriers contain clonal viral DNA integrations raised interest in the potential role of integration in hepatic oncogenesis [11,12]. In HBV carriers, integrated DNA is usually the sole form of viral DNA remaining in tumor tissue [11], whereas in woodchucks, episomal as well as integrated WHV DNA is present in some tumors [12]. Initially, it was hoped that by cloning the integrations from HCCs, it would be possible to identify cellular genes that function in association with viral DNA as hepatic oncogenes. The first integrations to be cloned were from two woodchuck hepatomas, each of which contained a single WHV DNA integration [13]. Unique cellular DNA sequences flanking the WHV integration were used to screen a large panel of woodchuck HCCs for a common cellular integration site, but none was found (unpublished data). Subsequent cloning of additional HBV, WHV, and DHBV DNA integrations from HCCs have confirmed the initial observation that viral integrations do not occur at a common site in HCCs. Therefore, it is now generally accepted that insertional mutagenesis of a specific cellular gene is not the predominant mechanism by which hepadna virus integrations function in hepatic oncogenesis.

In a few cases, hepadna virus integration adjacent to genes involved in cellular growth control has been reported [14,15]. Cases in which WHV DNA integration is in or near the c-*myc* gene [14], and one case in which HBV is integrated in a cellular retinoic acid receptor gene [15], have been reported. The mechanism by which these integrations alter the transcription of the respective genes or modify the function of the viral proteins leading to tumorigenesis has not yet been clarified.

Sequencing of viral integrations and restriction mapping of cellular integration sites have provided clues as to how such integrations occur and the mechanism(s) by which integrated viral sequences function in oncogenesis. Two general groups of integrations, those with linear integrated sequences and those with rearranged viral DNA sequences, have been described (Figure 30.2). The linear integrations show a strong preference to have one viral-cell junction within or immediately adjacent to one of two 11 base-pair, directly repeated sequences in the viral genome, designated DR1 and DR2 (Figure 30.2) [16,17]. The DR1 sequence is the site of initiation of minus strand DNA synthesis and is present in a unique triple strand region in virion DNA [4]. The other viral-cell junction of linear integrations apparently occurs randomly in viral DNA. Short homologies of two to five base pairs between viral and cellular DNA are common but not essential at the second viral-cell junction. The integration described in Figure 30.2**B** may be considered a prototype linear integration that exhibits both of the above characteristics.

Since the DR sequences are "hot spots" for viral integration, these sequences were examined for homology with cleavage sites for cellular enzymes known to be important in DNA recombination and repair mechanisms. Topoisomerase I (Topo I) has been shown in vitro to cleave integrated SV40 DNA specifically at sites where the virus is spontaneously excised in vivo in somatic cell hybrids [18]. Similar preferred Topo I cleavage sites were observed in the DR sequences of hepadna viruses, and a recent report [19] has shown that Topo 1 specifically cleaves WHV virion DNA at two places in the immediate vicinity of the DR sequences. Recent work has identified Topo I cleavage sites at the 3' end of the virion DNA minus strand, which also is a preferred viral integration site (Figure 30.2C). Since Topo I can link heterologous DNAs at cleavage sites, a possible role of Topo I in viral integration is suspected. Double-stranded viral DNA cleaved at the DR1 site or single-stranded replicative intermediates could function as substrates for Topo I-mediated integration. Since not all rearrangement sites are in the DR sequences or at Topo I cleavage sites, this represents only one of several mechanisms involved in generating integration and rearrangement of viral DNA.

The most common rearrangements of viral sequences in integrations are inverted duplications with deletion of viral sequences between the inversion [20]. Direct duplications or short deletions are also common [20]. Rearrangements of viral DNA are usually accompanied by rearrangement of flanking cellular DNA. The first clearly characterized rearrangement of cellular DNA at an HBV integration site was a large deletion (≥ 13 kb) that occurred at chromosome 11p13 [21]. Inverted duplication of HBV and cellular sequences has also been reported in a HBV clone from a Japanese patient [22]. Another integration involved an inverted duplication of HBV sequences in which the inversion served as the focal point for a chromosome 17q22:18q11 translocation [23]. Interestingly, one of the HBV sequences at the viral DNA inversion point was immediately adjacent to the DR1 sequence, at the exact position of the 3' end of virion minus strand DNA and a preferred Topo I cleavage site (Figure 30.2C). When the inverted duplication of HBV DNA from this clone was introduced into transgenic mice, the HBV sequences were rearranged in the mice at a high frequency [24]. Whether this instability is a specific property of the inverted HBV sequence, or is common to any inverted duplication of foreign DNA in transgenic mice, has yet to be determined. Several additional HBV integrations, cloned from HCCs, have been localized to chromosome translocations, each on a different chromosome [20].

In summary, initial studies on the role of HBV integrations in HCC have focused primarily on their potential to activate cellular oncogenes. These experiments have demonstrated that the location of HBV integrations in tumors is not specific for any oncogene and occurs on many chromosomes. In addition, in cases in which cellular genome has been mapped, major rearrangements of cellular DNA have often been observed. Reports that have compared viral DNA integrations in primary tumors to the integrations present in cell lines established from these tumors have shown that deletion and rearrangement of integrations can occur during the establishment of cell lines [25]. These postintegration rearrangements are also quite likely to occur during chronic active hepatitis as a consequence of continuous cellular regeneration. The occurrence of integrations that have linear viral sequences versus those with varying degrees of rearrangement fits a general model in which postintegration rearrangements are derived from integrations that initially had a specific linear structure.

The above model is consistent with results obtained from sequencing of five integrations from the liver and tumor of two young Japanese children [17]. The common features of these integrations were (1) that they were all comprised of linear viral DNA sequences without internal rearrangements and (2) that one viral-cell junction was in or adjacent to the DR 1 sequence in four of the cases. One explanation for these results, in the context of the above model, is that the relatively short period of chronic infection in the children (relative to the 20- to 40-year carrier state before HCC usually occurs in adults) did not provide sufficient cell generations for rearrangement of viral integrations to occur.

Three consequences of postintegration rearrangements that are significant with regard to cancer are as follows: (1) Such rearrangements, whether they involve excision, transposition, or translocation of viral and/or cellular DNA, result in the gradual accumulation of mutations in the cellular genome. Accumulation of mutations will invariably cause a cell to become predisposed to malignant transformation or may directly cause a mutation that results in transformation. (2) They may cause cellular DNA mutations that are no longer linked to viral DNA. This was clearly the case for the viral integration characterized at a chromosome translocation (Figure 30.2C), in which the reciprocal chromosome arm (18q11-centromere) contained a deletion of cellular DNA and no detectable viral DNA [23]. Thus, it is clear that, in addition to the translocation, viral DNA integration can generate chromosome aberrations by a "hit-and-run" mechanism. (3) The integration site in a tumor is not necessarily the cellular site at which the integration originated or caused a mutation predisposing the cell to malignant transformation. The apparent random distribution of HBV and WHV DNA integration sites in HCCs is consistent with this conclusion. Furthermore, only a few selected integrations would be expected to be associated with genes known to participate in oncogenesis. Experiments to trace the "footprints" (sites of excision) of integrations in the genome and possibly work backwards to identify cellular DNA rearrangements that are no longer associated with viral DNA, but are common to many HCCs, would be of interest. This may lead to the identification of common cellular mutation sites in HCCs. In light of the complications associated with utilizing HBV integrations to study genes and chromosomes involved in HCC, additional methods that do not depend on cloning viral integrations have been applied to this problem. These methods, described below, can detect chromosome defects that have relevance to mechanisms of cancer involving the loss of tumor suppressor genes.

Possible Role of Recessive Oncogenes (Tumor Suppressor Genes) in HCC

Since it has become clear that dominant oncogenes present only a partial picture of tumorigenesis mechanisms in vivo, the recessive oncogene hypothesis, originally proposed by Knudson [26], has attracted considerable attention. According to this hypothesis, a cell must lose or otherwise inactivate both copies of a recessive oncogene (another term is *tumor suppressor gene*) before cancer can develop. This hypothesis has gained strong experimental support through the study of such genetic tumors as Wilm's tumor [26,27] and retinoblastoma [28]. In these tumors, an individual who inherits a chromosome in which one copy of the tumor suppressor gene has been inactivated will be predisposed to cancer because inactivation of the remaining allele for the gene will result in the complete loss of function of the gene. Tumors arise at high rates in individuals who inherit chromosomes that predispose them to the tumorigenic consequences of such a "second" hit. Tumor suppressor genes have been localized on chromosome site 11p13 in Wilm's tumor [27,29] and 13q14 in retinoblastoma [28].

One molecular test of the Knudson hypothesis, which has been successfully applied to Wilm's tumor [30–33] and retinoblastoma [28], is that tumor tissues should exhibit the specific loss of alleles (or loss of heterozygosity) in regions of chromosomes in which tumor suppressor genes reside. Restriction fragment length polymorphisms (RFLPs) are normal variations in DNA sequence in the human population. Using probes for genes that exhibit RFLPs, it is possible to distinguish between the maternal or paternal copies of a gene and to detect the specific loss of one allele of a gene in tumor tissue. The only requirement of this analysis is that the initially normal tissue must be heterozygous for the RFLP to detect loss of one of the alleles in tumor tissue.

Since HCC is a solid tumor of epithelial origin, as are Wilm's tumors and retinoblastoma, HCCs from HBV carriers have been examined to determine whether they also exhibit specific chromosome losses particularly associated with chromosomes containing tumor suppressor genes, using RFLP loss as the genetic marker. Normal and HCC tumor tissues were screened with polymorphic gene probes from 20 chromosomes using Southern blotting. This study revealed a high frequency of loss of alleles from chromosomes 11p (45%) and 13q (50%) in tumor DNAs [34]. Nine of 14 tumors had lost one allele from either chromosome 11p, 13q, or, in some cases, both 11p and 13q, as opposed to random losses from other chromosomes. By analyzing tumor DNAs with probes spanning the region of chromosome 11p15 to 11p13, it was possible to localize deletions to either the distal end of 11p or internally at 11p13 (Figure 30.3). The data are consistent with previous reports for Wilm's tumor, in which some deletions were limited to the 11p15 region and others to 11p13 [30–33]. Both chromosomes 13q and 11p are believed to contain tumor suppressor genes.

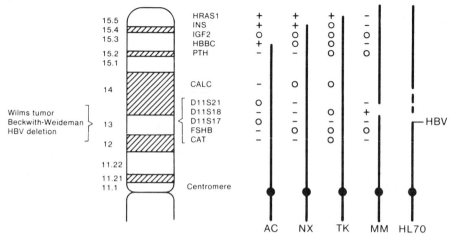

Figure 30.3. Diagram of deletions in chromosome 11p that caused the loss of gene heterozygosity on chromosome 11p in primary hepatocellular carcinomas from five HBV carriers. **Left** Diagram of chromosome 11p with chromosome bands numbered. The location of the tumor suppressor gene involved in Wilms' tumor is noted to the left; the gene probes used for restriction fragment length polymorphism (RFLP) analysis of tumor DNA are noted to the right of the banded chromosome diagram. **Right** Stick diagrams illustrating the results of RFLP analysis using chromosome 11p gene probes. Deletions of chromosome 11p are noted by a shorter stick diagram or a break in the line. The RFLP results for each probe are presented to the left of each stick diagram for chromosome 11p. The code for interpretation of RFLP results is as follows: +, normal DNA was heterozygous for this locus and tumor DNA was hemizygous, meaning that there was a deletion of one allele; 0, normal DNA was homozygous at the locus, no conclusion was possible; −, normal DNA was heterozygous, and tumor DNA was also heterozygous (no deletion at this locus).

In the HCCs examined, deletions occurred in 13q, 11p, or both 11p and 13q, raising the possibility that the tumor suppressor genes on these chromosomes may affect a common third site. Specifically, loss of alleles at 11p and/or 13q may serve to remove suppressor genes, which then allows expression of a recessive oncogene located elsewhere in the genome.

In each of the HCCs studied, the deletion of an allele for a gene resulted in a hemizygous condition in which the remaining allele of that gene was not reduplicated in the tumor. In this respect, the HCC data contrast with the data of Wilm's tumor and retinoblastoma, in which the predominant mechanism involves reduplication of the remaining allele to establish a homozygous condition of the mutant allele [28,30–33]. In embryonal tumors, there is probably a strong selection against monosomy, and therefore loss of chromosomes by mitotic nondisjunction is accompanied by reduplication of the remaining chromosome. In adult tumors, it is likely that the action of a hit-and-run carcinogenic agent, such as an HBV integration, would result in deletions of chromosome segments without the need to reduplicate the remaining chromosomal counterpart.

Transcriptional Activation of Insulin-Like Growth Factor II in HCC

One aspect of the tumor suppressor theory is that loss of a suppressor gene should allow expression of a recessive cellular oncogene. In addition to the Wilm's tumor gene at chromosome 11p13, several other genes involved in endocrine and autocrine control of growth are located on chromosome 11p, including c-H-Ras, insulin, insulin-like growth factor II (IGF-II), and parathyroid hormone at 11p15 and follicle-stimulating hormone and calcitonin in the 11p13 region. Since IGF-II is a fetal growth factor highly expressed in liver [35,36], its expression was studied in human HCC cell lines. Three out of the five HCC cell lines examined contained high levels of IGF-II RNA, and the IGF-II RNAs were characteristic of transcripts present in fetal human liver. Studies of primary human HCCs have shown a fetal liver pattern of IGF-II RNA in most cases [37]. Since IGF-II is a fetal growth factor, it might be expected to come under the control of suppressor genes that function to maintain the adult program of differentiation in hepatocytes. Derepression of alpha fetoprotein in HCCs is another example of a fetal gene that is derepressed in most HCCs.

Studies of IGF-II transcription in woodchucks have shown that IGF-II RNA is also highly elevated in most woodchuck HCCs [38]. A series of IGF-II transcripts identified in woodchuck HCCs were not detectable in normal liver. By analogy with the rat IGF-II gene, several of these RNAs are probably initiated from a promoter that is inactive in adult tissue and active in fetal rat liver [39]. In situ hybridization has demonstrated early cancer nodules with high levels of IGF-II RNA, and precancerous nodules also contain high levels of IGF-II RNA [38]. Therefore, IGF-II activation occurs at an early stage in cancer development. A two- or threefold elevation of serum IGF-II levels was also detected in woodchucks with chronic active hepatitis in the precancerous stages [38]. The biological activity of IGF-II in liver has not been directly determined. However, since IGF-II is expressed in fetal liver and is a potent mitogen in cell culture [40], it is likely that it has either autocrine or paracrine functions to stimulate regeneration of liver cells during chronic hepatitis. It is, therefore, not unreasonable that IGF-II may remain active in tumors originating from cells in which the gene is activated.

Model for Tumorigenesis in WHV Carriers

A schematic model of the series of events in the progression from initial WHV infection in woodchucks to HCC is presented in Figure 30.4. The model proposes that, once WHV persistent infection is established, portal hepatitis (PH) ensues and several pathological reactions occur on a self-perpetuating basis. These include (1) the death and regeneration of hepatocytes, (2) an inflammatory reaction primarily limited to the portal tracts, and (3) mild proliferation of bile ducts (Figure 30.4, line 2). As the disease progresses to chronic active hepatitis (CAH), involvement of the rest of the hepatic lobule becomes evident (lobular hepatitis) (Figure 30.4, line 3). The inflammatory reaction, composed primarily of lymphocytes with some macrophages and plasma cells, spills over the limiting plate and disrupts the lobular architecture. Clusters of dyplastic nodules arise in the liver at this precancerous stage. As a result of derepression of specific cellular gene expression, IGF-II transcription is activated, and elevated IGF-II is also detected in the serum of woodchucks with CAH. In addition, WHV DNA integrations can be cloned from infected liver cells at this stage [6].

Within the cellular milieu maintained by CAH, HCCs develop with distinct phenotypic and biochemical characteristics (Figure 30.4, line 4), including some tumors with and others without IGF-II activation. The role of the virus in this process includes maintenance of a selective pressure for

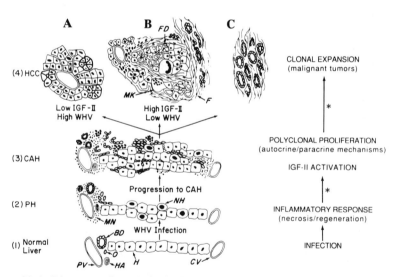

Figure 30.4. Diagram of pathological changes in woodchuck livers during the progression from initial persistent WHV infection to hepatocellular carcinoma (see text for full explanation). **Line (1),** normal liver; normal liver structural organization. **Line (2),** PH, portal hepatitis during early persistent infection. **Line (3),** CAH, progression to chronic active hepatitis. **Line (4),** HCC, fully malignant liver cells exhibiting heterogeneous gene expression. High IGF-II means high level of IGF-II RNA in HCC tissue; High WHV means high level of WHV RNA in tumor tissue. **Right side,** several factors important in the development of HCC at different stages of disease. **Asterisk** marks stages in the progression to HCC in which viral integration may function by mechanisms discussed in the text.

liver cell turnover and generation of chromosome aberrations that accumulate and predispose the cell to malignant transformation. Whether the virus plays a direct role in IGF-II activation or whether this activation is a consequence of proliferation of liver stem cells is a matter for further investigation. In any event, it is clear that polyclonal expansion during CAH generates a population of cells in the precancerous liver (regenerative nodules) that represent the first step to cancer. Mutational events within these cells eventually lead to the transformation of a single cell and that cell's clonal expansion to a fully malignant HCC.

References

1. Mason WS, Halpern MS, England JM, Seal G, Egan J, Coates L, Aldrich C, Summers J (1983) Experimental transmission of duck hepatitis B virus. Virology 131:575–584
2. Summers J, Mason WS (1982) Replication of the genome of a hepatitis B-like virus by reverse transcription of an RNA intermediate. Cell 29:403–415
3. Lien JM, Aldrich CE, Mason WS (1986) Evidence that a capped oligoribonucleotide is the primer for duck hepatitis B virus plus-strand DNA synthesis. J Virol 57:229–236
4. Ganem D, Varmus, HE (1987) The molecular biology of the hepatitis B virus. Ann Rev Biochem 56:651–693
5. Tuttleman JS, Pourcel C, Summers J (1986) Formation of the pool of covalently closed circular viral DNA in hepadnavirus infected cells. Cell 47:451–460
6. Rogler CE, Summers J (1984) Cloning and structural analysis of integrated woodchuck hepatitis virus sequences from a chronically infected liver. J Virol 50:832–837
7. Raimondo G, Burk RD, Lieberman HM, Muschel J, Hadziyannis SJ, Will H, Kew MC, Dusheiko JM, Shafritz DA (1988) Interrupted replication of hepatitis B virus in liver tissue of HBsAg carriers with hepatocellular carcinoma. Virology 165:103–112
8. Beasley RP, Hwang LY, Lin CC, Chien CS (1981) Hepatocellular carcinoma and hepatitis B virus. Lancet 3:1129–1133
9. Summers J, Smolec JM, Snyder R (1978) A virus similar to human hepatitis B virus associated with hepatitis and hepatomas in woodchucks. Proc Natl Acad Sci USA 75:4533–4537
10. Snyder RL, Summers J (1980) Woodchuck hepatitis virus and hepatocellular carcinoma. In Essex M, Todaro G, Zur Huasen H (eds) Viruses in Naturally Occurring Cancers. Cold Spring Harbor Conference of Cell Proliferation, vol 7. p 447–457
11. Shafritz DA, Shouval D, Sherman HI, Hadziyannis SJ, Kew MC (1981) Integration of hepatitis B virus DNA into the genome of liver cells in chronic liver disease and hepatocellular carcinoma. New Engl J Med 305:1067–1073
12. Summers J, Smolec JM, Werner BG, Kelly TG Jr, Tyler GV, Snyder RL (1980) Hepatitis B virus and woodchuck hepatitis virus are members of a novel class of DNA viruses. In Essex M, Todaro G, Zur Hausen H (eds) Viruses in Naturally

Occurring Cancers. Cold Spring Harbor Conference on Cell Proliferation, vol 7. pp 459–470

13. Ogston CW, Jonak GJ, Rogler CE, Astrin SM, Summers J (1982) Cloning and structural analysis of integrated woodchuck hepatitis virus sequences from hepatocellular carcinomas of woodchucks. Cell 29:385–394

14. Hsu TY, Moroy T, Etiemble J, Louise A, Trepo C, Tiollas P, Buendia MA (1988) Activation of cMyc by woodchuck hepatitis virus insertion in hepatocellular carcinoma. Cell 55:627–635

15. Brand N, Patkovich M, Krust A, Chambon P, de Th'e H, Marchio A, Tiollais P, Dejean A (1988) Identification of a second human retinoic acid receptor. Nature 332:850–853

16. Dejean A, Sonigo P, Wain-Hobson S, Tiollais P (1984) Specific hepatitis B virus integration in hepatocellular carcinoma DNA through a 11 base-pair direct repeat. Proc Natl Acad Sci USA 81:5350–5354

17. Yaginuma K, Kobayashi H, Kobayashi M, Morishima T, Matsuyama K, Koike K (1987) Multiple integration sites of hepatitis B virus DNA in hepatocellur carcinoma and chronic active hepatitis tissues from children. J Virol 61:1808–1813

18. Bullock P, Champoux JJ, Botchan M (1985) Association of crossover points with topoisomerase I cleavage sites: A model for nonhomologous recombination. Science 230:954–958

19. Wang HP, Rogler CE (1987) Evidence for topoisomerase I cutting of WHV virion DNA in the cohesive overlap region. Presented at Cold Spring Harbor Hepatitis B Viruses Meeting, September 28–October 1, Abstr p 121

20. Nagaya T, Nakamura T, Tokino T, Tsurimoto M, Mayumi T, Kamino K, Yamamura K, Matsubara K (1987) The mode of hepatitis B virus DNA integration in chromosomes of human hepatocellular carcinoma. Genes and Devel 1:773–782

21. Rogler CE, Sherman M, Su CY, Shafritz DA, Summers J, Shows TB, Henderson A, Kew M (1985) Deletion in chromosome 11p associated with a hepatitis B integration site in hepatocellular carcinoma. Science 230:319–322

22. Mizusawa H, Taira M, Yaginuma K, Kobayashi H, Yoshida E, Koike K (1985) Inversely repeating integrated hepatitis B virus DNA and cellular flanking sequences in the human hepatoma-derived cell line huSP. Proc Natl Acad Sci USA 82:208–212

23. Hino O, Shows TB, Rogler CE (1986) Hepatitis B virus integration site in hepatocellular carcinoma at chromosome 17;18 translocation. Proc Natl Acad Sci USA 83:8338–8342

24. Hino O, Nomura K, Ohtahe K, Kitagawa T, Sugano H, Kimura S, Yokoyamo M, Katsuki M (1986) Rearrangement of integrated HBV DNA in descendants of transgenic mice. Proc Japan Acad 62 Ser B:1–4

25. Unora M, Kobayashi K, Fukuoka K, Matsushita F, Morimoto H, Oshima T, Kameko S, Hattori N, Murakami S, Yoshikawa H (1985) Establishment of a cell line from a woodchuck hepatocellular carcinoma. Hepatology 6:1106–1111

26. Knudson AG, Strong IC (1972) Mutation and cancer: A model for Wilm's tumor of the kidney. J Natl Cancer Inst 48:313–324

27. Francke U, Riccardi VM (1979) Aniridia–Wilms' tumor association: Evidence for specific deletion of 11p13. Cytogenet Cell Genet 23:185–192

28. Cavenee WK, Dryja TP, Phillips RA, Benedict WF, Godbout R, Gallie BO,

Murphree AL, Strong LC, White RL (1983) Expression of recessive alleles by chromosomal mechanisms in retinoblastoma. Nature 305:779–784

29. Glaser TG, Lewis WH, Bruns GAP, Watkins PC, Rogler CE, Shows TB, Powers VE, Willard HF, Goguen JM, Simola KOJ, Housman DE (1986) The β-subunit of follicle stimulating hormone is deleted in patients in aniridia and Wilms' tumor, allowing a further definition of the WAGR locus. Nature 321:882–887
30. Koufos A, Hansen MF, Lampkin BC, Workman ML, Copeland NG, Jenkins NA, Cavenee WK (1984) Loss of alleles at loci on human chromosome 11 during genesis of Wilms' tumor. Nature 309:170–172
31. Orkin SH, Goldman DS, Sallan SE (1984) Development of homozygosity for chromosome 11p markers in Wilms' tumor. Nature 309:172–174
32. Reeve AE, Housiaux PJ, Gardner RJM (1984) Loss of Harvey ras allele in sporadic Wilms' tumor. Nature 309:174–176
33. Fearon ER, Vogelstein B, Feinberg AP (1984) Somatic deletion and duplication of genes on chromosome 11 in Wilm's tumors. Nature 309:176–178
34. Wang HP, Rogler CE (1988) Deletions in chromosomes 11p and 13q in primary hepatocellular carcinomas. Cytogenet Cell Genet 48:72–78
35. Soares MB, Ishii DN, Efstratiadis A (1985) Developmental and tissue-specific expression of a family of transcripts related to rat insulin-like growth factor II mRNA. Nuc Acid Res 14:1119–1134
36. Frunzio R, Chiarotti L, Brown A, Graham DE, Rechler MW, Bruni CB (1986) Structure and expression of the rat insulin-like growth factor II (rIGF-II) gene. J Biol Chem 261:17138–17149
37. Lasserre C, Cariani E, Hamelin B, Brechot C (1987) Re-expression of insulin-like growth factor II fetal transcripts in human primary liver tumors. Presented at Cold Spring Harbor Hepatitis B Viruses Meeting, September 28–October 1, Abstr p 127
38. Fu XX, Su CY, Lee Y, Hintz R, Biempica L, Snyder R, Rogler CE (1988) Insulin-like growth factor II expression and oval cell proliferation associated with hepatocarcinogenesis in woodchuck hepatitis virus carriers. J Virol 62:3422–3430
39. Soares MB, Turken A, Ishii D, Mills L, Epeskopou V, Cotter S, Zeitlin S, Efstratiadis A (1986) Rat insulin-like growth factor II gene, a single gene with two promoters expressing a multitranscript family. J Mol Biol 192:737–753
40. Dulak NC, Temin HM (1973) Multiplication-stimulating activity for chicken embryo fibroblasts from rat liver cell conditioned medium. A family of small polypeptides. J Cell Physiol 82:161–170

CHAPTER 31

Molecular Basis of Rhabdovirus Pathogenicity

ROBERT R. WAGNER

The viruses of the family Rhabdoviridae are widely distributed in nature as disease-causing agents of plants, insects, fish, and mammals, including humans. A great deal is known about the structure, genetics, and molecular biology of rhabdoviruses, much of which are described in great detail in a book published in 1987 [1]. Most mammalian rhabdoviruses can be classified in two genera: Vesiculovirus, the prototype of which is vesicular stomatitis virus (VSV), and Lyssavirus, the major species being rabies virus. The rhabdovirus prototype VSV grows very rapidly to high titer and is cytopathogenic for almost all eucaryotic cells, thus providing an excellent model system for studying viral pathogenicity [2].

The VSV genome is an unsegmented single strand of RNA containing 11,162 nucleotides. The genomic RNA is of negative sense in that it cannot serve as messenger but requires its own endogenous RNA polymerase for transcription. The infectious virion is bullet-shaped, $\sim$180 X 65 nm in diameter, and contains a tightly wound ribonucleoprotein (RNP) core surrounded by a lipoprotein bilayer membrane [1,2]. In addition to five mRNAs, which are also capped and polyadenylated, the VSV polymerase transcribes a 47-nucleotide leader RNA that is neither capped nor adenylated and whose only known function is transcription initiation [1]. Ultraviolet mapping studies have shown that the gene order is 3'-leader-N-NS-M-G-L-5' [1,2].

Another important aspect of VSV replication is the almost invariable production of defective-interfering (DI) particles. These DI virions are one-quarter to one-half the length of the infectious B virions and contain virion RNA deleted by 50% to 80% that of the wild-type genome. Generally, these DI–particle RNAs retain only the 5' half of the L gene and a 46-nucleotide leader sequence, partially homologous to the wild-type leader

RNA sequence. These 5'-DI particles contain all the virion proteins, including the L/NS polymerase, but are incapable of synthesizing mRNA and can transcribe only the 46-nucleotide DI–leader RNA [1,2].

Cellular Responses to Infection

Although cells vary somewhat in their response to VSV infection, virtually all vertebrate cells that have been tested, and to some extent invertebrate cells, are susceptible. It is wise to remember that the times at which cytopathic effects (CPE) occur in cell cultures depend on the multiplicity of infection. At low doses of infection, VSV CPE is not evident for 24 hours or longer. However, at high multiplicities, cell rounding occurs in several hours, progressing rapidly to membrane permeability to supravital dyes and terminating in cell detachment from adherent surfaces. An even earlier event, within the first hour or so, is inhibition of cellular macromolecular synthesis, resulting in the progressive shutdown of cellular RNA, DNA, and protein synthesis [3]. The rapidity with which these CPEs occur, dubbed cytotoxicity, led to the hypothesis that these early cellular reactions to VSV infection are due to the input virus or its constituents, most logically structural proteins. Putative support for this hypothesis comes from experiments that appeared to show that ultraviolet-irradiated virus or metabolically inactive DI particles at very high multiplicity caused CPE and inhibition of cellular RNA synthesis similar to that caused by fully infectious virus [4]. It is now clear that enormous input multiplicities of DI particles have no effect on CPE or cellular RNA synthesis, provided that they are really free of contaminating infectious virions [3,5]. Moreover, very high levels of ultraviolet irradiation are required to render VSV incapable of inhibiting cellular nucleic acid synthesis [6]. It is now generally accepted that newly synthesized viral products, and not the preexisting components of the infecting virion, are responsible for CPE and inhibition of cellular macromolecular synthesis caused by VSV [2].

Inhibition of Cellular Macromolecular Synthesis

Cellular RNA Synthesis
Having ruled out the possibility that structural components of invading virions are responsible for shutting off cellular RNA synthesis, the only reasonable hypothesis was that newly synthesized RNA or protein coded by the viral genome could be the inhibitor. Investigators resorted to indirect methods, particularly the use of temperature-sensitive mutants and ultraviolet irradiation. Weck and Wagner [3] tested several ts mutants and found that only the L-gene mutant, tsG114(I), in which all primary transcription is restricted at 39°C, failed to shut off cellular RNA synthesis at the nonpermissive temperature. Further studies by Wu and Lucas-Lenard [7]

revealed that group II mutant *ts*G22, assigned to the gene coding for the NS protein, was also restricted in its capacity to inhibit cellular RNA synthesis at restrictive temperature. These and other studies strongly suggest that primary VSV transcription is required to shut off cellular RNA synthesis.

Weck and colleagues [8] used the technique of mapping by increasing doses of ultraviolet irradiation and reported that a dose of 72,000 ergs/mm^2 was required to reduce the capacity of VSV to shut off cellular RNA synthesis to 37% that of the unirradiated control virus; calculations of target size implicated the gene coding for leader RNA as the only viral genetic determinant that could be responsible for inhibiting cellular RNA synthesis. Subsequent studies [9] with more purified VSV showed that 12,000 ergs/mm^2 was the true ultraviolet dose for reducing the cellular-RNA inhibiting factor to 37% for VSV-Indiana and VSV-New Jersey serotypes in vitro and in vivo; these studies revealed a target size of 85 nucleotides rather than the actual 47 nucleotides for the leader. An independent study by Dunigan and Lucas-Lenard [10] showed a two-phase ultraviolet target of 42 and 373 nucleotides for the VSV gene products required to inhibit cellular protein synthesis; this would implicate both the leader RNA gene and the N gene products as inhibitors of cellular RNA and protein synthesis.

The *wt* leader RNA is synthesized in cells soon after infection with VSV and then migrates to the nucleus. This small leader RNA is rapidly bound by a cellular protein that reacts antigenically with anti-La lupus antibody [11], in some respects resembling a small nuclear RNP (Snurp). A correlation was reported [9] between the rate and degree of VSV leader RNA synthesis with that for inhibition of cellular RNA synthesis, suggesting that VSV leader RNA migrates to the nucleus of MPC-11 mouse myeloma cells where it down-regulates cellular RNA synthesis. However, Dunigan et al. [12] could find no correlation between the accumulation of plus strand leader RNA and the inhibition of protein or RNA synthesis in VSV-infected mouse L cells. Moreover, Whitaker-Dowling and Youngner [13] have presented contradictory evidence that the large doses of ultraviolet irradiation, which they acknowledge are required to inhibit synthesis of cellular RNA and protein [8,10], also directly affect the virion polymerase L and NS proteins as determined by gel electrophoresis and Western blotting. These studies do not of course necessarily rule out the role of VSV leader, which patently requires the VSV polymerase for its production.

McGowan et al. [14] set out to design an in vitro system for testing the hypothesis that the leader RNA, or some other VSV product, is the inhibitor of cellular RNA synthesis. At the same time, such a system could shed some light on the target for inhibition of cellular RNA synthesis. It had been shown that inhibition of RNA synthesis in MPC-11 and L cells was not a function of defective nucleoside uptake, nucleotide conversion, RNA precursor processing, or RNA degradation [3,15]. Further studies showed that all three RNA polymerases were affected in VSV-infected cells and that the block in cellular RNA synthesis is at the level of transcription initiation, not chain elongation [15]. The HeLa cell extract system was used to test the effects of VSV leader RNA on in vitro transcription of the adenovirus major-late promoter (MLP) and the virus-associated (VA) genes under the control of polymerase II (pol II) and polymerase III (pol III), respectively. We found that *wt* leader RNA, but not DI leader, inhibited both pol II and pol III transcription. The *wt* and DI leaders are relatively homologous except for nucleotides 18 to 30, the sequence of which is 5'-AUUAUUAUCAUUA-3' for the *wt* leader. DNA clones and synthetic oligodeoxynucleotides with this sequence also inhibited MLP and VA transcription, but they were more efficient if they also contained flanking sequences [16]. Synthetic, leader-like oligodeoxynucleotides with certain base substitutions lose their capacity to inhibit in vitro transcription [16,17].

The target for inhibition of transcription by leader RNA or homologous oligodeoxynucleotide is not known. The polymerases themselves, pol II or pol III, do not appear to be the targets. Nor are there apparent DNA template binding sites for transcription inhibition by leader RNAs or homologous DNAs [17]. The most likely target for leader inhibition is a cellular factor(s) that recognizes a gene promoter or enhancer. The only candidate to date for such a transcription factor that can be blocked by leader RNA is a 65-kd protein in HeLa cells that bound ^{32}P-labeled *wt* leader or homologous DNA but not DI leader or extraneous DNA [17]. Eluted from nitrocellulose, this 65-kd protein also reversed to some extent the transcription–inhibition activity of *wt* leader or homologous DNA [17]. Much more research is needed before this factor or other factors can be implicated as targets for transcription inhibition by VSV *wt* leader RNA.

Cellular DNA Synthesis
The effect of VSV infection on cellular DNA synthesis is strikingly similar to that of VSV inhibition of cellular RNA synthesis. A comparative study showed that 3' DI particles and mutant *ts*G114 did not shut off MPC-11 cellular DNA synthesis and that *wt* VSV did so by kinetics indistinguishable from that of RNA synthesis inhibition. Moreover, similar doses of ultra-violet irradiation were found to be required to suppress the capacity of VSV to shut off cellular DNA and RNA synthesis [6]. All these studies suggest that synthesis of VSV leader RNA is essential to inhibit cellular DNA, as well as RNA synthesis. As a logical extension of these studies, Remenick and McGowan [18] have shown that the isolated *wt* VSV leader RNA, but not the DI leader RNA, is capable of blocking adenovirus DNA replication in a cell-free system, which strongly suggests that the same viral leader RNA is responsible for inhibiting both transcription and replication.

In an extension of these studies, it was found that synthetic oligodeoxy-nucleotides homologous to those of *wt* leader RNA also inhibited adenovirus DNA replication in vitro [19]. The inhibition produced was found to be at the level of pre-terminal protein (pTP)–dCMP complex formation, the step that initiates adenovirus DNA replication. The pTP–adenovirus DNA polymerase complex, the adenovirus DNA-binding protein, and nuclear factor I are all required for initiation of adenovirus replication. Nuclear extracts of uninfected or adenovirus-infected cells were found to be capable of restoring in vitro adenovirus DNA replication under conditions of inhibition by *wt* leader RNA. The addition of purified, pTP–adenovirus polymerase or HeLa-cell–DNA polymerase α-primase also restored replicative activity to leader RNA-inhibited adenovirus DNA replication. These findings suggest that the target of leader RNA is the pTP–adenovirus polymerase complex, possibly because of the striking degree of homology between *wt* leader RNA and the 5' terminal sequence of adenovirus type 2. An extension of this hypothesis would be recognition by the VSV leader RNA sequences of pTP–adenovirus polymerase complex and nuclear factor I that are known to bind the homologous adenovirus replication–initiation site, which is highly conserved among many strains of adenovirus. Unpublished studies by Remenick and McGowan indicate that the *wt* leader RNA can associate with at least one protein present in an immunopurified HeLa cell–DNA polymerase α-primase complex. It would be of interest to determine whether similar nuclear factors are able to compete successfully with both the transcription–inhibition and replication–inhibition activities of the VSV leader RNA, particularly in view of evidence that transcription promoter/enhancer factors also can function as replication enhancers.

Cellular Protein Synthesis

VSV inhibition of cellular protein synthesis also requires active viral transcription as determined by the ineffectiveness of transcriptase-restricted *ts* mutants [20] or of DI particles [21]. Partial evidence for differences in the capacity of VSV to inhibit synthesis of cellular nucleic acids and proteins was provided by the dissimilar kinetics of the two reactions [6]. Support for the hypothesis that a different VSV gene product is required to inhibit cellular protein synthesis came from ultraviolet inaction data revealing a genome target equivalent to that required for transcription of the N gene mRNA. The most definitive studies are those of Dunigan and Lucas-Lenard [10], which provide evidence for a biphasic ultraviolet-inactivation curve, implicating both the N gene and the leader RNA transcripts as dual requirements for maximal shutoff of cellular protein synthesis. The postulate [22] that inhibition of cellular protein synthesis by *wt* VSV and not by transcription-restricted mutants is the result of viral mRNA successfully competing with cellular messengers has been refuted in several ways, most tellingly by evidence that marked suppression of *wt* VSV mRNA synthesis by coinfecting DI particles has no effect on *wt* VSV inhibition of cellular protein synthesis [21].

Only limited and rather inconclusive studies have been performed to identify the VSV gene product(s) that inhibit cellular protein synthesis or the target for this inhibition. It was found that VSV leader and mRNAs per se had no effect on in vitro protein synthesis, but double-stranded RNA associated with VSV nucleocapsids did significantly reduce translation of endogenous and exogenous globin mRNA in a reticulocyte system [23]. The possible target for translation inhibition was examined by comparing fractionated and reconstituted products from protein-synthesizing machinery of uninfected and VSV-infected mouse L cells [24]. The results indicated that a salt-wash fraction containing eIF-3 and eIF-4B from ribosomes of VSV-infected cells, but not from mock-infected cells, was defective in initiation of translation in a reconstituted system. Other investigators consider eIF-2 to be the ribosomal translation initiation factor that is altered by VSV infection [25].

In conclusion, the take-home lesson from all studies to date on VSV cytopathic effects is to interpret them with caution. Newer molecular biological techniques will, it is hoped, provide more conclusive evidence.

References

1. Wagner RR (1987) The Rhabdoviruses. Plenum Publishing, New York
2. Wagner RR, Thomas JR, McGowan JJ (1984) Rhabdovirus cytopathology: Effects on cellular macromolecular synthesis. In Fraenkel-Conrat H, Wagner RR (eds) Comprehensive Virology, vol. 19. Plenum Publishing, New York, pp 223–295
3. Weck PK, Wagner RR (1979) Transcription of vesicular stomatitis virus is required to shut off cellular RNA synthesis. J Virol 30:410–413
4. Huang AS, Wagner RR (1965) Inhibition of cellular RNA synthesis by nonreplicating vesicular stomatitis virus. Proc Natl Acad Sci USA 54:579–583
5. Marcus PI, Sekellick MJ (1974) Cell killing by viruses. I. Comparison of cell killing by plaque-forming and defective-interfering particles of vesicular stomatitis virus. Virology 57:321–338
6. McGowan JJ, Wagner RR (1981) Inhibition of cellular DNA synthesis by vesicular stomatitis virus. J Virol 38:356–367
7. Wu FS, Lucas-Lenard JM (1980) Inhibition of ribonucleic acid accumulation in mouse L cells infected with vesicular stomatitis virus requires nucleic acid transcription. Biochemistry 19:804–810
8. Weck PK, Carroll AR, Shattuck DM, Wagner RR (1978) Use of UV irradiation to identify the genetic information of vesicular stomatitis virus responsible for shutting off cellular RNA synthesis. J Virol 30:746–753
9. Grinnell BW, Wagner RR (1983) Comparative inhibition of cellular transcription by vesicular stomatitis virus serotypes New Jersey and Indiana: Role of viral leader RNA. J Virol 48:88–101
10. Dunigan DD, Lucas-Lenard JM (1983) Two transcription products of the vesicular stomatitis genome may control L-cell protein synthesis. J Virol 45:618–626
11. Kurilla MG, Keene JD (1983) The leader RNA of vesicular stomatitis virus is bound by a cellular protein reactive with anti-La lupus antibody. Cell 34:837–845
12. Dunigan DD, Baird S, Lucas-Lenard J (1986) Lack of correlation between the accumulation of plus-strand leader RNA and its inhibition of protein and RNA

synthesis in vesicular stomatitis virus infected mouse L cells. Virology 150:231–246

13. Whitaker-Dowling P, Youngner JS (1988) Alteration of vesicular stomatitis L and NS proteins by UV irradiation: Implications for the mechanism of host cell shut-off. Virology 164:171–175

14. McGowan JJ, Emerson SU, Wagner RR (1982) The plus-strand leader RNA of VSV inhibits DNA-dependent transcription of adenovirus and SV40 genes in a soluble whole cell extract. Cell 28:325–333

15. Weck PK, Wagner RR (1979) Vesicular stomatitis virus infection reduces the number of active DNA-dependent RNA polymerases in myeloma cells. J Biol Chem 254:5430–5434

16. Grinnell BW, Wagner RR (1984) Nucleotide sequence and secondary structure of VSV leader RNA and homologous DNA involved in inhibition of DNA-dependent transcription. Cell 36:533–543

17. Grinnell BW, Wagner RR (1985) Inhibition of DNA-dependent transcription by the leader RNA of vesicular stomatitis virus: role of specific nucleotide sequences and cell protein binding. Mol Cell Biol 5:2502–2513

18. Remenick J, McGowan JJ (1986) A small RNA transcript of vesicular stomatitis virus inhibits the initiation of adenovirus replication in vitro. J Virol 59:660–668

19. Remenick J, Kenny MK, McGowan JJ (1988) Inhibition of adenovirus DNA replication by vesicular stomatitis virus leader RNA. J Virol 62:1286–1292

20. McAllister PE, Wagner RR (1976) Differential inhibition of host protein synthesis in cells infected with RNA⁻ temperature-sensitive mutants of vesicular stomatitis virus. J Virol 18:550–558

21. Schnitzlein WM, O'Banion MK, Poirot MK, Reichmann ME (1983) Effect of intracellular vesicular stomatitis virus mRNA concentration on the inhibition of host cell protein synthesis. J Virol 45:206–214

22. Lodish HF, Porter M (1980) Translational control of protein synthesis after infection by vesicular stomatitis virus. J Virol 36:719–733

23. Thomas JR, Wagner RR (1982) Evidence that vesicular stomatitis virus produces double-stranded RNA that inhibits protein synthesis in a reticulocyte lysate. J Virol 44:189–198

24. Thomas JR, Wagner RR (1983) Inhibition of translation in lysates of mouse L cells infected with vesicular stomatitis virus. Presence of defective ribosome-associated factor. Biochemistry 22:1540–1546

25. Jaye MC, Godchaux W, Lucas-Lenard J (1982) Further studies on the inhibition of cellular protein synthesis by vesicular stomatitis virus. Virology 116:148–162

CHAPTER 32
Gene Functions Directing Advenovirus Pathogenesis

HAROLD S. GINSBERG AND GREGORY PRINCE

The pathogenic potentials of adenoviruses are great. Thus, of the 41 distinct types of adenoviruses that infect humans, the diseases produced vary from mild upper respiratory infections to severe pneumonia, from mild conjunctivitis to epidemic keratoconjunctivitis, from esophagitis to epidemic gastroenteritis, and even to hemorrhagic cystitis [1]. It is noteworthy that these widely varied clinical syndromes produced by adenoviruses are not due to host range variations resulting from differences in reactions with different receptors. All or most adenoviruses replicate well in the gastrointestinal tract, but types 40 and 41 are essentially solely responsible for epidemics of infantile gastroenteritis. Many types infect and replicate in the respiratory tract, but relatively few types are responsible for producing epidemics (e.g., types 3, 4, and 7). In recruits in the Armed Forces (ARD), Ad4 produces extensive epidemics of acute respiratory diseases, including pneumonia, but this virus is rarely detected in the civilian population. Many types of adenoviruses infect the eye, but only a few types produce epidemic keratoconjunctivitis (e.g., Ad8 and 19). Clearly, the pathogenic potentials of adenoviruses do not reside predominately in their host range (i.e., infection of cells with appropriate receptors), which is commonly used to define virulence, but rather they depend on other gene functions that are responsible for producing the pathological changes.

Since the initial isolations of adenoviruses by Rowe, Huebner, and their colleagues [2] and by Hilleman and Werner [3], elegant studies have described the virion structure, events in viral replication and assembly, the genome's nucleotide sequence and the gene regions contained therein, and the gene products produced [4]. Mutants in most genes have been isolated and characterized, and these mutants have helped to identify many of the functions of both early and late gene products [5]. Despite the extensive data available, there is essentially no information relating to the mechanisms of pathogenesis by which adenoviruses produce these diseases.

Two critical factors, however, have made it possible to investigate the molecular mechanisms of pathogenesis, that is, the gene functions that are central to producing the pathological changes that are the basis of disease: (1) the discovery of an animal model for adenovirus pneumonia, the cotton rat [6]; and (2) the availability of mutants containing defects in almost every viral gene, as well as the technology to construct a mutant at any site desired [5]. Using these tools, it has been possible to begin to determine the viral gene functions required to produce adenovirus pneumonia and the mechanisms by which they induce disease (at least pneumonia in this animal model). The results of these early studies suggest that pathogenesis depends on mechanisms not previously considered in acute viral diseases: an early elaboration of cytokine(s) and the induced subsequent inflammatory reaction followed by a cytotoxic T cell response.

The Animal Model

Two strains of cotton rats were found to be susceptible to adenovirus infections: *Sigmadon hispidus* (inbred at the National Institutes of Health) and *Sigmadon fulviventer*. Following intranasal inoculation, virus replicates well in the lungs and nares of both species, but the pathological lesions are more extensive in *S. hispidus,* and therefore, most of the studies have been carried out in these animals. Viral replication can be detected within one day of inoculation of infectious material varying from 10^6 to 10^{10} PFU/animal, but the maximum titers obtained and the time at which maximum virus is attained vary directly with the size of the inoculum: That is, the larger the inoculum the higher the titer and the faster it is obtained. Consequently, the greater the infectious dose of virus used the earlier the inflammatory response appears and the more extensive the pneumonia that develops [7]. These findings vary considerably from data obtained when the pulmonary infection of mice is produced with mouse-adapted strains of influenza viruses: Over a relatively wide range of infectious doses, the maximum titers are similar and the extent of the pneumonia is comparable, although the kinetics directly depend on the infectious dose employed [8].

When 10^8 to 10^9 PFU/animal is employed, peak titers are reached on day 2 or 3, and maximum pneumonia develops by 5 to 7 days after infection. Immunofluorescence analysis detects viral replication occuring only in the epithelial cells lining the bronchioles. As in cell cultures, intranuclear inclusion bodies are present in many cells, and electron microscopic examination reveals crystal-like arrays of viral particles within nuclei of these cells. Thus, evidence of adenovirus replication in the cotton rat lungs is similar to that seen in pathological human lesions and in infected cells in culture [1,4,6].

The inflammatory response to the viral infection, that is, the pneumonia, can first be detected 1 to 2 days after infection, depending on the infecting dose and the viral strain used. The initial response consists of peribronchial lymphocytic infiltration and the appearance of alveolar macrophages. The lymphocytic, monocyte/macrophage infiltration continuously increases, extending into the intraalveolar septae, eventually invading the outer bronchial walls and even extending into the epithelial cell lining. It is striking, however, that lysis of the infected epithelial cells is never evident and that the walls of the bronchi and bronchioles are not denuded. Early in the infiltration process, small foci of polymorphonuclear leukocytes appear, particularly when very large viral inocula are used (e.g., 10^9 to 10^{10} PFU). Maximum infiltration develops 5 to 7 days postinfection, and then gradual resolution occurs so that the lungs appear normal by 14 days after infection.

Genes Effecting the Inflammatory Response

Early studies in cell cultures showed that only about 10% of the newly synthesized viral DNA and structural proteins was assembled into virions and that the excess viral products accumulated in the infected cell nuclei, forming inclusion bodies in addition to the virion crystals [9]. It was assumed, therefore, that a similar accumulation of virions and viral structural proteins and DNA would be the basic mechanism inducing the adenovirus pneumonia described. Experiments devised to test this hypothesis, however, yielded surprising results. Thus, the use of conditionally lethal, temperature-sensitive mutants unable to replicate the viral DNA at the nonpermissive temperature of the cotton rat's lungs (about 39.2° C), conditions that do not permit synthesis of late viral structural proteins, suggested that the above hypothesis is incorrect: Infection with H5ts125 or H5ts149, DNA-minus temperature-sensitive mutants [10] produced peribronchial lymphocytic infiltration affecting the majority of the bronchi and bronchioles, although there was little or no production of infectious virus or even late viral proteins. The infiltration was not as extensive as that caused by wild-type virus, since there was only a single cycle of early events, but qualitatively the pneumonias were comparable. Hence, early viral gene products appear to be responsible for the pathogenic reactions.

Mutants in other early gene regions were therefore used to determine which of the early genes products were critical for the activation of the molecular events basic to this pathogenic process of adenoviruses. Deletion and insertion mutants in early region 1A (E1A) were unsatisfactory for these studies, since it could not be determined whether their failure to produce a pathological response was due to the inability of the mutants to replicate and to enhance transcriptional activation of genes in other early regions or whether one or more E1A gene products are essential for the inflammatory reactions. However, when mutants defective in the early region 1B (E1B) gene encoding the 496 amino-acid protein (e.g., H5d1110) were used, very little pathology was seen, although the virus multiplied almost as well as Ad5 wild-type virus. The E1B 496 amino-acid protein is required to shut off host protein synthesis during productive infection [11], and it is possible that this reaction is critical for the development of the pathogenic process. However, it is also possible that this early protein is expressed on the surface of the infected cell, serving as an antigenic marker or as a signal to induce other reactions, which will be discussed.

The early region 3 (E3) has been largely neglected in the elegant functional and genetic studies of adenoviruses, since early investigations revealed that numerous mutants with large E3 deletions exist in nature and that the deletions do not diminish the ability of the mutated viruses to replicate or to transform cells. Indeed, the E3 region was unfortunately termed *nonessential* for these reasons. It seemed unlikely, however, that as much as 9.5% of a relatively small viral genome would have been evolutionarily maintained if it did not have an important function. Moreover, the E3 region contains at least nine open reading frames (ORF), of which three proteins have thus far been identified [12], and it also seems unlikely that these proteins are without functions. Several mutants containing large deletions of E3 have been isolated from stocks of wild-type virus. Two of these, H2d1801 [13] and H5d1327 (T. Shenk, personal communication), which replicate like wild-type viruses in cell cultures in vitro and in cotton rat lung in vivo, have been used to determine whether the gene products encoded in this genome region play any role in the pathogenic process. The DNA in each of these mutants is deleted between map units 76.5 and 83.5, and the phenotypes of both mutants are similar in the cotton rat model and in KB cells in vitro. Since the viral mutants previously studied generally were able to attenuate the pathogenic potentials of the virus, the finding that these mutations markedly increased their virulence was unexpected. In cotton rats infected with either H2d1801 or H5d1327, the pulmonary inflammatory response appeared sooner, involvement of 100% of the bronchi and bronchioles occurred earlier, the extent of the lymphocyte, monocyte/macrophage infiltration was more extensive, and there were more numerous, larger foci of polymorphonuclear leukocytes. Concomitant with the beginning of this investigation, Andersson and colleagues reported that adenovirus infection reduces expression of the class I major histocompatibility (MHC) antigen on the surface of infected cells and that a 19-kd glycoprotein (gp19k) encoded in the E3 region complexes with the class I MHC Ag to reduce its transport to the infected cell's surface [14]. Using H2d1801 and H5d1327 and their parental wild-type viruses, we confirmed the finding that a viral gene product encoded in the E3 region inhibited expression of the class I MHC Ag on the infected cell surface. Using the E3 mutant containing a small deletion that only abrogated the expression of gp19 (H5d1754), in collaboration with Wold (W.S.M. Wold, unpublished data), we confirmed that this E3 glycoprotein affected class I MHC Ag expression. And H5d1754 is now being studied in cotton rats to determine whether the gp19 gene product alone is responsible for the markedly increased pulmonary infiltration resulting from infection with the large E3 deletion mutants as compared with wild-type viruses.

The 3′ region of the E3 (E3B) region contains three ORFs, the gene products of which may also play a role in pathogenesis. H5sub304 [15] has a deletion between 83.2 and 85.1 mu and, like other E3 mutants, multiplies without restriction. When inoculated intranasally into cotton rats, H5-sub304 elicits an inflammatory response that is quantitatively like that produced by Ad5 wild-type virus; however, more foci of polymorphonuclear leukocytes are present than in the inflammation produced by wild-type virus. Indeed, collections of polymorphonuclear leukocytes are similar to those seen following the injection of tumor necrosis factor (TNF) (cachectin) into animals [16]. Although there is no evidence that TNF is elaborated during this adenovirus infection of cotton rats, it is noteworthy that the E3B region encodes a 14.7-kd protein that protects adenovirus-infected cells from TNF lysis [17].

Implications

Adenoviruses, at least types 2 and 5, produce acute pneumonias in cotton rats that are pathologically similar to those observed in humans. The mechanism by which adenoviruses produce the pneumonia, however, appears to be unique and totally different from that predicted [9]. Indeed, although maximum pneumonia requires complete viral replication to infect many susceptible cells, only early viral functions are necessary to initiate the usual initial peribronchial inflammatory infiltration of lymphocytes and monocytes/macrophages. The inflammatory response thus does not require the accumulation of unassembled capsid proteins and viral DNA, as previously hypothesized [9], but the early viral nonstructural proteins are critical for the induction of the inflammatory response. The E1B 55-kd protein plays a central role either via its function in shutting off host protein synthesis during productive infections [11] or possibly by being expressed on the infected cells' surfaces to serve as a signal for a cytotoxic T cell attack and/or for inducing secretion of cytokines. Indeed, the adenovirus-infected bronchial epithelial cells may themselves secrete cytokines and/or activate other cells, for example, macrophages, to elaborate such cytokines as interferons, interleukin-1, or TNF.

Gene products of early region 3 (E3) must play a particularly important role in the pathogenic process. Normally, two proteins encoded in this region probably function to protect infected cells and to permit establishment of persistently infected cells. The 19-kd glycoprotein (gp19kd) encoded in and expressed from this region markedly reduces transport of the class I antigen of the infected cell surface [14], apparently to protect the infected cells from cytotoxic T cell (CTL) attack. When the region of E3 encoding the gp19k is deleted, these mutants (e.g., H2d1801) induce a significantly increased pulmonic inflammatory response [7]. The first stage of the increased lymphocyte, monocyte/macrophage infiltration appears too soon after infection for it to be due to a CTL response. It is possible that the increased expression of the class I MHC antigen on the surfaces of infected cells also serves to signal the increased induction of cytokines, which call forth the inflammatory cell invasion around the infected bronchi and bronchioles.

A gene encoded in the 3′ end of the E3 region (so-called E3B), which expresses a 14.7-kd protein, has been reported [17] to protect infected cells from lysis by TNF. It must, however, be emphasized again that the infected bronchial bronchiolar epithelial cells are not lysed during development of the adenovirus pneumonia. Rather, when the region of E3B that encodes the 14.7-kd protein is deleted (e.g., H5sub304), the extent of pulmonary infiltration is not increased, but foci of polymorphonuclear leukocytes (PMNs) are amplified. These foci of PMNs resemble those reported when TNF is injected into mice [16], which suggests that the E3 14.7-kd protein either reduces the ability of another early viral protein to induce TNF or inhibits the infected cell signaling another cell to produce this cytokine.

Studies using cotton rats to explore the molecular mechanism(s) by which adenoviruses produce disease are just beginning. However, they have already demonstrated the potential strength of this approach by suggesting that production of at least some acute viral diseases depends on the action of cytokines as well as on cellular immune responses. This approach may also permit an understanding at a molecular level as to why some adenoviruses commonly produce infections but relatively few clinical diseases (e.g., types 2 and 5), whereas others more commonly initiate clinical diseases and even epidemics (e.g., types 3, 4, and 7) [1]. An understanding of the reactions directing the pathogenesis of viral diseases at a molecular level appears to be a possibility and to present an exciting future.

References

1. Straus SE (1984) Adenovirus infection in humans. *In* Ginsberg HS (ed) The Adenoviruses. Plenum Publishing, New York, pp 451–496
2. Rowe WP, Huebner RJ, Gilmore LK, Parrott RN, Ward TG (1953) Isolation of a cytopathogenic agent from human adenoids undergoing spontaneous degeneration in tissue culture. Proc Soc Exp Biol Med 84:570–573
3. Hilleman MR, Werner JR (1954) Recovery of a new agent from patients with acute respiratory illness. Proc Soc Exp Biol Med 85:183–188
4. Ginsberg HS (ed) (1984) The Adenoviruses. Plenum Publishing, New York
5. Young CSH, Shenk T, Ginsberg HS (1984) The Genetic System. *In* Ginsberg HS (ed) The Adenoviruses. Plenum Publishing, New York, pp 125–172
6. Pacini DL, Dubovi EJ, Clyde WA Jr (1984) A new animal model for human respiratory tract disease due to adenoviruses. J Infect Dis 150:92–97

7. Ginsberg HS, Lundholm-Beauchamp U, Horswood RL, Pernis B, Wold WSM, Chanock RM, Prince GA (1988) Role of early region 3 (E3) in pathogenesis of adenovirus disease. Proc Natl Acad Sci, USA (in press)

8. Ginsberg HS, Horsfall FL Jr (1952) Quantitative aspects of the multiplication of influenza A virus in the mouse lung. Relation between the degree of viral multiplication and the extent of pneumonia. J Exp Med 95:135–145

9. Ginsberg HS (1969) The biochemical basis of adenovirus cytopathology. *In* Mudd S (ed) Infectious Agents and Host Reactions. Saunders, Philadelphia, pp 466–486

10. Ensinger MJ, Ginsberg HS (1972) Selection and preliminary characterization of temperature-sensitive mutants of type 5 adenovirus. J Virol 10:328–339

11. Babiss LE, Ginsberg HS (1984) Adenovirus type 5 early region 16 gene product is required for efficient shut-off of host protein synthesis. J Virol 50:202–212

12. Bhat BM, Wold WSM (1986) Genetic analysis of mRNA synthesis in adenovirus region E3 at different stages of productive infection by RNA-processing mutants. J Virol 60:54–63

13. Challberg SS, Ketner G (1981) Deletion mutants of adenovirus 2: Isolation and initial characterization of virus carrying mutations near the right end of the viral genome. Virology 114:196–209

14. Andersson M, Paabo S, Nilsson T, Peterson PA (1985) Impaired intracellular transport of class I MHC antigens as a possible means for adenoviruses to evade immune surveillance. Cell 43:215–222

15. Jones N, Shenk T (1978) Isolation of deletion and substitution mutants of adenovirus type 5. Cell 13:181–188

16. Beutler B, Cerami A (1987) Cachectin: More than a tumor necrosis factor. New Engl J Med 316:379–385

17. Gooding LR, Elmore LW, Tollefson AE, Brody HA, Wold WSM (1988) A 14,700 MW protein from the E3 region of adenovirus inhibits cytolysis by tumor necrosis factor. Cell (in press)

CHAPTER 33
Enterovirus-Induced Cardiomyopathy

REINHARD KANDOLF AND PETER HANS HOFSCHNEIDER

Enteroviruses of the human Picornaviridae, such as the group B coxsackie-viruses (types 1 to 5), are generally considered to be the most common agents of viral myocarditis [1–4]. Various coxsackie A viruses and echoviruses have also been associated with human viral heart disease. These agents appear to be capable of producing dilated cardiomyopathy of acute onset, or they can lead to life-threatening arrhythmias. Particularly intriguing is the concept of a chronically dilated cardiomyopathy evolving from acute or subacute infections of the human heart.

The structure and molecular genetics of the enteroviruses are well understood, chiefly by analogy with the thoroughly studied poliovirus [5,6]. The genomic single-stranded RNA of enteroviruses consists of about 7,500 nucleotides of positive polarity, covalently linked at the 5′ end to a small virus-encoded protein, VPg. The viral RNA, which is polyadenylated at the 3′ end, functions as a messenger in the cytoplasm of infected cells. It is translated into a polyprotein that is subsequently cleaved by viral proteases into the mature viral proteins. The four structural proteins VP1–VP4 (1A–1D; for a systematic nomenclature of picornaviruses, cf. Rueckert and Wimmer [7]) are generated from the amino-terminal part of the polyprotein, whereas the nonstructural proteins like the polymerase (3D), VPg (3B), two proteases (2A, 3C), and several polypeptides of as yet unknown function are released from the carboxy-terminal part.

The difficulty of establishing an unequivocal diagnosis of viral heart disease is a major problem in clinical cardiology. Confirmation of the clinical suspicion of enterovirus-induced cardiomyopathy demands the demonstration of replicating virus inside myocardial cells, which is exceedingly difficult by conventional methods [8,9]. Molecular genetic techniques have now provided investigators with powerful tools to define the role of enteroviruses in the induction of the disease [10–14] as well as to study the molecular basis of pathogenicity. This chapter will focus on the development of an enterovirus, group-specific, in situ hybridization approach [13], which is undoubtedly the method of choice for diagnosing enteroviral heart disease using endomyocardial biopsy samples. In addition, antisera raised against bacterially expressed coxsackievirus B3 (CVB3) coat proteins are described, and they reveal a broad spectrum of cross-reactivity within the enteroviruses [15]. Visualizing both nucleic acid and distinct proteins on the same section will be enormously useful in future studies, extending the application of the in situ hybridization technique in the study of acute and persistent enterovirus infection. Finally, in the absence of effective antiviral therapy, in vitro experiments are reviewed, demonstrating the high antiviral activity of human natural fibroblast interferon (IFN-β) in cultured human heart cells.

Cloned Coxsackievirus B3 cDNA as an Enterovirus Diagnostic Reagent

One major prerequisite for the introduction of in situ hybridization as a diagnostic tool in suspected enterovirus-induced cardiomyopathy was the molecular cloning and characterization of the single-stranded genomic RNA of the cardiotropic CVB3 [12]. Full-length, reverse-transcribed CVB3 cDNA generated replication-competent CVB3 upon transfection of recombinant viral cDNA into mammalian cells, offering unique opportunities for the genetic analysis of this virus. From the diagnostic point of view, cloned CVB3 cDNA offers the unique possibility for a group-specific diagnosis of enterovirus infections [10–13]. Because of the high degree of nucleic acid sequence identity among the numerous serotypes of the human enterovirus group, including the group A and B coxsackieviruses and the echoviruses, detection of these various agents commonly implicated in human viral heart disease is possible in a single hybridization assay. This broad detection spectrum of cloned CVB3 cDNA greatly facilitates diagnosis of enterovirus heart disease, since, from the clinical point of view, the antigenic typing of an etiologically implicated enterovirus serotype appears to be of secondary importance and can be carried out later—for example, by standard virological techniques or by hybridization with serotype-specific cDNA fragments available to date only for a limited number of enterovirus serotypes.

Highly specific hybridization conditions have been established for the detection of enteroviruses. When radioactively labeled cloned CVB3 cDNA corresponding to 95.4% of the viral genome (nucleotides 66 to 7.128) was hybridized to electrophoretically resolved CVB3 RNA and to total RNA from cultured human heart cells, specific hybridization was found for the viral RNA but not for the human myocardial RNA [12].

In Situ Detection of Enterovirus RNA in Infected Cells by Nucleic Acid Hybridization

The feasibility of using the in situ hybridization technique [16,17] to detect enterovirus RNA was first established in cell culture systems and then applied to myocardial tissue of athymic mice persistently infected with CVB3 [13]. Uninfected Vero cells exhibited essentially no grains when hybridized to the ^{3}H-labeled or ^{35}S-labeled CVB3 cDNA probe. By contrast, highly significant labeling was achieved in Vero cells infected with various enteroviruses—for example, CVB1, CVB3, CVB5, or echovirus 11.

Using myocardial tissue sections from CVB3-infected mice, in situ hybridization proved to be a powerful tool not only with respect to establishing an unequivocal diagnosis of myocardial infection but also with respect to understanding its pathogenesis (Figure 33.1). The autoradiographic silver grains, which indicate hybridization between viral RNA and the radiolabeled CVB3 cDNA probe, are clearly related to distinct infected myocytes (Figure 33.1**A**). These cells are easily identified by interference contrast microscopy in unstained sections because of their characteristic size and morphology. In this model system of enteroviral heart disease, myocardial involvement was found to be multifocal and randomly distributed in the heart muscle (Figure 33.1**B**). Myocardial cross sections revealed a transmural disseminated infection of the myocardium as demonstrated in Figure 33.1**C** for the left ventricle. Infected myocytes were often found in clusters within areas of severe myocardial lesions and fibrosis (Figure 33.1**D**). Furthermore, progression of the infection could be observed from areas with myocardial fibrosis to as yet uninfected myocytes (Figure 33.1**E,F**), indicating the possible cell-to-cell spread of the virus. Viral RNA, however, was also found in isolated myocytes in apparently normal myocardial tissue; this was primarily observed in the early stage of the infection (Figure 33.1**G**). In addition, viral RNA also appeared to be located within the small interstitial myocardial cells (Figure 33.1**B,E**). Labeled myocardial cells were not found when myocardial sections were probed with the radiolabeled plasmid p2732B control DNA [12] or with the cloned EcoRI J fragment of the genetically unrelated cytomegalovirus [18]. In addition, no labeled myocardial cells were found when myocardial tissue sections of uninfected mice were probed with radiolabeled CVB3 cDNA (Figure 33.1**H**).

A high sensitivity of detection was made possible by several improvements in methodology, including optimized hybridization conditions to prevent nonspecific binding and the use of radiolabeled cloned cDNA fragments about 100 nucleotides in length [13]. Quantification of CVB3 copy numbers in infected Vero cells by RNA blot analysis and comparison with in situ hybridization of cells from the same culture indicated that as few as 20 viral copies are easily detectable within 2 weeks of autoradiographic exposure. Clearly positive hybridization signals with infected mouse myo-

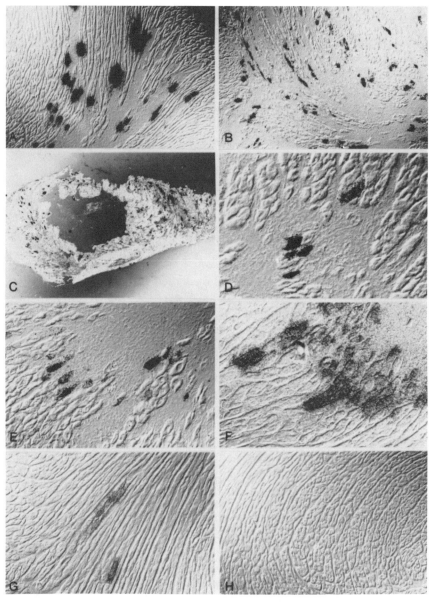

Figure 33.1. Autoradiographs of CVB3-infected (**A–G**) and uninfected (**H**) mouse myocardial tissue hybridized in situ with ^{3}H-labeled (**B–E**) or ^{35}S-labeled (**A, F–H**) cloned CVB3 cDNA. Days after infection of athymic mice were 56 for **A,** 23 for **B–E,** 42 for **F,** and 4 for **G.** Exposure times were 9 days for **A** and **H,** 6 weeks for **B–E,** 4 days for **F,** and 2 days for **G.** Note that since silver grains are positioned at various levels within the photoemulsion, some grains are not observed and appear out of focus in the photographs. Interference contrast microscopy of unstained sections; × 200 (**A,G,** and **H**), × 100 (**B**), × 25 (**C**), × 400 (**D** and **E**), × 650 (**F**). (Reprinted with permission from Kandolf et al., Proc. Natl. Acad. Sci. USA 84:6272–6276, 1987.)

cardial tissue were observed after only 2 days of exposure to the ^{35}S-labeled cDNA probe, indicating a high copy number of replicating viral genomes in myocardial cells (Figure 33.1**G**). Furthermore, overexposed myocardial slide preparations showed extremely low background signals, which confirmed the high specificity of in situ hybridization for the detection of enteroviral RNA.

In Situ Detection of Enterovirus RNA in the Failing and Failed Human Heart

To date, in situ hybridization has already been proved to be a valuable tool to assess the presence of enterovirus RNA in endomyocardial biopsy samples obtained from patients with etiologically unexplained heart failure [9,19]. Replicating enterovirus RNA was found in 18 of 78 patients with a clinical suspicion of viral myocarditis or dilated cardiomyopathy, the latter being a condition that in its end-stage can be treated only by heart transplantation. All 25 patients with other specific heart muscle diseases (e.g., ischemic, hypertrophic, or metabolic cardiomyopathies) were negative when myocardial tissue was examined using the in situ hybridization approach.

An important finding is the observation that enterovirus RNA was not only detected in myocytes but also in small interstitial cells, presumably fibroblasts, which agrees with the previous in vitro findings in cultured human heart cells and persistently infected, human myocardial fibroblasts [20].

The number of infected myocardial cells appears to be related to the severity of clinical symptoms. In patients with mild perimyocarditis or healing myocarditis, only a few myocardial cells were found to express enterovirus RNA, but numerous infected cells were found in patients with severe dilated-type cardiomyopathy. Furthermore, enterovirus RNA was not only present at an early stage of the disease but also in late endomyocardial biopsies (e.g., 12 months after onset of dilated cardiomyopathy), indicating persistence of the virus in the human heart. Consequently, enteroviral RNA could also be demonstrated in explanted hearts of a significant number of patients. To date, 4 of 19 patients who underwent heart transplantation because of end-stage dilated cardiomyopathy were found to have enteroviral heart disease. The concept of enterovirus persistence in dilated cardiomyopathy, evolving from acute or subacute infections, is further substantiated by the presence of enterovirus RNA in subsequent biopsies obtained from patients with ongoing disease. Nonetheless, the possibility has to be considered that in some cases a preexisting cardiomyopathy might also potentiate the growth of the virus as a secondary event. However, this does not exclude the possibility that an initial, perhaps symptomless infection may trigger the train of events that lead to dilated cardiomyopathy.

Generation of Enterovius Group-Specific Antisera

Although a number of cross-relationships exist between different enteroviruses [8,21,22], the applicability of group-specific antigen detection has met with only modest success in the identification of cardiotropic enteroviruses because of the antigenic heterogeneity among and within the serotypes. Compared with hyperimmune antisera obtained by imunization of rabbits with purified enteroviral virions [23], antisera raised against bacterially synthesized CVB3 proteins revealed a much broader spectrum of cross-reactivity within the enteroviruses [15]. Distinct subgenomic fagments from infectious recombinant CVB3 cDNA were inserted into the expression plasmid pPLc24 [24,25] within the coding sequence of the replicase gene of the MS2-phage, thereby achieving a fusion of both reading frames. The expressed viral fusion proteins contain MS2-replicase-specific amino-acid sequences at the N terminus. In this approach, plasmids were constructed to express either the structural proteins VP4, VP2, and VP3 (p1A-1C) or the structural protein VP1 (p1D) of CVB3. Polyclonal antisera raised in rabbits against the purified expression products offer the unique possibility for enterovirus group-specific identification of various commonly implicated agents of human viral heart disease, including group A and group B coxsackieviruses and echoviruses [15]. This broad detection spectrum is expected to facilitate identification of enterovirus infections by antigen detection. The use of these antibodies in combination with in situ hybridization is currently being pursued. With a double-labeling assay [26], simultaneous in situ detection of enteroviral RNA and demonstration of distinct viral proteins are now possible at the single-cell level. This approach will allow the question of whether restricted virus replication is implicated in persistent forms of enterovirus-induced cardiomyopathy to be answered.

Antiviral Activity of Human Fibroblast Interferon in Cultured Human Heart Cells

Effective antiviral therapy has not yet been established in the treatment of viral heart disease. The optimal goal of antiviral treatment, to restore function to the infected cell, usually appears unattainable. A realistic goal would be to inhibit viral replication and thus prevent spread to as yet uninfected cells. This is achieved, in part, in many natural infections, by the endogenous interferon (IFN) system. With respect to the potential clinical application of exogenous IFN in viral heart disease, one major prerequisite is the demonstration that the virus to be treated is susceptible in vitro to the action of a given type of IFN in the specific host cell.

The protective role of natural human fibroblast interferon (IFN-β) in CVB3-infected, enriched, fetal human myocytes has been established [20]. Myocytes protected by IFN-β continued to beat rhythmically as long as IFN-β was given, with exchanges of medium, every 2 days. By contrast, virus replication in unprotected cultures was accompanied by loss of spontaneous contractility within 9 hours, followed by complete lysis of myocytes within 20 hours of infection.

The discovery of persistently infected, CVB3 carrier cultures of human myocardial fibroblasts [20] provided another useful test system for studying the activity of antiviral agents. In this type of infection, only a small proportion of the cell population is productively infected. Virus titers of 10^7 plaque-forming units of infectious CVB3/ml culture medium are obtained. Treatment of these cultures with IFN-β at a dose of 300 U/ml culture medium every 24 hours eliminated infectious CVB3 from the culture medium within 11 to 13 days. An important finding is the observation that this potent antiviral activity of IFN-β is completely blocked in the presence of 10 μM prednisolone [27].

Comment

Enterovirus research is in a very dynamic phase. Recombinant DNA techniques have not only allowed an important breakthrough in the study of enterovirus genetics but have already contributed significantly to an improved diagnosis of enterovirus infections. To date, there is firm evidence from the nucleic acid hybridization approach that enterovirus infections of the human heart lead to a significant number of patients presenting with clinical signs and symptoms of myocarditis or dilated cardiomyopathy. In patients with dilated cardiomyopathy of recent onset, the most dramatic manifestation of myocarditis, the incidence of enterovirus infections was found to be approximately 30%. The concept of enterovirus persistence in chronic dilated cardiomyopathy was substantiated by the presence of enterovirus RNA in follow-up biopsies of patients with ongoing cardiac disease. Presumably, enterovirus persistence in myocardial cells is the result of continued synthesis of the viral precursor polyprotein and correct processing, at least of the virus-encoded polymerase. However, a lack of correct processing of other enterovirus gene products, for example, coat proteins, could restrict replication. Restricted replication could explain the common failure to isolate infectious virus from patients with chronic cardiomyopathies. If cloned CVB3 cDNA are used in combination with enterovirus group-specific antisera, a simultaneous in situ detection of viral RNA and proteins will provide a powerful means to address questions concerning the molecular basis of persistent enterovirus infections of the human heart.

The most appealing treatment for enterovirus-induced cardiomyopathy, which in some cases may conceivably prevent the onset of dilated cardiomyopathy, is antiviral therapy. Although the relationship between the in vitro effects of IFN-β in cultured human heart cells and its potential in vivo activities remains to be determined, the in vitro observations are an important prerequisite for correct therapeutic use of exogenous IFN-β in severe enterovirus infections. When given at an early stage of the disease, IFN-β would probably reduce viral spread in the myocardium by protecting as yet uninfected cells, and thereby limit the disease.

Acknowledgments

The contributions of Dr. A. Canu, Dr. P. Kirschner, Dr. D. Ameis, Dr. W. Klump, S. Werner, R. Widmer, and H. Schönke are gratefully acknowledged. Endomyocardial biopsies were obtained from Dr. E. Erdmann and Dr. H. P. Schultheiß, Department of Internal Medicine I, University of Munich. Myocardial tissue from explanted hearts was obtained from Dr. B. Kemkes, Department of Cardiac Surgery, University of Munich. This work was supported in part by Grant Ka 593/2-1 from the Deutsche Forschungsgemeinschaft and by Grant 321-7291-BCT-0370 from the German Ministry for Research and Technology. R. K. is a Hermann and Lilly Schilling Professor for Experimental Medicine.

References

1. Abelmann WH (1973) Viral myocarditis and its sequelae. Ann Rev Med 24:145–152
2. Woodruff JF (1980) Viral myocarditis—A review. Am J Pathol 101:427–479
3. Bolte HD (ed) (1984) Viral Heart Disease. Springer-Verlag, New York
4. Reyes MP, Lerner AM (1985) Coxsackievirus myocarditis—With special reference to acute and chronic effects. Prog Cardiovasc Dis 27:373–394
5. Melnick JL (1985) Enteroviruses: Polioviruses, coxsackieviruses, echoviruses, and newer enteroviruses. In Fields BN (ed) Virology. Raven Press, New York, pp 739–794
6. Wimmer E, Kuhn RJ, Pincus S, Yang CF, Toyoda H, Nicklin MJH, Takeda N (1987) Molecular events leading to picornavirus genome replication. J Cell Sci (suppl 7):251–276
7. Rueckert RR, Wimmer E (1984) Systematic nomenclature of picornavirus proteins. J Virol 50:957–959
8. Melnick JL, Wenner HA, Phillips CA (1979) Enteroviruses. In Lennette EH, Schmidt NJ (eds) Diagnostic Procedures for Viral, Rickettsial and Chlamydial Infections, 5th ed. American Public Health Association, Washington, DC, pp 471–534
9. Kandolf R (1988) The impact of recombinant DNA technology on the study of enterovirus heart disease. In Bendinelli M, Friedman H (eds) Coxsackieviruses—A General Update. Plenum Publishing, New York, pp 293–318
10. Hyppiä T, Stålhandske P, Vainionpää R, Petterson U (1984) Detection of enteroviruses by spot hybridization. J Clin Microbiol 19:436–438
11. Tracy S (1984) A comparison of genomic homologies among the coxsackievirus B group: Use of fragments of the cloned coxsackievirus B3 genome as probes. J Gen Virol 65:2167–2172
12. Kandolf R, Hofschneider PH (1985) Molecular cloning of the genome of a cardiotropic coxsackie B3 virus: Full-length reverse-transcribed recombinant

cDNA generates infectious virus in mammalian cells. Proc Natl Acad Sci USA 82:4818–4822

13. Kandolf R, Ameis D, Kirschner P, Canu A, Hofschneider PH (1987) In situ detection of enteroviral genomes in myocardial cells by nucleic acid hybridization: An approach to the diagnosis of viral heart disease. Proc Natl Acad Sci USA 84:6272–6276

14. Bowles NE, Richardson PJ, Olsen EGJ, Archard LC (1986) Detection of coxsackie-B-virus-specific RNA sequences in myocardial biopsy samples from patients with myocarditis and dilated cardiomyopathy. Lancet i:1120–1123

15. Werner S, Klump WM, Schönke H, Hofschneider PH, Kandolf R (1988) Expression of coxsackievirus B3 capsid proteins in Escherichia coli and generation of virus-specific antisera. DNA 7:307–316

16. Wolf H, zur Hausen H, Becker V (1973) EB viral genomes in epithelial nasopharyngeal carcinoma cells. Nature New Biol 244:245–247

17. Haase A, Brahic M, Stowring L, Blum H (1984) Detection of viral nucleic acids by in situ hybridization. In Maramorosh K, Koprowski H (eds) Methods in Virology, vol 7. Academic Press, New York, pp 189–226

18. Nelson JA, Fleckenstein B, Galloway DA, McDougall JK (1982) Transformation of NIH 3T3 cells with cloned fragments of human cytomegalovirus strain AD169. J Virol 43:83–91

19. Kandolf R, Kirschner P, Ameis D, Müller BC, Erdmann E, Schultheiss HP, Kemkes B, Hofschneider PH (1987) Diagnosis of enteroviral heart disease by in situ nucleic acid hybridization. Circulation 76 (suppl IV):262 (Abstr)

20. Kandolf R, Canu A, Hofschneider PH (1985) Coxsackie B3 virus can replicate in cultured human foetal heart cells and is inhibited by interferon. J Mol Cell Cardiol 17:167–181

21. Herrmann JE, Hendry RM, Collins MF (1979) Factors involved in enzyme-linked immunoassay of viruses and evaluation of the method for identification of enteroviruses. J Clin Mirobiol 10:210–217

22. Yolken RH, Torsch VM (1981) Enzyme-linked immunosorbent assay for detection and identification of coxsackieviruses A. Infect Immun 31:742–750

23. Mertens T, Pika U, Eggers HJ (1983) Cross antigenicity among enteroviruses as revealed by immunoblot technique. Virology 129:431–442

24. Remaut E, Tsao H, Fiers W (1983) Improved plasmid vectors with a thermoinducible expression and temperature-regulated runaway replication. Gene 22:103–133

25. Klump W, Marquardt O, Hofschneider PH (1984) Biologically active protease of foot-and-mouth disease virus is expressed from cloned cDNA in Escherichia coli. Proc Natl Acad Sci USA 81:3351–3355

26. Brahic M, Haase AT, Cash E (1984) Simultaneous in situ detection of viral RNA and antigens. Proc Natl Acad Sci USA 81:5445–5448

27. Kandolf R, Kirschner P, Ameis D, Canu A, Hofschneider PH (1987) Cultured human heart cells: A model system for the study of the antiviral activity of interferons. Eur Heart J 8 (suppl J):453–456

CHAPTER 34

Hantavirus Infection: A Newly Recognized Viral Enzootic of Commensal and Wild Rodents in the United States

RICHARD YANAGIHARA AND D. CARLETON GAJDUSEK

Viruses belonging to the newly defined Hantavirus genus in the family Bunyaviridae possess a segmented (tripartite), single-stranded RNA genome of negative polarity. Each RNA segment has a conserved 3' terminal nucleotide sequence (3' AUCAUCAUCUG), which differs from the consensus sequences reported for the four established genera of Bunyaviridae [1]. These viruses cause hemorrhagic fever with renal syndrome (HFRS) or muroid virus nephropathy, an acute interstitial nephropathy that varies greatly in clinical severity. The severe form of disease, caused by Hantaan virus, the prototype virus of HFRS, is characterized by high fever, renal failure, and varying degrees of hemorrhage [2]. A milder form of disease, typically without hemorrhage, called nephropathia epidemica, is caused by Puumala virus. Though occasionally severe, infections caused by strains of Seoul virus isolated from peridomestic rats in urban centers of Far Eastern Asia, are usually mild, simulating a flu-like illness with pyrexia, myalgia, flank pain, headache, and proteinuria, with minimal or no hemorrhagic manifestations.

The conjecture, originally made by Gajdusek in 1953 [3], that Korean hemorrhagic fever is related to milder forms of hemorrhagic fever occurring in European Russia and Scandinavia and that these diseases are caused by the same virus or closely related viruses has now been verified. Serological and virological studies indicate that the epidemic hemorrhagic fever in the People's Republic of China and the Far Eastern hemorrhagic nephroso-nephritis in Eastern Siberia are caused by viruses that are indistinguishable from Hantaan virus and that the HFRS of European Russia and the nephropathia epidemica of Sweden, Finland, and Norway are caused by viruses that are antigenically similar to, but distinct from, Hantaan virus.

In regions in which HFRS is endemic, which span much of the Eurasian landmass and Scandinavia, small, circumscribed foci or microfoci are typically found, prompting the concept of a "place" disease [4]. Epidemic outbreaks usually occur during widescale invasion and upheaval of rodent habitats or following irruptions of reservoir rodent populations. Prolonged shedding of virus in saliva, urine, and feces, long suspected to be the source of infection and mode of virus maintenance among rodents in endemic areas, has been demonstrated in *Apodemus agrarius* and in *Clethrionomys glareolus* experimentally infected with Hantaan and Puumala virus, respectively [5,6]. Exposure to nesting material heavily contaminated with infectious secretions and excretions, grooming behavior, and intraspecies wounding by biting are important factors in the maintenance of the enzootic cycle. Infection results in a short-lived viremia and virus dissemination to various tissues, particularly lung, salivary glands, and kidney [5,6]. Virus excretion persists unabated, despite the brisk development of high-titer, neutralizing antibodies. Unlike arenavirus infection in their reservoir rodent hosts, vertical transmission does not occur in hantavirus-infected rodents.

The respiratory droplet, or airborne, route of infection is highly efficient and constitutes the principal mode of virus transmission to humans, as evidenced by laboratory outbreaks of HFRS, such as those occurring in 1961 at the Gamaleya Institute in Moscow, following the introduction of wild rodents caught in enzootic foci, and sporadic cases of HFRS, involving very brief, seemingly trivial exposures to rodent-infested habitats. Thus, exposure to the excretions and secretions of persistently infected rodents is a primary risk factor. Not surprisingly then, individuals involved in such occupations as forestry and farming have traditionally been the groups at highest risk, and as a consequence, the disease occurs predominantly in men, usually between the ages of 20 and 50 years. Women and children are exposed during the colder months, when small rodents, seeking food and shelter, invade hay-barns, granaries, and houses.

Epizootiology of Hantavirus Infection in the United States

The importance of wild rodents as reservoirs of the etiological agents of HFRS was established by the investigations of Soviet and Japanese scientists 50 years ago. Serological diagnosis of infection, since made possible by the isolation of Hantaan virus from the striped field mouse (*Apodemus agrarius*) [7], indicates that hantaviruses are widespread in several genera of rodents and that the geographical distribution of infection extends far beyond the previously recognized areas of endemicity [2].

Rodents classified in the superfamily Muroidea and the family Muridae (genera: *Apodemus* and *Rattus*) and the family Arvicolidae (genera: *Cleth-*

rionomys and *Microtus*) serve as the major reservoirs of hantaviruses. Hantaan virus isolated from *Apodemus agrarius,* Seoul virus from *Rattus norvegicus,* and Puumala virus from *Clethrionomys glareolus* cause HFRS. Prospect Hill virus [8], a Hantaan-related virus isolated from meadow voles (*Microtus pennsylvanicus*), an indigenous arvicolid (microtine) rodent species that is widely distributed in the continental United States and Canada, is not known to cause disease in humans, although experimentally infected cynomolgus monkeys develop transient proteinuria and azotemia [9].

Apodemus species, the murid rodent reservoir of prototype Hantaan virus, do not exist in the Americas, but *Rattus norvegicus* and several species of *Clethrionomys* do. Serological evidence for hantavirus infection has been found in several species of indigenous arvicolid and cricetid rodents, including *Microtus californicus, Clethrionomys gapperi, Peromyscus californicus, Peromyscus difficilis, Peromyscus leucopus, Peromyscus maniculatus, Peromyscus truei, Neotoma cinerea,* and *Neotoma mexicana,* captured in Maryland, West Virginia, Minnesota, Colorado, New Mexico, California, and Alaska [8,10–12]. Preliminary data indicate that these indigenous rodents are infected with viruses that are antigenically more closely related to Prospect Hill virus than to Hantaan or Seoul virus [11,12].

Viruses serologically indistinguishable from disease-causing *Rattus*-derived hantaviruses isolated in Far Eastern Asia have been recoverd from *Rattus norvegicus* captured in New Orleans, Houston, Philadelphia, and Baltimore [10,13,14]. High prevalence rates of hantavirus infection have also been found in rats trapped in other inland cities that lack commercial waterways and in residential districts of many port cities in the United States [10,14]. Longitudinal and mark–release–recapture studies of *Rattus* populations in Baltimore, Maryland, indicate an age-dependent acquisition of infection [12,15].

Another hantavirus has been isolated recently from a common house mouse (*Mus musculus*) captured in Texas [16]. The new isolate, designated Leakey virus, is antigenically distinct from the four known serotypes of hantavirus [16,17]. The circulation of three, and possibly more, hantaviruses in rodent populations in the United States augments our concepts of the epizootiology of hantavirus infection in commensal rodents. It raises important questions about the prevalence of infection in feral and laboratory mice and about whether the worldwide distribution of Leakey virus in *Mus* populations resembles that of Seoul virus in *Rattus* populations. Finally, since *Microtus pennsylvanicus, Rattus norvegicus, Mus musculus,* and cricetid rodents, such as *Peromyscus leucopus,* are sympatric and synchronistic in some habitats [8,12], the possibility exists for genetic reassortment among hantaviruses, which may account for differences in virulence.

Nonmuroid Reservoirs of Hantaviruses

Other small mammals, including the peridomestic musk shrew (*Suncus murinus*) and the domestic cat (*Felis catus*), have recently been incriminated in cases of HFRS in the People's Republic of China. In the United States, cats, particularly stray cats living in alleys inhabited by seropositive rats, have also been found to be infected with hantaviruses; low-titered fluorescent antibodies against Hantaan virus, however, have been found rarely in shrews (*Blarina brevicauda*) and weasels (*Mustela frenata*) [8]. The extent to which these small mammals are involved in the maintenance and transmission of hantaviruses is not known, but like humans, they are probably end-stage hosts and, therefore, unimportant.

Seroepidemiology of Hantavirus Infection in Humans in the United States

Many rodent species harboring hantaviruses are probably ancillary hosts and are rarely, if ever, involved in disease transmission to humans under normal conditions. Alternatively, certain rodent species may be infected with hantavirus strains that are nonpathogenic for humans. In this regard, although neutralizing antibodies against Hantaan and Seoul viruses have been found in Americans [10,18,19], cases of HFRS have not been recognized in the United States, or in Brazil, Colombia, and Bolivia, where antibodies against hantavirus have also been found in humans. Also, serological evidence of Prospect Hill virus infection has been demonstrated in American mammalogists with considerable field exposure to arvicolid rodents [20].

Generally, the clinical severity of HFRS in a specified area can be predicted by the ecology of the predominant reservoir rodent species. Severe disease is the rule when *Apodemus agrarius* is the principal reservoir, and mild disease is expected in areas where virus-infected *Clethrionomys glareolus* abound. In such geographical regions as eastern Europe, where both *Apodemus* and *Clethrionomys* species serve as reservoirs, both severe and mild forms of HFRS can be anticipated. Serological and virological evidence for the coexistence of Hantaan and Puumala viruses in Eastern Europe and Europen Russia is consistent with the ecology of *Apodemus* mice and *Clethrionomys* voles. The absence of *Apodemus* species in the Americas would seem to preclude the severe form of HFRS, but severe disease has been reported occasionally with *Rattus*-derived hantaviruses in Far Eastern Asia and Europe. It is unlikely that American clinicians are misdiagnosing the severe form of HFRS. More likely, hantavirus infection in humans in the United States is subclinical, mild, or atypical. Studies to define the clinical features of human infection with the American hantaviruses are needed.

The obvious issues posed by these newly discovered bunyaviruses involve concerns about whether hantaviruses, like the plague bacillus, constitute a potential threat to the public health of the United States, particularly in areas with high human population densities, and where living conditions are poor or contacts between humans and rodents are common. Since commensal rats are the principal reservoirs of hantaviruses in urban centers, the potential exists for worldwide dissemination of rats harboring disease-causing hantavirus strains via international shipping. To guard against this possibility, rodent control measures need to be maintained or instituted in major port facilities. Also, to safeguard animal handlers and laboratory investigators engaged in research requiring rodents, every precaution should be taken to prevent introducing infected rodents into breeding colonies or research institutions.

The widespread existence of hantaviruses in commensal rats and indigenous wild rodents in the United States, in the absence of human disease, offers a unique opportunity to study the biology and ecology of these viruses to learn whether the absence of disease results from differences in pathogenicity or virulence of the American strains or from differences in social structure or cultural practices that make Americans, as a group, less likely to be in frequent contact with infected rodents. Molecular genetic comparisons between disease-causing hantaviruses and the American isolates may yield additional clues about pathogenicity.

Addendum. A full bibliography with a historical introduction to HFRS and the hantaviruses has been published recently [21]. It contains more than 3,000 references, many of which are in foreign languages.

References

1. Schmaljohn CS, Hasty SE, Dalrymple JM, LeDuc JW, Lee HW, von Bonsdorff C-H, Brummer-Korvenkontio M, Vaheri A, Tsai TF, Regnery HL, Goldgaber D, Lee P-W (1985) Antigenic and genetic properties of viruses linked to hemorrhagic fever with renal syndrome. Science 227:1041–1044
2. Yanagihara R, Gajdusek DC (1988) Hemorrhagic fever with renal syndrome: A historical perspective and review of recent advances. *In* Gear JHS (ed) CRC Handbook of Viral and Rickettsial Hemorrhagic Fevers. Boca Raton, FL: CRC Press, Inc, pp 151–188
3. Gajdusek DC (1953) Acute Infectious Hemorrhagic Fevers and Mycotixicoses in the Union of Soviet Socialist Republics. Medical Science Publication No 2, Army Medical Service Graduate School, Walter Reed Army Medical Center, Washington, DC, 140 pp
4. Smadel JE (1953) Epidemic hemorrhagic fever. Am J Public Health 43:1327–1330
5. Lee HW, Lee P-W, Baek LJ, Song CK, Seong IW (1981) Intraspecific transmission of Hantaan virus, the etiologic agent of Korean hemorrhagic fever, in the rodent *Apodemus agrarius*. Am J Trop Med Hyg 30:1106–1112
6. Yanagihara R, Amyx HL, Gajdusek DC (1985) Experimental infection with

Puumala virus, the etiologic agent of nephropathia epidemica, in bank voles (*Clethrionomys glareolus*). J Virol 55:34–38

7. Lee HW, Lee P-W, Johnson KM (1978) Isolation of the etiologic agent of Korean hemorrhagic fever. J Infect Dis 137:298–308

8. Lee P-W, Amyx HL, Yanagihara R, Gajdusek DC, Goldgaber D, Gibbs CJ Jr (1985) Partial characterization of Prospect Hill virus isolated from meadow voles in the United States. J Infect Dis 152:826–829

9. Yanagihara R, Amyx HL, Lee P-W, Asher DM, Gibbs CJ Jr, Gajdusek DC (1988) Experimental hantavirus infection in nonhuman primates. Arch Virol 101:125–130

10. Tsai TF, Bauer SP, Sasso DR, Whitfield SG, McCormick JB, Caraway TC, McFarland L, Bradford H, Kurata T (1985) Serological and virological evidence of a Hantaan virus-related enzootic in the United States. J Infect Dis 152:126–136

11. Yanagihara R, Daum CA, Lee P-W, Baek LJ, Amyx HL, Gajdusek DC, Gibbs CJ Jr (1987) Serological survey of Prospect Hill virus infection in indigenous wild rodents in the United States. Trans R Soc Trop Med Hyg 81:42–45

12. Childs JE, Glass GE, Korch GW, LeDuc JW (1987) Prospective seroepidemiology of hantaviruses and population dynamics of small mammal communities of Baltimore, Maryland. Am J Trop Med Hyg 37:648–662

13. LeDuc JW, Smith GA, Johnson KM (1984) Hantaan-like viruses from domestic rats captured in the United States. Am J Trop Med Hyg 33:992–998

14. Childs JE, Korch GW, Glass GE, LeDuc JW, Shah KV (1987) Epizootiology of *Hantavirus* infections in Baltimore: Isolation of a virus from Norway rats, and characteristics of infected rat populations. Am J Epidemiol 126:55–68

15. Childs JE, Korch GW, Smith GA, Terry AD, LeDuc JW (1985) Geographical distribution and age-related prevalence of Hantaan-like virus in rat populations of Baltimore, Maryland, U.S.A. Am J Trop Med Hyg 34:385–387

16. Baek LJ, Yanagihara R, Gibbs CJ Jr, Miyazaki M, Gajdusek DC (1988) Leakey virus: A new hantavirus isolated from *Mus musculus* in the United States. J Gen Virol 69:3129–3132

17. Lee P-W, Gibbs CJ Jr, Gajdusek DC, Yanagihara R (1985) Serotypic classification of hantaviruses by indirect immunofluorescent antibody and plaque reduction neutralization tests. J Clin Microbiol 22:940–944

18. Yanagihara R, Chin C-T, Weiss MB, Gajdusek DC, Diwan AR, Poland JB, Kleeman KT, Wilfert CM, Meiklejohn G, Glezen WP (1985) Serological evidence of Hantaan virus infection in the United States. Am J Trop Med Hyg 34:396–399

19. Childs JE, Glass GE, Korch GW, Arthur RR, Shah KV, Glasser D, Rossi C, LeDuc JW (1988) Evidence of human infection with a rat-associated *Hantavirus* in Baltimore, Maryland. Am J Epidemiol 127:875–878

20. Yanagihara R, Gajdusek DC, Gibbs CJ Jr, Traub R (1984) Prospect Hill virus: Serological evidence for infection in mammalogists. N Engl J Med 310:1325–1326

21. Gajdusek DC, Goldfarb LG, Goldgaber D (1987) Bibliography of Hemorrhagic Fever with Renal Syndrome, Second Edition (NIH Publication No. 88-2603). U.S. Department of Health and Human Services, Public Health Service, National Institutes of Health, Bethesda, MD, 290 pp

CHAPTER 35
Virus-Induced, Cell-Mediated Autoimmunity

VOLKER TER MEULEN

Recovery from a viral infection is the result of very complex interactions between specific and nonspecific immune reactions and the infectious agent. A variety of immune mechanisms, including viral antibodies, cell-mediated immunity (CMI), complement, phagocytic processes, and interferon, are undoubtedly important factors in this event and operate together in overcoming the infectious process. However, despite much available information about these viral defense mechanisms, it has proved remarkably difficult to assign a determinative role in vivo to any single antiviral immunological mechanism in recovery from any single viral disease, particularly since the immune response to the virus itself may frequently contribute to the pathology of the disease. In this context, delayed type hypersensitivity (DTH), immune complexes, and immune cytolysis are examples of immune pathological reactions [1]. Furthermore, if virus-induced immune responses are also directed against normal host components, this may set the stage for an autoimmunity that may be of pathogenetic significance. In acute viral infection, this phenomenon may merely lead to increased tissued damage, whereas in chronic infection, autoimmune reactions may perpetuate the disease process.

Although many important human diseases reveal an autoimmune component involving organ- or nonorgan-specific antigens, the etiology is unknown for the majority of these disorders. Many theories, ranging from immunogenetic abnormalities to exogenous insults, have been proposed [2]. With respect to the latter mechanism, it is clear that viruses have the potential to induce an immune response against self-antigens, as seen in experimental animal infections as well as in human infections. The majority of these reports describe the occurrence of autoantibodies in the course of RNA and DNA virus infections [3], whereas research is just beginning to explore the potential of virus infection to induce cell-mediated autoimmune (CMAI) responses. The evidence for CMAI reactions in relation to viruses is reviewed here and the mechanisms by which viruses may alter host reactivity against self-antigens are discussed.

Virus-Induced CMAI Reactions in Experimental Animals

Inoculation of rats with the murine coronavirus JHM or measles virus usually lead to a persistent central nervous system (CNS) infection. Recent studies have also revealed the presence of a cell-mediated immune response against host antigens in these animals [4,5].

Infection of rats with JHM virus at the age of 21 to 35 days leads to the development of a subacute demyelinating encephalomyelitis (SDE), with an incubation period of 2 weeks to 8 months [6–8]. The rats display a paresis of the hindlegs and an ataxic gait leading to paralysis and tetraplegia. The most prominent neuropathological changes are lesions of primary demyelination disturbed in the white matter, mainly of the spinal cord and midbrain. The demyelinating plaques reveal well-preserved axons and neurons and are associated with perivascular cuffing of lymphocytes and mononuclear cells. In addition, cellular infiltrations are found within demyelinating areas; these cells are mainly macrophages. Infectious virus can be reisolated from diseased rats by conventional methods independent of the duration of incubation period, and viral antigens are found primarily in glial cells in the neighborhood of demyelinating plaques.

Despite these serious histopathological changes, many of the SDE animals survive and recover. However, some of these surviving rats may succumb to a second attack of disease in 2 to 3 months [9]. Histopathologically, fresh demyelinating lesions with infiltration of mononuclear cells are found next to remyelinated areas, often in the absence of viral information.

A similar disease is induced with measles virus in weanling Lewis rats [10]. After an incubation period of weeks to months, infected animals come down with a subacute measles encephalomyelitis (SAME) that, histologically, consists of an intense inflammatory infiltrate within both the white and gray matter of the CNS without apparent demyelination. However, in contrast to coronavirus-infected rats, no relapses occur in the survivors of SAME. Moreover, no infectious measles virus can be recovered from brain tissue, which is probably the result of a restricted measles virus gene expression in infected brain cells.

Analysis of the humoral- and cell-mediated immune responses of diseased Lewis rats in both animal models reveals an autoimmune reaction. In SDE as well as in SAME rats, the T lymphocytes present react not only with the virus but also with myelin basic protein (MBP). Lymphocytes collected from spleen, thymus, and peripheral blood and cultured in the presence of MBP show a proliferative response to this antigen [4,5]. Furthermore, adoptive transfer of lymphocytes derived from these diseased rats to normal rats induces the clinical symptoms of experimental allergic encephalomyelitis (EAE) 5 days after transfer. Histologically, the recipient animals reveal perivascular mononuclear cell infiltrations in the dorsal area of the white matter of the spinal cord and in the pons, cerebellar white matter, and thalamus. Moreover, in MHV–JHM–infected Lewis rats, an intrathecal virus-specific immune response is only occasionally found, despite the presence of oligoclonal IgG. Thus, these immunoglobulins of restricted heterogeneity are not directed against viral antigens but rather against antigens of host origin [11,12].

These observations indicate that in JHM virus-infected Lewis rats, as well as in measles virus-infected Lewis rats with a persistent infection of the CNS, development of a cellular autoimmune reaction may be of pathogenetic significance.

Virus-Induced CMAI Reaction in Viral Infection in Man

The possibility of a CMAI reaction as a result of viral infection has been discussed for CNS complications in the course of measles, mumps, vaccinia, rubella, and herpes zoster virus infections. In these postinfectious encephalomyelitides, recovery of the infectious agent in question from brain tissue or cerebrospinal fluid (CSF) is rare, and attempts at identification of viral footprints in diseased brain areas with immunohistological or in situ hybridization techniques have usually failed. However, the neuropathological changes observed in encephalomyelitides resemble those seen in EAE, and thus, an autoimmune component to these diseases may be implicated. To date this interpretation is only indirectly supported by circumstantial evidence. In patients with acute measles encephalomyelitis, a significant proliferative response of isolated peripheral lymphocytes against human MBP was found, which was recorded not only during the clinical symptoms but also as late as 30 days after the onset of disease [13]. In addition, in CSF specimens from such patients, MBP protein was detected as a consequence of myelin breakdown.

Such MBP-specific lymphoproliferative responses have not only been seen after measles infection, they have also been seen in patients with postinfectious encephalomyelitis following rubella, varicella, and respiratory infection and in patients with complications of postexposure rabies immunization [14]. The latter disorder is probably the human equivalent of EAE, since such patients received rabies vaccine prepared in brain tissue. By analogy to EAE, it is not surprising, therefore, that the finding of an MBP-specific lymphoproliferative response in these virus infections is considered to be of pathogenetic importance.

Possible Mechanisms of Virus-Induced CMAI Reactions

The most important question concerns the mechanisms by which viruses induce a T lymphocyte response against host antigens. This problem is still unresolved, but a number of possibilities exists:

1. The fact that viruses only multiply in living cells has major consequences for the cell as well as for the host. It has been proposed that, during replication, the virus may incorporate host antigens into its envelope and insert, modify, or expose cellular antigen on the cell surface. Such newly exposed antigen could be recognized as foreign by the host and could elicit a reaction in the same manner as any other previously unencountered protein [15]. So far, definitive proof of this hypothesis has not been presented, although for certain virus infections, this mechanism may induce autoimmune reactions [3].

2. The virus interacts with the immune regulatory system by destroying subpopulations of lymphocytes or stimulating generation of lymphocyte clones that are autoreactive. In general, many viruses are lymphotropic; one of the prime examples is the Epstein–Barr virus, which infects and transforms human B lymphocytes. Such immortalized cells may secrete, under certain conditions, autoantibodies that can react with cellular constituents [16]. In the animal models described above, the murine corona virus JHM or the neurotropic measles virus do not seem to be lymphotrophic, since B or T lymphocytes of rats cannot be infected. However, it is well known that measles infection in man does alter host immune functions (see discussion above) as well as antibody production, responsive to both mitogens, and allogeneic stimulus and that the disease is sometimes followed by such serious complications as an exacerbation of tuberculosis [17].

3. Another mechanism could be molecular mimicry by which an immune response is raised against certain viral antigens that may cross-react with normal host cell antigens [18]. By computer analysis of a variety of viral sequences, it was recently found that a number of viruses contains the sequence of MBP in the viral genome [19]. Moreover, the immunization of a rabbit with a synthesized peptide from such a sequence of hepatitis B virus polymerase led to the induction of EAE lesions in rabbits [20]. So far, a rat MBP sequence has not been found in the viral genome of JHM or of measles virus, and measles virus-specific T cell lines isolated from infected Lewis rats do not proliferate in the presence of MBP [5]. However, this mechanism could play an important role in the pathogenesis of autoimmune reactions if an infectious agent and a host-cell protein share an antigenic site.

4. A final possibility for the development of an immune pathological reaction in the course of a CNS viral infection is the induction of class II antigen on brain cells by the virus and a subsequent pathological response, such as a delayed type sensitivity (DTH) reaction, in genetically susceptible hosts. It has been established that major histocompatibility (MHC) antigens are expressed only at low levels or not at all on the bulk of cells in the CNS [21]. It is evident, however, that a number of glial cells may be induced to express (MHC) antigens by treatment with such agents as interferon gamma (IFN-γ) in vitro [22] or as a result of an inflammatory reaction in vivo, presumably due to the release of IFN-γ by infiltrating T cells [23]. However, the problem remains as to the events that may initiate such a reaction in the CNS, since without the presence of MHC antigens, it is undoubtedly very difficult for T cells to recognize antigen, become activated, and release lymphokines. Therefore, in a viral infection of the CNS, other mechanisms must operate to induce MHC antigen expression in order for the T lymphocyte to find its target cell.

With respect to corona virus JHM or measles virus, it has been shown that exposure of astrocytes in culture to these two viruses leads directly to the expression of MHC class II antigens that can be further enhanced by the addition of tumor necrosis factor (TNF) [24,25]. Moreover, it has been shown that astrocytes expressing class II antigen can present MBP to CD4+ T cells [26]. Thus, if one extrapolates these observations to the in vivo situation, it is possible that viral infection may create a situation in which cells that previously were not recognizable by components of the host's immune system because of lack of the appropriate restriction elements may now interact with and perhaps even present antigen to autoreactive T-cells. Once this has occurred, subsequent induction of MHC antigens may be mediated by products of the activated T cell. This mechanism could potentially play a role in the JHM virus-induced chronic demyelinating disease of Lewis rats, since the hyperexpression of MHC class II molecules on astrocytes after exposure to JHM viral particles in vitro is genetically regulated [27,28].

Conclusion

The development of cell-mediated autoimmune reactions as a consequence of a viral infection is probably of pathogenetic importance for a number of subacute and chronic diseases in animals and man. At present one cannot point to any one defect that may be responsible for autoimmune disorders, in general, and it is conceivable that such disorders have a multifactorial basis. Indeed, it may transpire that autoimmune phenomena are the sequela of a number of different virus-induced changes, each one a relatively common event.

References

1. Oldstone M (1984) Virus-induced immune complex formation and disease: Definition, regulation, importance. *In* Notkins AL, Oldstone MBA (eds) Concepts in Viral Pathogenesis. Springer Verlag, New York, pp 201–209
2. Yamamura T, Tada T (1984) Progress in Immunology V. Academic Press, Tokyo
3. Notkins AL, Onodera T, Prabhaker BS (1984) Virus-induced autoimmunity. *In* Notkins AL, Oldstone MBA (eds) Concepts in Viral Pathogenesis. Springer-Verlag, New York, pp 210–215
4. Watanabe R, Wege H, ter Meulen V (1983) Adoptive transfer of EAE-like lesions by BMP stimulated lymphocytes from rats with coronavirus-induced demyelinating encephalomyelitis. Nature 305:150–153
5. Liebert UG, Linington C, ter Meulen V (1988) Induction of autoimmune reactions to myelin basic protein in measles virus encephalitis in Lewis rats. J Neuroimmunol 17:103–118
6. Nagashima K, Wege H, Meyermann R, ter Meulen V (1978) Corona virus induced subacute demyelinating encephalomyelitis in rats: A morphological analysis. Acta Neuropathol (Berl) 44:63–70
7. Nagashima K, Wege H, Meyermann R, ter Meulen V (1979) Demyelinating encephalomyelitis induced by a long-term corona virus infection in rats. Acta Neuropathol (Berl) 45:205–213
8. Sörensen O, Percy D, Dales S (1980) In vivo and in vitro models of demyelinating diseases. III. JHM virus infection of rats. Arch Neurol 37:478–484
9. Wege H, Watanabe R, ter Meulen V (1984) Relapsing subacute demyelinating encephalomyelitis in rats in the course of coronavirus JHM infection. J Neuroimmunol 6:325–336
10. Liebert UG, ter Meulen V (1987) Virological aspects of measles virus induced encephalomyelitis in Lewis and BN rats. J Gen Virol 68:1715–1722
11. Dörries R, Watanabe R, Wege H, ter Meulen V (1986) Murine corona-virus induced encephalomyelitides in rats: Analysis of immunoglobulins and virus-specific antibodies in serum and cerebrospinal fluid. J Neuroimmunol 12:131–142
12. Dörries R, Watanabe R, Wege H, ter Meulen V (1987) Analysis of the intrathecal humoral immune response in Brown Norway (BN) rats, infected with the murine coronavirus JHM. J Neuroimmunol 14:305–316
13. Johnson RT, Griffin DE, Hirsch RL, Wolinsky JS, Roedenbeck S, DeSoriano IL (1984) Measles encephalomyelitis: Clinical and immunological studies. N Engl J Med 310:137–141
14. Johnson RT, Griffin D (1986) Virus-induced autoimmune demyelinating disease of the CNS. *In* Notkins AL, Oldstone MBA (eds) Concepts in Viral Pathogenesis II. Springer-Verlag, New York, pp 203–209
15. Hirsch MS, Proffitt MR (1975) Autoimmunity in viral infection. *In* Notkins AL (ed) Viral Immunology and Immunopathology. Academic Pess, New York, pp 419–434
16. Rosen A, Gergely P, Jondal M, Klein G, Britton S (1977) Polyclonal Ig production after Epstein–Barr virus infection of human lymphocytes in vitro. Nature 267:52–54
17. Cherry JD (1981) Measles *In* Feigin RD, Cherry JD, Saunders WB (eds) Textbook of Pediatric Infectious Diseases, Vol II. Philadelphia, London, Toronto, Mexico City, Rio de Janeiro, Sydney, Tokyo, pp 1210–1230

18. Oldstone MBA, Notkins AL (1986) Molecular mimicry. *In* Notkins AL, Oldstone MBA (eds) Concepts in Viral Pathogenesis II. Springer-Verlag, New York, pp 195–202.

19. Jahnke U, Fischer EH, Alvord EC (1985) Sequence homology between certain viral proteins and proteins related to encephalomyelitis and neuritis. Science 229:282–284

20. Fujinami RS, Oldstone MBA (1985) Amino acid homology and immune responses between the encephalitogenic site of myelin basic protein and virus: A mechanism for autoimmunity. Science 230:1043–1045

21. Hart DNJ, Fabre JW (1981) Demonstration and characterization of Ia-positive dendritic cells in the institial tissues of rat heart and other tissues, but not brain. J Exp Med 154:347–361

22. Wong GHW, Bartlett PF, Clark-Lewis I, McKimm-Breschkin JL, Schrader JW (1985) Interferon-gamma induces the expression of H-2 and Ia antigens on brain cells. J Neuroimmunol 7:255–278

23. Traugott U, Scheinberg LC, Raine CS (1985) On the presence of Ia-positive endothelial cells and astrocytes in multiple sclerosis lesions and its relevance to antigen presentation. J Neuroimmunol 8:1–14

24. Massa PT, Dörries R, ter Meulen V (1986) Viral particles induce Ia antigen expression on astrocytes. Nature 320:543–546

25. Massa PT, Schimpl, A, Wecker E, ter Meulen V (1987). Tumor necrosis factor amplifies measles virus-mediated Ia induction on astrocytes. Proc Natl Acad Sci USA 84:7242–7245

26. Fontana A, Fierz W, Wekerle H (1984) Astrocytes present myelin basic protein to encephalitogenic T-cell lines. Nature 307:273–276

27. Massa PT, ter Meulen V, Fontana A (1987) Hyperinducibility of Ia antigen on astrocytes correlates with strain-specific susceptibility to experimental autoimmune encephalomyelitis. Proc Natl Acad Sci USA 84:4219–4223

28. Massa PT, Brinkmann R, ter Meulen V (1987) Inducibility of Ia antigen on astrocytes by murine coronavirus JHM is rat strain dependent. J Exp Med 166:259–264

CHAPTER 36
Immunological Tolerance in Viral Infections

RAFI AHMED

The immune system is programmed to respond against foreign antigens and to remain unresponsive to self-antigens. This self–nonself discrimination is a fundamental aspect of immunology. It is believed that tolerance to an antigen is acquired during the development of the immune system; antigens encountered during embryonic life are considered self and do not evoke an immune response, whereas antigens seen after the development of the immune system are considered foreign and elicit responses [1]. In this chapter I will review what is known about immune responses to viruses that initiate infection at the fetal or neonatal stage, and consider the biological and medical significance of such congenitally or perinatally acquired infections.

Immune Response in Viral Infections of the Fetus and Neonate

B Cell Responses
In the 1930s Traub [2] reported that mice infected in utero with lymphocytic choriomeningitis virus (LCMV) carried high levels of virus in their blood and tissues throughout life without any detectable complement-fixing or neutralizing antibody, whereas mice first infected as adults made a potent virus-specific antibody response. These studies on LCMV, along with Owen's [3] observations on chimeric cattle, formed the basis on which Burnet and Fenner [4] predicted that an antigen introduced into the body during embryonic life would be mistaken for self, and Burnet [5] subsequently proposed the theory of clonal deletion of self-reactive clones. The notion that congenitally (or perinatally) acquired viral infections do not elicit an immune response was challenged by the work of Oldstone and Dixon [6]

who showed that LCMV carrier mice do make antibodies but that this antiviral antibody combines with LCMV antigen in the circulation to form virus–antibody immune complexes that are deposited in various tissues. This was an important finding for two reasons: First, it showed unequivocally that exposure to a virus during ontogeny does not necessarily result in complete tolerance at the B cell level; second, it demonstrated the presence of circulating antigen–antibody complexes with the potential to cause immune complex disease in a chronically infected host. Both these observations have now been confirmed in a large number of congenitally/perinatally acquired viral infections of humans and animals (Table 36.1) [7]. In addition, similar findings have been reported for bacterial and parasitic infections [7].

Although complete B cell tolerance is not seen in most viral infections of the fetus and neonate, there is evidence of both qualitative and quantitative changes in the antibody response. For example, marked differences have been noted in the isotype profiles of LCMV-specific antibody present in neonatally infected mice compared to the profiles for mice infected as adults [8]. The majority (~70%) of the antiviral antibody in mice first infected as adults is of the IgG2a subclass. In contrast, most (>95%) of the LCMV-specific antibody in adult carriers infected at birth is of the IgG1 subclass (A.R. Salmi, R. Ahmed, and M.B.A. Oldstone, unpublished data). There are also indications of quantitative differences between the numbers of spleen cells producing LCMV-specific antibodies during acute infection compared to persistent infection for certain mouse strains [9]. The antibody response appears to be lower in LCMV carrier mice, especially those infected in utero as opposed to neonatally infected mice (R. Ahmed and B.D. Jamieson,

Table 36.1. Representative congenital and perinatal viral infections in which virus-specific antibody responses have been documented.[a]

Host	Virus
Human	Rubella
	Cytomegalovirus
	Human immunodeficiency virus
	Herpes simplex
	Hepatitis B
	Parvovirus
	Coxsackieviruses B
Mouse	Lymphocytic choriomeningitis virus
	Lactic dehydrogenase virus
Mink	Aleutian disease virus
Swine	Porcine cytomegalovirus
Cat	Feline leukemia virus
Horse	Equine infectious anemia virus
Goat	Caprine arthritis–encephalitis virus

[a] This table is not comprehensive, and only a few examples are listed.

unpublished data). Some other examples of congenitally acquired infections with virus-specific B cell hyporesponsiveness are avian leukosis virus (chicken), bovine virus diarrhea virus (cattle), murine leukemia virus (mouse), and Border disease virus (sheep) [10]. Transplacental infections in humans with rubella virus usually result in an antibody response, but there are a few instances of B cell defects in congenital rubella [11]. However, it is difficult to assess the contribution of specific immunological tolerance in these cases because such individuals have a generalized immunosuppression and exhibit diminished B and T cell responses to a variety of unrelated antigens. Similar findings have recently been reported in transplacental and perinatal human immunodeficiency virus (HIV) infections [12]. Although antibody against HIV and hypergammaglobulinemia is present in most pediatric cases, there are a few examples of HIV-infected infants with severe hypogammaglobulinemia and no detectable antibodies against HIV. As in the few cases of B cell suppression in congenital rubella, it is not clear whether the HIV-specific unresponsiveness is due to immunological tolerance or generalized immunosuppression.

T Cell Responses
Congenital infection of mice with LCMV is the best-studied model of T cell tolerance in a viral infection [13]. Mice infected with LCMV at birth or in utero become lifelong carriers, with high levels of virus in most of their organs, including the thymus. Absence of virus-specific T cells is a hallmark of chronic LCMV infection [14,15]. These carrier mice show minimal or no detectable cytotoxic T lymphocyte (CTL) or delayed-type hypersensitivity responses against LCMV; they also have few or no T cells that will proliferate when stimulated with the virus. Attempts to induce a CTL response in carrier mice after rechallenge with homologous and heterologous strains of LCMV in vivo, or after stimulation of carrier lymphocytes with free virus or infected macrophages in vitro, have been unsuccessful [16]. No virus-specific CTL are detected even after addition of lymphokines (purified interleukin 2 or concanavalin A supernatants) to cultures of carrier spleen and lymph node cells. LCMV carrier mice do not exhibit a generalized T cell suppression; they are fully capable of generating allogeneic CTL responses, they can reject skin grafts, and they are not susceptible to opportunistic infections. Thus, there is convincing evidence of a highly specific T cell unresponsiveness in LCMV carrier mice. This T cell defect is the primary reason why these carrier mice are unable to eliminate the virus [13–18]. Although other viral systems have not been as extensively studied, considerable data suggest a decreased virus-specific T cell response in several congenitally and perinatally acquired viral infections [7,10].

Is the virus-specific T cell unresponsiveness seen in a congenitally acquired chronic viral infection reversible? Jamieson and Ahmed [16] have recently documented virus-specific T cell competence in mice cured of congenitally acquired chronic LCMV infection. In these studies the persistent infection was eliminated by transfer of CD8+ T cells from LCMV immune mice. To determine whether these cured carriers could generate their own LCMV-specific CTL response, mice congenic at the Thy-1 locus (Thy1.1 and Thy1.2) were used in the adoptive transfer experiments. Host-derived T cell responses were checked after treating the cured carriers with a monoclonal antibody to deplete the immune donor T cells. Such cured carrier mice were able to generate a host-derived, virus-specific CTL response and resisted a second LCMV challenge in the absence of any donor T cells. These results demonstrate unequivocally that exposure to a virus during fetal life does not result in permanent deletion of specific T cells and that the unresponsive state is reversible. It will be of interest to determine when the carrier mice acquired T cell competence, whether LCMV-specific CTL were present throughout the life span of these mice in a "nonfunctional" state, or whether these CTL differentiated from the bone marrow after the elimination of viral antigen (i.e., export of new T cells from the thymus).

Tolerance Mechanisms for Self-Antigens and Viruses

Three possible mechanisms have been proposed to explain tolerance to self-antigens: (1) the elimination of autoreactive lymphocytes during ontogeny (clonal deletion), (2) their functional inactivation by exposure to antigen at an early stage of development (clonal anergy), and (3) their suppression by antigen-specific suppressor T cells or the idiotypic regulatory network [1]. These mechanisms are not mutually exclusive, and it is possible that immunological tolerance to self is mediated by several distinct mechanisms. The elegant experiments of Kappler et al. [19] utilizing monoclonal antibodies against specific T cell receptor chains to monitor the fate of self-reactive lymphocytes have provided direct evidence for clonal deletion of autoreactive T cells within the thymus. Studies using transgenic mice have also demonstrated the elimination of self-reactive T cells during differentiation in the thymus [20]. Studies by Goodnow et al. [21] examining the mechanism of B cell tolerance in double transgenic mice expressing the genes for hen egg lysozyme and for a high-affinity, anti-lysozyme antibody have shown that the majority of lysozyme-specific B cells did not undergo clonal deletion but were present in a nonfunctional stage (i.e., unable to secrete anti-lysozyme antibody). These recent studies are beginning to provide a much better understanding of self–nonself discrimination and suggest a possible dichotomy in the mechanisms of self tolerance for B cells and T cells.

Are the rules and mechanisms for induction of tolerance to viruses different than those for self-antigens? Probably not. Then why do viral infections of the fetus often elicit an immune response? It should be pointed out that the mere presence of an antigen during ontogeny may not be sufficient to induce tolerance. Efficient tolerance induction depends not only on the presence of antigen during embryonic life but also on its tissue site of expression, structure, concentration, and continued persistence [1,16,22]. It is therefore not altogether surprising that transplacental infections do not necessarily result in tolerance to the virus. Finally, it is worth considering if viral infections can somehow interfere with the process of tolerance induction. Clonal deletion of self-reactive T cells in the thymus requires expression of such accessory molecules as CD8 and CD4, in addition to the antigen-specific T cell receptor. Recent studies have shown that the deletion process spares "autoreactive" T cells that lack or have low levels of these accessory molecules [20,23]. Since viruses can up- or down-regulate the expression of cellular proteins, it is conceivable that a viral infection may affect the process of clonal deletion by suppressing the expression of critical cellular molecules. It is also possible that viral infections may affect the generation of "second" signals, the presence or absence of which may constitute the difference between activation and tolerance of lymphocytes [1]. Thus, it is likely that studies on congenital viral infections, along with the use of transgenic animals, will continue to provide insights into the mechanisms of tolerance and self–nonself discrimination.

Medical Importance of Congenitally and Neonatally Acquired Viral Infections

An important consequence of transplacental and perinatal exposure to a virus is the increased likelihood of developing a persistent infection. This is best illustrated by hepatitis B virus (HBV) infections in humans [24]. The HBV infections acquired as adults are usually acute and resolve within a few months, whereas exposure to HBV during infancy results in persistence. In certain parts of Africa and Asia where mother-to-offspring transmission is common, up to 10 to 15% of the population are HBV carriers. Examples of other viruses that can persist in humans following infection of the fetus or neonate include rubella, herpes simplex, cytomegalovirus, HTLV-1, B19 parvovirus, and HIV—the causative agent of acquired immunodeficiency syndrome [7,12]. Perinatal infection is increasingly becoming a significant mode of HIV transmission, especially in some cities in Africa where 5 to 15% of pregnant women are infected [25].

It is clear that transplacental and neonatal viral infections are an important factor in chronic infections of humans. It is also clear that chronic virus infections now constitute a major health problem. In this context it is worth noting that complete B cell tolerance is rarely observed in fetal/neonatal infections and that the T cell unresponsiveness is reversible [6,16]. The latter finding is of particular significance, since failure of T cell immunity often leads to chronic infections. Research with the LCMV model shows that a chronic infection acquired in utero can be eliminated and that the host T cell response becomes functional and provides protection against reinfection [16]. These results demonstrating T cell competence in a previously unresponsive, persistently infected host have implications for the treatment of chronic infections. They suggest that even when an infection has been acquired congenitally, or as an infant, reduction of the virus load by an appropriate therapeutic protocol may restore the potential of the host to become immunocompetent.

Acknowledgments

This work was supported by U.S. Public Health Service Grant NS-21496 and an award from the University of California Systemwide Task Force on AIDS. The author is a Harry Weaver Neuroscience Scholar of the National Multiple Sclerosis Society.

References

1. Nossal GJV (1983) Cellular mechanisms of immunological tolerance. Ann Rev Immunol 1:33–62
2. Traub E (1938) Factors influencing the persistence of lymphocytic choriomeningitis virus in the blood of mice after clinical recovery. J Exp Med 68:229–250
3. Owen RD (1945) Immunogenetic consequences of vascular anastomoses between bovine twins. Science 102:400–401
4. Burnet FM, Fenner F (1949) The Production of Antibodies. Macmillan Publishing Co, New York
5. Burnet FM (1957) A modification of Jerne's theory of antibody production using the concept of clonal selection. Aust J Sci 20:67–69
6. Oldstone MBA, Dixon FJ (1967) Lymphocytic choriomeningitis: Production of antibody by "tolerant" infected mice. Science 158:1193–1195
7. Mims CA (1987) The Pathogenesis of Infectious Disease, 3rd ed. Academic Press, New York
8. Thomson AR, Volkert M, Marker O (1985) Different isotype profiles of virus-specific antibodies in acute and persistent lymphocytic choriomeningitis virus infections in mice. Immunology 55:213–223
9. Moskophidis D, Lehmann-Grube F (1984) The immune response of the mouse to lymphocytic choriomeningitis virus. IV. Enumeration of antibody-producing cells in spleens during acute and persistent infection. J Immunol 133:3366–3370
10. Fenner F, Bachmann PA, Gibbs EPJ, Murphy FA, Studdert MJ, White DO (1987) Veterinary Virology. Academic Press, New York
11. Hancock MP, Huntley CC, Sever JL (1968) Congenital rubella syndrome with immunoglobulin disorder. J Pediatr 72:636–645
12. Pahwa R, Good RA, Pahwa S (1987) Prematurity, hypogammaglobulinemia, and neuropathology with human immunodeficiency virus (HIV) infection. Proc Natl Acad Sci USA 84:3826–3820

13. Oldstone MBA, Ahmed R, Byrne J, Buchmeier MJ, Rivere Y, Southern P (1985) Virus and immune responses: Lymphocytic choriomeningitis virus as a prototype model of viral pathogenesis. Br Med Bull 41:70–74

14. Zinkernagel RM, Doherty PC (1979) MHC-restricted cytotoxic T cells: Studies on the biological role of polymorphic major transplantation antigens determining T-cell restriction, specificity, function, and responsiveness. Adv Immunol 77:51–177

15. Ahmed R, Salmi A, Butler LD, Chiller JM, Oldstone MBA (1984) Selection of genetic variants of lymphocytic choriomeningitis virus in spleens of persistently infected mice: Role in suppression of cytotoxic T lymphocyte response and viral persistence. J Exp Med 60:521–540

16. Jamieson BD, Ahmed R (1988) T cell tolerance: Exposure to virus in utero does not cause a permanent deletion of specific T cells. Proc Natl Acad Sci USA 85:2265–2268

17. Volkert M (1963) Studies on immunological tolerance to LCM virus: Treatment of virus carrier mice by adoptive immunization. Acta Pathol Microbiol Scand 57:465–487

18. Ahmed R, Jamieson BD, Porter DD (1987) Immune therapy of a persistent and disseminated viral infection. J Virol 61:3920–3929

19. Kappler JW, Roehm N, Marrack P (1987) T cell tolerance by clonal elimination in the thymus. Cell 49:273–280

20. Kisielow P, Bluthmann H, Staerz UD, Steinmetz M, von Boehmer H (1988) Tolerance in T-cell-receptor transgenic mice involves deletion of nonmature CD4$^+$8$^+$ thymocytes. Nature 333:742–746

21. Goodnow CC, Crosbie J, Adelstein S, Lavoie TB, Smith-Gill SJ, Brink RA, Pritchard-Briscoe H, Wotherspoon JS, Loblay RH, Raphael K, Trent RJ, Basten A (1988) Altered immunoglobulin expression and functional silencing of self-reactive B lymphocytes in transgenic mice. Nature 334:676–682

22. Morecki S, Lesham B, Eid A, Slavin S (1987) Alloantigen persistence in induction and maintenance of transplantation tolerance. J Exp Med 165:1468–1480

23. Fowlkes BJ, Schwartz RH, Pardoll DM (1988) Deletion of self-reactive thymocytes occurs at a CD4$^+$8$^+$ precursor stage. Nature 334:610–625

24. Marion PL, Robinson WS (1983) Hepadna viruses: Hepatitis B and related viruses. Curr Top Microbiol Immunol 105:99–121

25. Mann JM, Chin J (1988) AIDS: A global perspective. N Engl J Med 319:302–303

Control of Viral Diseases

CHAPTER 37
Designing Peptides to Immunologically Distinguish Specific Human Immunodeficiency Virus Isolates

JOHN W. GNANN, JR, AND MICHAEL B.A. OLDSTONE

Synthetic peptides representing defined segments of viral proteins have become increasingly useful both as reagents in epidemiological and clinical analyses and in the design of vaccines. Among the advantages of synthetic peptide reagents are ready availability in large quantities, high purity, defined specificity, and lack of infective potential. In addition, peptides can be designed to avoid potentially detrimental sequences that may act either as cross-reactive immunogens with host proteins or suppress immune functions. Applications of synthetic peptides can be grouped into two general categories. First, a peptide is used as an immunogen to generate an immune response able to cross-react with the native viral protein from which the peptide was derived. The ultimate application of synthetic immunogens is as vaccines to elicit a specific protective immune response. Peptides in the second category are designed as antigens to detect B and T lymphocyte responses generated during infection against the native viral protein. In this scenario, a synthetic peptide can serve either as an antigen to detect humoral immune responses or as a specific mitogen to induce T lymphocyte proliferative responses or to map cytotoxic T lymphocyte recognition epitopes. In this chapter we describe how synthetic peptides can be designed to descriminate between infections caused by human immunodeficency virus type 1 and type 2 (HIV-1, HIV-2) and to segregate antibody responses among HIV-1 strains.

Selection of Protein Regions for Synthesis

A given viral protein usually contains multiple immunogenic sites. Such a site may be either a continuous amino-acid sequence or discontinuous determinants that are presented by protein folding. The challenge for the investigator is to select for synthesis those linear amino-acid sequences, usually 10 to 20 residues in length, that have the highest probability of functioning as B cell epitopes. A number of "rules" have been proposed to aid in the selection of protein regions for synthesis, although exceptions to these rules abound. Characteristics that may indicate a B cell determinant include high local average hydrophilicity [1], positions near protein chain termini [2], and high flexibility and segmental mobility [3]. In general, these criteria enable one to select sequences that are likely to be exposed on the surface of a protein rather than buried in folds. Selection of sequences that are conserved among various strains of the virus is also important, particularly for HIV, which has several hypervariable regions in the envelope glycoproteins [4]. For a synthetic peptide to be useful as a diagnostic reagent or as a vaccine, it should represent a sequence that is highly conserved among many HIV strains.

Computer programs have been devised that use these parameters to predict possible antigenic determinants on HIV proteins [4–6]. Although these predictive schemes have proved useful to some extent, they still provide only rough guesses. For example, none of these programs predicted the major immunodominant epitope of HIV that has been identified near the amino end of the HIV transmembrane glycoprotein. This failure indicates that our understanding of the characteristics that define a B cell determinant is incomplete. However, the alternative to selecting potential determinants for synthesis is to synthesize sequential overlapping peptides covering the entire length of the protein. For a large protein, this can obviously be a laborious and expensive undertaking, although innovative methods for expediting the process have been described [7,8].

Envelope Glycoprotein Determinants of HIV-1

Serological studies of infected humans have shown the envelope components to be the most consistently immunogenic HIV proteins. In addition, epitopes that elicit protective neutralizing antibody to murine and feline retroviruses were identified on the envelope glycoproteins of those viruses. Consequently, initial efforts at identifying B cell epitopes of HIV have focused on the envelope glycoproteins.

The envelope of HIV type 1 contains two proteins, a 120kd external membrane glycoprotein (gp120) and a 41kd transmembrane glycoprotein (gp41), encoded by the *env* gene. Antibodies to both of these proteins, as well as to the 160kd precursor protein (gp160), are readily detectable in the sera of virtually all infected individuals.

To identify immunoreactive envelope determinants, investigators synthesized peptides representing potential epitopes and tested the peptides for reactivity with sera from HIV-1-infected individuals by enzyme-linked immunosorbent assay (ELISA) or radioimmunoassay (RIA). With these techniques, three distinct B cell epitopes were identified. One was located at the carboxy terminus of gp120, adjacent to the proteolytic cleavage site between gp120 and gp41 [9]. This region (env amino acids 504–518 of strain HTLV-III_B) was moderately hydrophilic, had a high surface probability, and was highly conserved among isolates of HIV-1 [4]. Absorption of reactive human sera with this synthetic peptide removed a high percentage of anti-gp120 antibody reactivity, indicating that this is a major recognition epitope [9]. However, only about 45% of HIV-1-infected individuals had detectable antibody against this peptide. Antibody bound by this peptide in an affinity column reacted specifically with gp120 by immunoblotting and RIA but did not exhibit HIV-1 neutralizing activity in vitro [9].

A second immunoreactive determinant was identified near the carboxy terminus of gp41 [10]. This region (env amino acids 735–752 of strain HTLV-III_B) has high predicted values for hydrophilicity, flexibility, and surface probability, and the residues are highly conserved among HIV-1 strains [4]. When used as an antigen in an ELISA, this peptide reacted with sera from a small number of HIV-1-infected patients [10]. Rabbit antibody raised against this synthetic peptide immunoprecipitated gp160, labeled HIV-1 infected cells in an immunofluorescence assay, and neutralized HIV-1 in vitro [10].

The best-studied HIV-1 B cell determinant lies between env amino acids 584–618 (strain LAV-1_BRU) near the amino terminus of gp41 [11–13]. This region is largely hydrophobic, and the only characteristics suggesting a potential B cell epitope are a high degree of amino-acid conservation (Table 37.1) and a postulated position on the exterior side of the virus membrane [4]. Peptides that represent various portions of this 35 amino-acid sequence showed reactivity against more than 99% of sera from HIV-1-infected individuals by ELISA [11–13].

Characterization of the Immunodominant gp41 Epitope (amino acids 598–609)

By testing a series of overlapping peptides extending across gp41 amino acids 584–618, an immunodominant 12 amino-acid B cell epitope was identified. A synthetic peptide analogous to env amino acids 598–609 (LeuGlyLeuTrpGlyCysSerGlyLysLeuIleCys—Table 37.1) reacted with more than 99% of serum specimens from HIV-1-infected patients from the United States and did not react with normal control sera [11].

To determine precisely the residues essential for antibody recognition, a nested set of peptides was synthesized with single amino acids sequentially deleted from the amino terminus of the immunoreactive 12 amino-acid peptide [23]. The smallest peptide that retained a degree of reactivity was a

Table 37.1. Amino-acid conservation in the transmembrane glycoproteins of HIV-1 and HIV-2.

Isolate name	Source	Amino-acid sequence[a,b]	References
HIV-1 (LAV-1$_{BRU}$; HTLV-III$_B$; HTLV-III$_{RF}$; WMJ-1; ARV-2; Zr6)	France, USA, Zaire, Haiti	598 609 Leu-Gly-Ile-[Trp-Gly-Cys-Ser-Gly-Lys-Leu-Ile]-Cys	14–18
HIV-1 (CDC-451)	USA	Leu-Gly-_Phe_-Trp-Gly-Cys-Ser-Gly-Lys-Leu-Ile-Cys	19
HIV-1 Z3	Zaire	Leu-Gly-_Leu_-Trp-Gly-Cys-Ser-Gly-Lys-Leu-Ile-Cys	20
HIV-1 (LAV-1$_{ELI}$)	Zaire	Leu-Gly-Ile-Trp-Gly-Cys-Ser-Gly-Lys-_His_-Ile-Cys	21
HIV-1 (LAV-1$_{MAL}$)	Zaire	Leu-Gly-_Met_-Trp-Gly-Cys-Ser-Gly-Lys-_His_-Ile-Cys	21
HIV-2 (HIV-2$_{ROD}$)	Cape Verde Islands	592 603 Leu-Asn-Ser-[Trp-Gly-Cys-Ala-Phe-Arg-Gln-Val]-Cys	22

a The amino-acid position numbers for HIV-1 are based on the sequence of LAV-1$_{BRU}$ [14].

b Residues in the HIV-1 sequences that differ from the sequence of the prototype isolate are underlined. The boxed areas show residues conserved between HIV-1 and HIV-2.

Adapted with permission from Gnann et al. [26]. Copyright 1987 by the AAAS.

7 amino-acid sequence (*env* amino acids 603–609) containing a cysteine (Cys) residue at each end that reacted with 48% of positive area. Deletion of either Cys markedly reduced the reactivity of the peptide, suggesting that disulfide bond formation is critical for the antigenic conformation of the peptide. Substitution of a serine (Ser) residue for the carboxy-terminal Cys of the 12 amino-acid peptide also substantially reduced the reactivity, thus confirming the importance of the two Cys residues [23]. The immunodominance of this gp41 epitope was confirmed by studies in which a human monoclonal antibody was produced from spleen cells from an HIV-1-infected individual [24]. When the monoclonal antibody was characterized, it was found to be directed to precisely the same gp41 epitope previously identified by peptide mapping [23].

By testing serum specimens obtained serially after acute HIV-1 infection, the kinetics of IgM and IgG antibody responses directed against this gp41 epitope were followed. A transient low-titered IgM response occurred early in the course of infection, whereas the IgG response was high-titered and persistent [25]. To establish whether the antibody response to this epitope was T cell dependent, homozygous nude (athymic) mice and their immuno-competent heterozygous littermates were immunized with the 12 amino-acid peptide. Both groups of mice produced transient IgM responses, but only the mice with intact T cell function produced antipeptide IgG, indicating that the IgG response to this epitope is T cell dependent [25].

Detection of Strain-Specific Antibody

The gp41 peptide was reactive with more than 99% of sera from HIV-1-infected American patients. However, when the peptide was tested against a panel of well-characterized HIV-1 positive sera from Zaire, only 87% of the sera reacted [26]. Although the gp41 region from which the peptide was derived is highly conserved among the HIV-1 isolates from Europe and North America (Table 37.1), sequencing studies revealed at least two nonconservative amino-acid substitutions in this region in isolates from Zaire [21]. This suggested the possibility that sera from Zairian patients infected with a different HIV-1 strain might not bind to the test peptide but might bind to peptides specifically designed to incorporate the amino-acid substitution(s). To test this hypothesis, peptides corresponding to the analogous gp41 sequences of Zairian isolates, LAV-1$_{MAL}$ and LAV-1$_{ELI}$, were synthesized (Table 37.1), and the Central African sera that failed to react with the original peptide reacted strongly with these newly designed peptides [26]. Most of the Zairian sera that reacted with the original peptide also bound the substituted peptides, although a few sera lost reactivity. These data indicate that synthetic peptide antigens can be constructed to detect strain-specific antibody responses and that immune recognition can be altered by substitution of a single amino acid in a critical position.

Immunogenic Epitope From HIV-2

HIV-1 and HIV-2, a West African immunodeficiency virus, are related viruses, but are clearly distinct at both genetic and antigenic levels. The transmembrane glycoproteins of the two viruses share identical amino acids at approximately 44% of positions [22]. There is some partial antigenic cross-reactivity between the more conserved core proteins of HIV-1 and HIV-2, but cross-reactivity between envelope proteins of the two viruses is reported to be minimal.

The HIV-2 transmembrane glycoprotein region (*env* amino acids 592–603 of strain HIV-2$_{ROD}$) that corresponds to the immunodominant gp41 epitope of HIV-1 has identical amino acids at 5 of the 12 positions (Table 37.1). The two Cys residues shown to be critical for the antigenicity of the HIV-1 epitope are conserved in the HIV-2 sequence. A peptide based on this region of HIV-2 was synthesized and tested by enzyme immunoassay against HIV-positive and -negative sera. The peptide reacted with all serum specimens from HIV-2-infected West African patients, but not with sera from individuals infected with HIV-1 only or with sera from uninfected individuals [26]. Thus, the similar regions of the transmembrane glycoproteins of both HIV-1 and HIV-2 serve as highly reactive B cell epitopes. A likely prediction is that as new HIV isolates are identified they will share this general characteristic of an immunodominant domain near the amino terminus of gp41 containing two Cys residues and presented as a cyclic epitope.

Conclusions

Studies with synthetic peptides that represent portions of HIV proteins have identified several immunoreactive viral epitopes. The best characterized is an immunodominant epitope located near the amino end of the transmembrane glycoprotein that elicits an antibody response in virtually every HIV-infected individual. The minimal determinant for antibody binding is a seven amino-acid sequence containing two essential cysteine residues. Single amino-acid substitutions can be made in this domain of HIV-1 to create peptide antigens that are increasingly or decreasingly strain specific. This reactive epitope is partially conserved in the transmembrane glycoprotein of HIV-2. Immunoassays employing these specific HIV-1 and HIV-2 peptide antigens can thus be used to distinguish the two infections.

Designer peptides also have been applied to the study of a variety of other human viral pathogens including Epstein–Barr virus, hepatitis A virus, hepatitis B virus, and respiratory syncytial virus. Data accumulating from all these studies demonstrate the utility of custom-designed peptides in biological research and clinical medicine.

Acknowledgments

This is Publication Number 5389-IMM from the Department of Immunology, Scripps Clinic and Research Foundation, La Jolla, CA 92037. This work was supported in part by U.S. Public Health Service Grants AI-07007 and NS-12428, and by Training Grant T32-NS-07078 (JWG) from the National Institutes of Health and the U.S. Army Medical Research and Development Command under Contract Number DAMD17-88-C-8103. Opinions, interpretations, conclusions, and recommendations are those of the author and are not necessarily endorsed by the U.S. Army.

References

1. Hopp TP, Woods KR (1981) Prediction of protein antigenic determinants from amino acid sequences. Proc Natl Acad Sci USA 78:3824–3828
2. Walter G, Scheidtmann K-H, Carbone A, Laudana AP, Doolittle RF (1980) Antibodies specific for the carboxy- and amino-terminal regions of simian virus 40 large tumor antigen. Proc Natl Acad Sci USA 77:5197–5200
3. Westhof E, Altschuh D, Moras D, Bloomer AC, Mondragon A, Klug A, Van Regenmortel MHV (1984) Correlation between segmental mobility and the location of antigenic determinants in proteins. Nature 311:123–126
4. Modrow S, Hahn BH, Shaw GM, Gallo RC, Wong-Staal F, Wolf H (1987) Computer-assisted analysis of envelope protein sequences of seven human immunodeficiency virus isolates: Prediction of antigenic epitopes in conserved and variable regions. J Virol 61:570–578
5. Pauletti D, Simmonds R, Dreesman GR, Kennedy RC (1985) Application of a modified computer algorithm in determining potential antigenic determinants associated with the AIDS virus glycoprotein. Anal Biochem 151:540–546
6. Sternberg MJ, Barton GJ, Zvelebil MJ, Cookson J, Coates ARM (1987) Prediction of antigenic determinants and secondary structures of the major AIDS virus proteins. FEBS 218:231–237
7. Geysen HM, Rodda ST, Mason TJ, Tribbick G, Schoofs PG (1987) Strategies for epitope analysis using peptide synthesis. J Immunol Methods 102:259–274
8. Houghten RA (1985) General method for the rapid solid-phase synthesis of large numbers of peptides: Specificity of antigen-antibody interaction at the level of individual amino acids. Proc Natl Acad Sci USA 82:5131–5135
9. Palker TJ, Matthews TJ, Clark ME, Cianciolo GJ, Randall RR, Langlois AJ, White GC, Safai B, Snyderman R, Bolognesi DP, Haynes BF (1987) A conserved region at the COOH terminus of human immunodeficiency virus gp120 envelope protein contains an immunodominant epitope. Proc Natl Acad Sci USA 84:2479–2483
10. Kennedy RC, Henkel RD, Pauletti D, Allan JS, Lee TH, Essex M, Dreesman GR (1986) Antiserum to a synthetic peptide recognizes the HTLV-III envelope glycoprotein. Science 231:1556–1559
11. Gnann JW, Schwimmbeck PL, Nelson JA, Truax AB, Oldstone MBA (1987) Diagnosis of AIDS using a 12 amino acid peptide representing an immunodominant epitope of human immunodeficiency virus. J Infect Dis 156:261–267
12. Wang JJG, Steel S, Wisniewolski R, Wang CY (1986) Detection of antibodies to human T-lymphotropic virus type III by using a synthetic peptide of 21 amino

acid residues corresponding to a highly antigenic segment of gp41 envelope protein. Proc Natl Acad Sci USA 83:6159–6163

13. Smith RS, Naso RB, Rosen J, Whalley A, Hom Y-L, Hoey K, Kennedy CJ, McCutchan JA, Spector SA, Richman DD (1987) Antibody to a synthetic oligopeptide in subjects at risk for human immunodeficiency virus infection. J Clin Microbiol 25:1498–1504

14. Wain-Hobson S, Sonigo P, Danos O, Cole S, Alizon M (1985) Nucleotide sequence of the AIDS virus, LAV. Cell 40:9–17

15. Ratner L, Haseltine W, Patarca R, Livak KJ, Starich B, Josephs SF, Doran FR, Rafalski JA, Whitehorn EA, Baumeister K, Ivanoff L, Petteway SR, Pearson ML, Lautenberger JA, Papas TS, Ghrayeb J, Chang NT, Gallo RC, Wong-Staal F (1985) Complete nucleotide sequence of the AIDS virus, HTLV-III. Nature 313:277–284

16. Starcich BR, Hahn BH, Shaw GM, McNeely PD, Modrow W, Wolf H, Parks ES, Parks WP, Josephs SF, Gallo RC, Wong-Staal F (1986) Identification and characterization of conserved and variable regions in the envelope gene of HTLV-III/LAV, the retrovirus of AIDS. Cell 45:637–648

17. Sanchez-Pescador R, Power MD, Barr PJ, Steimer KS, Stempien MM, Brown-Shimer SL, Gee WW, Renard A, Randolph A, Levy JA, Dina D, Luciw PA (1985) Nucleotide sequence and expression of an AIDS-associated retrovirus (ARV-2). Science 227:484–492

18. Srinivasan A, Anand R, York D, Ranganathan P, Feorina P, Schochetman G, Curran J, Kalyanaraman VS, Luciw PA, Sanchez-Pescador R (1987) Molecular characterization of human immunodeficiency virus from Zaire: Nucleotide sequence analysis identifies conserved and variable domains in the envelope gene. Gene 52:71–82

19. Desai SM, Kalyanaraman VS, Casey JM, Srinivasan A, Andersen PR, Devare SG (1986) Molecular cloning and primary nucleotide sequence analysis of a distinct human immunodeficiency virus isolate reveal significant divergence in its genomic sequences. Proc Natl Acad Sci USA 83:8380–8384

20. Willey RL, Rutledge RA, Dias S, Folks T, Theodore T, Buckler CE, Martin MA (1986) Identification of conserved and divergent domains within the envelope gene of the acquired immunodeficiency syndrome retrovirus. Proc Natl Acad Sci USA 83:5038–5042

21. Alizon M, Wain-Hobson S, Montagnier L, Sonigo P (1986) Genetic variability of the AIDS virus: Nucleotide sequence analysis of two isolates from African patients. Cell 46:63–74

22. Guyader M, Emerman M, Sonigo P, Clavel F, Montagnier L, Alizon M (1987) Genome organization and transactivation of the human immunodeficiency virus type 2. Nature 326:662–669

23. Gnann JW, Nelson JA, Oldstone MBA (1987) Fine mapping of an immuno-dominant domain in the transmembrane glycoprotein of human immunodeficiency virus. J Virol 61:2639–2641

24. Banapour B, Rosenthal K, Rabin L, Sharma V, Young L, Fernandez J, Engelman E, McGrath M, Reyes G, Lifson J (1987) Characterization and epitope mapping of a human monoclonal antibody reactive with the envelope glycoprotein of human immunodeficiency virus. J Immunol 139:4027–4033

25. Schrier RD, Gnann JW, Langlois AJ, Shriver K, Nelson JA, Oldstone MBA

(1988) B and T lymphocyte responses to an immunodominant epitope of human immunodeficiency virus. J Virol 62:2531–2536

26. Gnann JW, McCormick JB, Mitchell S, Nelson JA, Oldstone MBA (1987) Synthetic peptide immunoassay distinguishes HIV type 1 and type 2 infections. Science 237:1346–1349

CHAPTER 38
Control of Hepatitis B by Vaccination

COLIN R. HOWARD

Hepatitis B infection of man is a global public health problem, there being an estimated 280 million asymptomatic carriers worldwide. This reservoir of infection is maintained largely by transmission of the hepatitis B virus (HBV) to infants born to carrier mothers. There are thus two different requirements of a hepatitis B vaccine: first, the prevention of chronic persistent infections that occur most frequently in young children; and second, the prevention of clinical disease in populations at high risk of coming into contact with HBV carriers. In many parts of the developed world, for example, Europe and the United States, the carrier rate is less than 1% of the adult population. Here the emphasis is on the availability of a vaccine for the protection of groups and individuals at high risk of contracting an acute infection by virtue of employment or close contact with known hepatitis B carriers. Vaccination of health care workers who may be exposed to contaminated blood or body fluids, hemophiliacs, and others treated with blood products is highly recommended (Table 38.1). Transmission of HBV is by parenteral routes, with no evidence of aerosol spread.

In contrast, acute hepatitis is rarely seen in children infected with HBV; in endemic areas the carrier rate may exceed 10% of the adult population, and as a result of continued HBV genome presence within infected hepatocytes, carriers are predisposed to the development of hepatocellular carcinoma (HCC) later in life [1]. Only by the application of mass immunization programs within these regions will the numbers of asymptomatic carriers be controlled and, subsequently, the mortality rate associated with HCC.

Table 38.1. Recommended groups for hepatitis B immunization.

1. Health care personnel at high risk of contact with contaminated blood, blood products, and body fluids
2. Patients on hemodialysis and in closed institutions
3. Contacts of hepatitis B patients
4. Infants born to mothers infected with HBV
5. Staff of immigration centers
6. Male homosexuals and prostitutes
7. Selected groups at lower risk, for example, police officers and paramedical staff

Immunogenic Proteins

Hepatitis B infection of the liver is characterized by the production of excess envelope protein assembled into 22-nm surface antigen-positive (HBsAg) particles. The major constituent is the S protein (MW 25,400) (gp25) and its glycosylated form (gp30) cross-linked via disulfide bonds [2]. Antibody to the a determinants on these particles confers protection, whereas antibodies to the subtype determinants d, y, w, and r do not. In recent years considerable interest has been generated by the discovery of additional antigenic specificities expressed as HBV transcripts initiated upstream to the S gene. These pre-S determinants are present in a much higher proportion on the surface of 42-nm HBV particles coating the HBcAg-reactive nucleocapsid. Several studies comparing human antibody responses to vaccines with and without pre-$S2$ antigens have led to the suggestion that pre-$S2$ antibodies are neutralizing antibodies [3]. However, the collective experience since 1976 in over 15 million recipients of vaccine containing S protein alone has clearly demonstrated the efficacy of these products in healthy individuals.

The release of 22-nm HBsAg particles into the circulation of infected persons has allowed both the development of sensitive serological tests for diagnosis and the use of plasma from asymptomatic carriers as a source of immunogen. These "first-generation" vaccines are prepared by the initial separation of HBsAg particles from virions present in plasma and then by rigorous chemical treatment to ensure the biological safety of the final product. The 10 or more commercially available vaccines differ in the extent and nature of this treatment. In one instance, this includes disruption of the 22-nm particle form with 8 M urea followed by reaggregation and exposure to pepsin. However, such treatment removes pre-S antigens, and this may reduce vaccine potency in certain target populations. Formaldehyde inactivation is a final step common to all the first-generation, plasma-derived products. Although the plasma is obtained from the same group of persons at risk of acquiring HIV, there is no evidence of AIDS ever having developed in recipients of these plasma-derived products. The efficacy and safety of these vaccines are now universally accepted.

Immune Responses in Vaccinees

Protection against HBV is correlated, at least in part, with the appearance of antibody to the outer surface antigen of the virus (anti-HBs) at a level of 10 mIU/ml or greater. This value is based on long-term, follow-up studies of homosexual men immunized with three doses of a plasma-derived HBsAg product [4]. The desirability of a minimum level of antibody is reinforced by epidemiological data showing that prevention of asymptomatic infections is equally important, since the chronic persistent carrier state develops more often from anicteric infections. The predominant subclass of antibody in man is IgG1, and this subclass shows the highest affinity for HBsAg [5].

The classic study of Szmuness et al. [6] reported that 96% of a large group of male homosexuals developed a high titer of antibodies upon completion of the immunization regime, with a notably reduced attack rate of HBV infection 18 months later. Subsequent studies have shown that seroconversion rates vary between 85% and 100% in healthy adults, with a decline in response with increase in age. Weight also appears to be a critical factor, with diminished responses in obese individuals under the age of 30. There is no evidence that those who do not respond are more likely to develop a chronic infection upon subsequent infection. The longevity of the anti-HBs response is related to the level of antibodies present in the serum 4 to 6 weeks after the final vaccine dose with circulating antibody found 3 years or more in successfully vaccinated persons. There is a marked increase in antibody titer after the final dose, compared to preceding doses [6]. Completion of the vaccination regime is thus critical in order to confer full immunity, particularly as there is evidence that 1% to 3% of individuals at high risk show serological evidence of asymptomatic HBV infection during the course of immunization. Attempts to accelerate B cell immunity involve the use of an additional third dose of vaccine at around 12 weeks, followed by a booster 6 to 12 months later. The need for a 6-month course of immunization is clearly disadvantageous for the protection of nonimmune persons newly introduced to a high risk environment. Immunization with HBsAg of one subtype appears to protect immunized adults against heterologous subtypes, reinforcing the concept that antibodies directed against the *a* determinants are important in protection [7].

Nonresponsiveness to hepatitis B vaccines is considerably higher in patients with acquired or inherited immunodeficiency states. In particular, the seroconversion rates in patients on maintenance dialysis is reduced to below 63% with many showing anti-HBs titers less than the desired

10 mIU/ml level [8]. However, this may be improved considerably by the use of vaccine containing additional pre-*S* antigenic specificities [3]. Similarly, individuals infected with HIV respond equally poorly to the vaccine. There is some suggestion that individuals with HLA-D7, who lack D1, respond less well than others [9]. Few data are available, however, as to the nature of the deficiency in nonresponding vaccine recipients, and indeed the role of cellular immunity in generating both B and T cell protective responses is not clear. Approximately one-third of such individuals do respond to a later booster dose, albeit with only low levels of antibody. This is in marked contrast to individuals who respond to the initial vaccination; they also respond promptly to revaccination with anti-HBs levels in excess of those obtained by primary immunization [10]. This boosting effect is observed even in persons whose antibody level has dropped below the level of detection for several years, demonstrating the presence of immunological memory to the vaccine over this period and indicating the likelihood of protection upon subsequent exposure to the virus.

Infants born to carrier mothers are at high risk of HBV infection during the perinatal period, particulaly if the mother is positive for HBeAg, a serological marker of elevated HBV activity. Such infants have a greater than 90% probability of infection, and more than 85% of these infections become persistent. Passive immunization with hepatitis B–specific immunoglobulin merely delays the onset of infection to later in the first year of life. Young infants respond well to hepatitis B vaccines, even if they are administered immediately after birth. Seroconversion rates approach 100% regardless of vaccine type. The incidence of chronic infections in these immunized infants drops to below 20%, particularly if the vaccine is used in combination with a specific immunoglobulin [11]. The reason for the apparent failure of prophylaxis in one-fifth of this group is not clear, but most likely these infants were incubating the infection prior to immunization, possibly having acquired the infection *in utero*. There is the further possibility in a few cases that heterologous immunity is ineffective in infants [11]. There is therefore a case that vaccines designed for mass infant immunization should contain a mixture of the relevant subtypes endemic in the region in which it is to be used.

Experimental work in animals has indicated that the potency of hepatitis B vaccines may be improved by inclusion of pre-S antigens in addition to the major *S* protein of HBsAg. Although such data show that mouse strains normally unresponsive to S determinants produce anti-*a* antibodies if given HBsAg containing pre-S determinants, there is as yet little convincing evidence that this improvement can be directly elicited in humans. Indeed, similar results may be obtained by the use of *S* protein alone, reformulated into immune-stimulating complexes (iscoms), suggesting that antigen presentation is more important than inclusion of pre-*S* determinants *per se* (C.R. Howard, unpublished observations). The exception is the use of pre-S2-containing vaccine in hemodialysis patients [8].

Newer Vaccine Products

Cloning of the HBV genome has allowed the inclusion of the S gene in a number of expression systems (Table 38.2). In particular, the cloned S gene has been introduced into yeast (*Saccharomyces cerevisiae*, Figure 38.1), and the expressed HBsAg particles were shown to elicit anti-HBs antibodies in man indistinguishable in specificity from those induced by the plasma-derived products [12]. Unlike plasma-derived HBsAg, this product is not glycosylated, which suggests that the carbohydrate moiety does not play a significant role in immunogenicity. The vaccine is as efficacious as the plasma-derived products, although the addition of pre-S antigens is proving difficult because of the general instability of the pre-S antigens thought to be located externally on the virus surface [13].

An alternative approach is the insertion of the cloned S gene into a live virus vector. The expression of the S gene has been achieved in both vaccinia [14,15] and adenovirus 5 [16]. In addition to the general reservations concerning the reintroduction of vaccinia into vaccination programs, laboratory studies show that the level of immunity induced in chimpanzees does not always give complete protection [15]. There is also concern as to

Table 38.2. Current status of hepatitis B vaccines.

Vaccine	Antigenic composition	Status
First generation		
Plasma-derived	S only, S + pre-$S2$	Licenced
Second generation		
1. Recombinant products, expressed in eukaryote cells		
Yeast-derived	S only	Licenced
Mammalian-cell-derived	S only	Experimental
2. Recombinant products, expressed in live virus vectors		
Vaccinia	S only	Clinical trials
Adenovirus 5	S only, S + pre-S	Experimental
3. Reformulations of existing materials for immune enhancement		
Micelles	S only	Clinical trials
Iscoms	S only	Experimental
Third generation		
1. Synthetic peptides	Selected epitopes of S or pre-$S2$	Experimental
2. Anti-idiotype antibodies	Internal images of a epitopes	Experimental

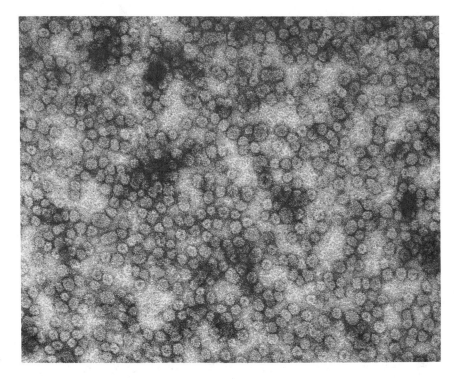

Figure 38.1. Electron micrograph of recombinant (yeast-derived) hepatitis B vaccine (kindly provided by Dr. P.R. Young).

the potential altered virulence for humans of these vectors, and markers of attenuation are urgently required, together with further modification to increase the general level of expression of the inserted *S* gene. Adenovirus is similar in that comparatively large quantities of foreign information can be introduced without impairing the ability of the vector to replicate. In contrast to vaccinia, the use of adenovirus is attractive, since it may be administered orally.

Two further approaches to hepatitis B vaccination may be mentioned. These are the use of anti-idiotype antibodies bearing internal images of the relevant *a* epitopes and synthetic peptides mimicking similar short amino-acid sequences. Both offer the opportunity to use molecules that represent single, defined epitopes able to substitute for antigen (i.e., surrogate antigens). Although anti-idiotypes may stimulate B cells of higher affinity to HBsAg, these antibodies elicit a more restricted range of B cell reactivities compared to synthetic peptides. Such a surrogate hepatitis B vaccine would consist of either a single peptide containing a number of B cell epitopes together with appropriate T helper cell cites or a cluster of monoclonal internal image anti-idiotypes [17].

Other Considerations

There is evidence for the existence of a second serotype of hepatitis B, HBV2 [18]. Immunity to HBV induced by vaccination does not protect against HBV2, although HBsAg particles of both viruses have common antigenic determinants. However, HBV2 appears to contain a serologically distinct nucleocapsid reminiscent of the hepatitis delta virus. Although the epidemiological significance of HBV2 infections has yet to be determined, the message is clear. Existing hepatitis B vaccines may have a more limited potential in protecting certain populations against hepatitis B-like disease than has hitherto been recognized. It may therefore prove desirable for vaccines in the future to have a wider antigenic repertoire than they have at present. The use of chimeric molecules containing both HBsAg and nucleocapsid antigens could be used to exploit the strong T helper cell stimulatory properties of HBcAg sequences [19], although the role of anti-HBc antibodies in disease prevention needs careful evaluation.

References

1. Zukerman AJ (1982) Primary hepatocellular carcinoma and hepatitis B virus. Trans R Soc Trop Med Hyg 76:711–718
2. Howard CR (1986) The biology of hepadnaviruses. J Gen Virol 67:1215:1235
3. Neurath AR, Kent SBH, Parker K, Prince AM, Strick N, Brotman B, Sproul P (1986) Antibodies to a synthetic peptide from the preS 120-145 region of the hepatitis B virus envelope are virus-neutralizing. Vaccine 4:35–37
4. Hadler SC, Francis DP, Maynard JE, Thompson SE, Judson FN, Echenberg DF, Ostrow DG, O'Malley PM, Penley KA, Altman NL, Braff E, Shipman GF, Coleman PJ, Mandel EJ (1986) Long-term immunogenicity and efficacy of hepatitis B vaccine in homosexual men. New Engl J Med 315:209–214
5. Persson MAA, Brown SE, Steward MW, Hammarstrom L, Smith CIE, Howard CR, Wahl M, Rynnel-Dagoo B, Lefranc G, Carbonara AO (1988) IgG subclass-associated affinity differences of specific antibodies in humans. J Immunol 140:3875–3879
6. Szmuness W, Stevens CE, Harley EJ, Zang EA, Olesko WR, William DC, Sadovsky R, Morrison JM, Kellner A (1980) Hepatitis B vaccine: Demonstration of efficacy in a controlled clinical trial in a high risk population in the United States. New Engl J Med 303:833–841
7. Szmuness W, Stevens C, Harley E, Zang E, Alter H, Taylor P, DeVera A, Chen T, Kellner A (1982) Hepatitis B vaccine in medical staff of hemodialysis units—Efficacy and subtype cross-protection. New Engl J Med 307:1481–1486
8. Stevens CE, Alter HJ, Taylor PE, Zang EA, Harley EJ, Szmuness W. (1984) Hepatitis B vaccine in patients receiving hemodialysis. New Engl J Med 311:496–501
9. Dienstag JL (1984) Immunologic mechanisms in chronic hepatitis. *In* Vyas GN, Dienstag JL, Hoofnagle JH (eds) Viral Hepatitis and Liver Disease. Grune and Stratton, New York, pp 135–166

10. Jilg W, Schmidt M, Deinhardt F (1988) Immune response to hepatitis B revaccination. J Med Virol 24:377–384
11. Stevens CE, Taylor PE, Tong MJ, Toy PT, Vyas GN (1984) Hepatitis B vaccine: An overview. *In* Vyas GN, Dienstag JL, Hoofnagle JH (eds) Viral Hepatitis and Liver Disease. Grune and Stratton, New York, pp 275–291
12. Brown SE, Stanley C, Howard CR, Zuckerman AJ, Steward MW (1986) Antibody responses to recombinant and plasma derived hepatitis B vaccines. Brit Med J 292:159–161
13. Chen S-H, Howard CR (1988) Properties and stability of the pre-S region of hepatitis B virus. *In* Zuckerman AJ (ed) Viral Hepatitis and Liver Disease. Alan R. Liss, New York, pp 622–626
14. Paoletti E, Lipinskas BR, Samsonoff C, Mercer S, Panicali D (1984) Construction of live vaccines using genetically engineered poxviruses: Biological activity of vaccinia virus recombinants expressing the hepatitis B virus surface antigen and the herpes simplex virus glycoprotein D. Proc Natl Acad Sci USA 81:193–197
15. Smith GL, Mackett M, Moss B (1983) Infectious vaccinia virus recombinants that express hepatitis B virus surface antigen. Nature 302:490–495
16. Davies AR, Kostek B, Mason BB, Hsiao CL, Morin J, Dheer SK, Hung PP (1985) Expression of hepatitis B surface antigen with a recombinant adenovirus. Proc Natl Acad Sci USA 82:7560–7564
17. Thanavala YM, Brown SE, Howard CR, Roitt IM, Steward MW (1986) A surrogate hepatitis B virus antigenic epitope represented by a synthetic peptide and an internal image antiidiotype antibody. J Exp Med 164:227–236
18. Coursaget P, Yvonnet B, Bourdil C, Melvelec MN, Adamowicz P, Barres JL, Chotard J, N'Doye R, Diop Mar I, Chiron JP (1987) HBsAg positive reactivity in man not due to hepatitis B virus. Lancet ii:1354–1358
19. Clarke BE, Newton SE, Carroll AR, Francis MJ, Appleyard G, Syred AD, Highfield PE, Rowlands DJ, Brown F (1987) Improved immunogenicity of a peptide epitope after fusion to hepatitis B core protein. Nature 330:381–384

CHAPTER 39
Antiviral Compounds Bind to a Specific Site Within Human Rhinovirus

FRANK J. DUTKO, MARK A. MCKINLAY, AND MICHAEL G. ROSSMANN

The approximately 100 serotypes of human rhinovirus (HRV) [1], the most commonly isolated virus from those suffering from an upper respiratory infection (common cold), represent a formidable therapeutic challenge [2]. A number of structurally unrelated compounds (Figure 39.1) have been shown to inhibit HRV uncoating (i.e., the release of viral RNA into the cytosol) as a result of site-specific binding to virions. Recently, X-ray crystallographic analysis of disoxaril, [5-[7-[4-(4,5-dihydro-2-oxazolyl)phenoxy]heptyl]-3-methylisoxazole]=HRV(type 14) complexes has identified the specific binding site in the viral capsid [3,4]. In this chapter the molecular details and biological consequences of the binding of several antiviral compounds to HRV-14 are discussed.

Molecular Details of Binding

Picornavirus Structure and Function

The three-dimensional structures of HRV-14 [5], poliovirus type 2 [6], and mengovirus [7] have been solved using X-ray crystallography. The protein capsid of picornaviruses has a radius of approximately 300 Å and consists of 60 protomers in an icosahedral arrangement. Each protomer consists of one copy of each of the four virion proteins (i.e., VP1, VP2, VP3, and VP4).

One important function of the viral capsid is to prevent degradation of viral RNA in extracellular virus and to facilitate the early stages of virus replication, such as adsorption, penetration, and uncoating. Picornaviruses utilize the normal cellular process of endocytosis in order to enter cells and uncoat. For example, poliovirus adsorbs to specific receptors on the cell

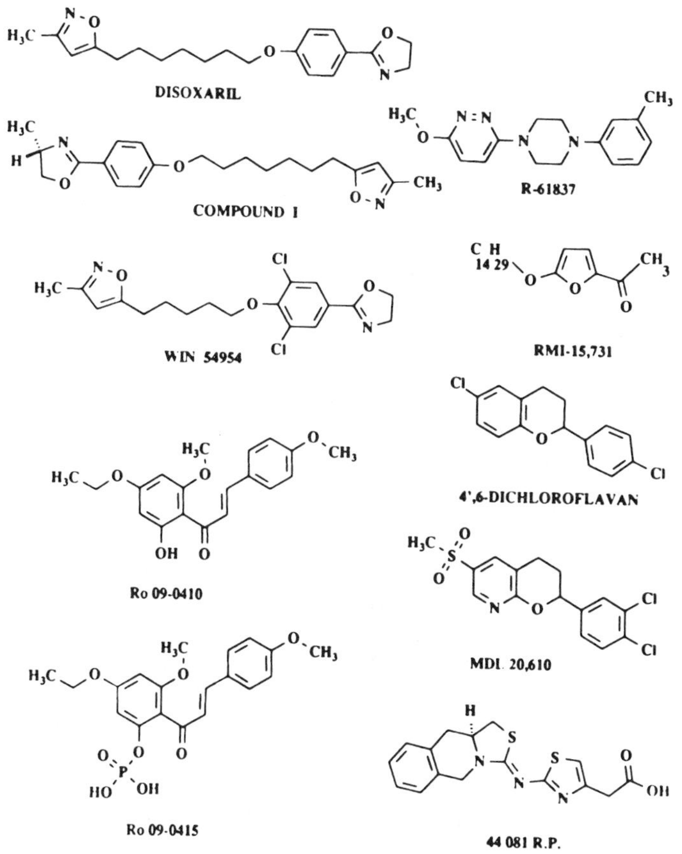

Figure 39.1. Chemical structures of capsid-binding compounds.

surface and is endocytosed by clathrin-coated pits and vesicles [8]. Acidification of endosomes and lysosomes is a key to the uncoating process because agents that raise the pH of endosomes and lysosomes inhibit the uncoating of viruses. The final result of the uncoating process is the translocation of the viral RNA through the capsid and the endosome or lysosome vesicle into the cytosol, where the viral RNA can then function as a viral mRNA.

Parameters of Binding

Disoxaril and several structurally related antiviral compounds bind within a hydrophobic pocket in the eight-stranded, anti-parallel β barrel in viral protein 1 (VP1) of HRV-14 (Figure 39.2). The interactions between the compounds and the viral protein are mostly hydrophobic. The compounds displace several water molecules from the hydrophobic pocket and alter the locations of other water molecules. Hydrophobic bonds occur between the

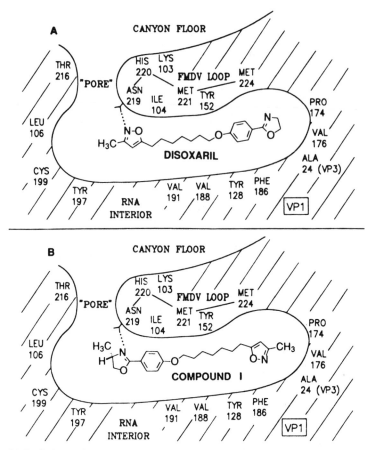

Figure 39.2. Schematic representation of the hydrophobic binding pocket of human rhinovirus type 14 within 3.6 Å of **A,** disoxaril, or **B,** compound I.

phenyl ring of compounds and tyrosine and valine side chains of the virion. The binding of the eight compounds examined to date to HRV-14 induce similar significant conformational changes in the virion structure [4]. In the most extreme case, the C_α positions of the residues from isoleucine (VP1; residue number 215) to methionine (VP1; residue number 224) move by distances of up to 4.5 Å [3,4]. The similar conformational changes are independent of the orientation of the compounds within the pocket, which may differ by 180° (i.e., the methylisoxazole and phenyl-oxazoline are flipped), or the antiviral potency of the compounds, which differ by 100-fold.

Despite the different orientations of the eight compounds within HRV-14, a hydrogen bond between the virion and the compounds appears to be conserved with all compounds. This hydrogen bond involves asparagine (Asn) (VP1; residue number 219) of HRV-14 and either the nitrogen atom of the methylisoxazole group of disoxaril or the nitrogen atom of the oxazoline group of compound 1 (Figure 39.2) [4]. Because this Asn is conserved in the amino-acid alignments of several HRV serotypes for which sequence data are available [9], this probable hydrogen bond could be an important factor in the binding of antiviral compounds to HRV-14.

Enantiomeric effects have been observed with homologs of disoxaril [10]. When substituents were introduced into the 4-position of the oxazoline group, the *S* isomers were approximately 10-fold more active against HRV-14 in plaque-reduction tests than the *R* isomers. Furthermore, X-ray crystallographic studies of a racemic mixture indicated selective binding of the *S* isomer. The enatiomeric effect on antiviral activity can be explained by energetically favored hydrophobic interactions of the *S*-substituent with a pocket formed by leucine (Leu) (VP1; residue number 106) and serine (VP1; residue number 107) [10].

Relationship Between Capsid-Binding Compounds

X-ray crystallographic studies have not been performed to demonstrate that the other classes of antiviral compounds in Figure 39.1 can bind to the same hydrophobic pocket in HRV-14 as disoxaril and its homologs. However, experiments using resistant viruses derived with one compound have demonstrated cross-resistance to other compounds. Approximately 50 viruses resistant to disoxaril or compound I have been sequenced by B. Heinz and R. Rueckert (University of Wisconsin). Viruses that were highly resistant to compound I had single amino-acid substitutions in only two positions: cysteine (VP1; residue number 199) or valine (VP1; residue number 188). One of these resistant viruses that contains Leu at Position 188 of VP1 has been well characterized. This virus was resistant to disoxaril or compound I as well as to Ro 09-0410 in plaque-reduction tests. In addition, rhinoviruses resistant to RMI 15,731 and dichloroflavan were cross-resistant to Ro 09-0410 [11]. Furthermore, the binding of radioactive Ro 09-0410 to HRV-2 was inhibited by unlabeled Ro 09-0410, as well as by dichloroflavan or by RMI 15,731 [12]. These results demonstrate that the binding site for several antiviral compounds within HRV is similar.

Biological Consequences of Binding

Mode of Action of Capsid-Binding Compounds

Several studies have demonstrated that the biological consequence of the binding of these compounds to virions is the prevention of uncoating (i.e., the release of viral RNA to the cytosol) [13–15]. As mentioned before, the uncoating process is pH mediated and must involve the unfolding and disassembly of virions in order to release RNA. Our hypothesis is that the compounds, through the hydrophobic interactions with amino-acid side chains in the pocket, act as "molecular glue" by decreasing the protein flexibility and therefore hold the virions together. This hypothesis is supported by studies in which polioviruses were protected from heat inactivation by disoxaril [13].

In Vivo Activity

Oral administration of antiviral, capsid-binding compounds has been shown to protect mice from challenge with enteroviruses (members of Picornavirus genus). Enterovirus models of infection are used to investigate the *in vivo* efficacy of compounds because HRVs can only infect primates and humans. The phenoxypyridinecarbonitrile and a related compound, 2-(3,4-dichloro-phenoxy)-5-(methylsulfonyl)pyridine, have been shown to protect mice from coxsackievirus A-21 infection [16,17]. Furthermore, disoxaril and a related compound, WIN 54954, are orally active in animal models of poliovirus, coxsackievirus A-9, and echovirus 9 infection [18]. These results demonstrate that compounds that inhibit the uncoating of picornaviruses *in vitro* are orally effective as antiviral agents in animal models of virus infection.

The potential clinical efficacy of the capsid-binding compounds has been realized with R-61837 [19]. In contrast to the ineffectiveness of dichlorofla-van (oral or intranasal), Ro 09-0410/Ro 09-0415 (oral or intranasal), RMI 15,731 (intranasal), and 44 081 R.P. (intranasal), intranasal administration of R-61837 prophylactically to volunteers challenged with a very sensitive HRV serotype reduced symptoms and nasal mucus weights [20]. These positive clinical findings may open up a new era of therapy of the rhinovirus common cold with capsid-binding drugs whose binding site is known at the atomic level.

Acknowledgments

We gratefully acknowledge the significant contributions of Guy Diana, Tom Bailey, Adi Treasurywala, Daniel Pevear, Dorothy Young, Martin Seipel, Pat Fox, Marilyn Fancher, Maureen Woods, Don Baright, Peter Furlani, Peter Felock, Wendy Shave, LoAnne Visosky, Alix Ehrenberg, and Valerie Csontos at Sterling-Winthrop Research Institute; of Tom Smith, John Badger, Sangsoo Kim, Ming Luo, Marcia Kremer, and Iwona Minor at Purdue University; of Beverly Heinz and Roland Rueckert at the University of Wisconsin; and of Heinz Zeichhardt at the Freie University in Berlin, West Germany.

References

1. A Collaborative Report (1987) Rhinoviruses—Extension of the numbering system from 89 to 100. Virology 159:191–192
2. Sperber SJ, Hayden FG (1988) Chemotherapy of rhinovirus colds. Antimicrob Agents Chemother 32:409–419

3. Smith TJ, Kremer MJ, Luo M, Vriend G, Arnold E, Kamer G, Rossmann MG, McKinlay MA, Diana GD, Otto MJ (1986) The site of attachment in human rhinovirus 14 for antiviral agents that inhibit uncoating. Science 233:1286–1293

4. Badger J, Minor I, Kremer MJ, Oliveira MA, Smith TJ, Griffith JP, Guerin DMA, Krishnaswamy S, Luo M, Rossmann MG, McKinlay MA, Diana GD, Dutko FJ, Fancher M, Rueckert RR, Heinz BA (1988) Structural analysis of a series of antiviral agents complexed with human rhinovirus 14. Proc. Natl Acad Sci USA 85:3304–3308

5. Rossmann MG, Arnold E, Erickson JW, Frankenberger EA, Griffith JP, Hecht H-J, Johnson JE, Kamer G, Luo M, Mosser AG, Rueckert RR, Sherry B, Vriend G (1985) Structure of a human common cold virus and functional relationship to other picornaviruses. Nature 317:145–153

6. Hogle JM, Chow M, Filman DJ (1985) Three-dimensional structure of poliovirus at 2.9 Å resolution. Science 229:1358–1365

7. Luo M, Vriend G, Kamer G, Minor I, Arnold E, Rossmann MG, Boege U, Scraba DG, Duke GM, Palmenberg AC (1987) The structure of Mengo virus at atomic resolution. Science 235:182–191

8. Zeichhardt H, Wetz K, Willingmann P, Habermehl K-O (1985) Entry of poliovirus type 1 and mouse Elberfeld (ME) virus into HEp-2 Cells: Receptor-mediated endocytosis and endosomal or lysosomal uncoating. J Gen Virol 66:483–492

9. Palmenberg AC (1988) Sequence alignments of picornaviral capsid proteins. International Conference on Molecular Biology of Picornaviruses, ICN–UCI, Abstr, Jan 1988

10. Diana GD, Otto MJ, Treasurywala AM, McKinlay MA, Oglesby RC, Maliski EG, Rossmann MG, Smith TJ (1988) Enantiomeric effects of homologues of disoxaril on the inhibitory activity against human rhinovirus-14. J Med Chem 31:540–544

11. Ishitsuka H, Ninomiya Y, Suhara Y (1986) Molecular basis of drug resistance to new antirhinovirus agents. J Antimicrob Chemother 18 (suppl B):11–18

12. Ninomiya Y, Aoyama M, Umeda I, Suhara Y, Ishitsuka H (1985) Comparative studies on the modes of action of the antirhinovirus agents Ro 09-0410, Ro 09-0179, RMI-15,731 4',6-dichloroflavan, and enviroxime. Antimicrob Agents Chemother 27:595–599

13. Fox MP, Otto MJ, McKinlay MA (1986) Prevention of rhinovirus and poliovirus uncoating by WIN 51711, a new antiviral drug. Antimicrob Agents Chemother 30:110–116

14. Alarcon B, Zerial A, Dupiol C, Carrasco L (1986) Antirhinovirus compound 44 081 R.P. inhibits virus uncoating. Antimicrob Agents Chemother 30:31–34

15. Zeichhardt H, Otto MJ, McKinlay MA, Willingmann P, Habermehl K-O (1987) Inhibition of poliovirus uncoating by disoxaril (WIN 51711). Virology 160:281–285

16. Kenny MT, Dulworth JK, Torney HL (1985) In vitro and in vivo antipicornavirus activity of some phenoxypyridinecarbonitriles. Antimicrob Agents Chemother 28:745–750

17. Kenny MT, Dulworth JK, Torney HL (1986) In vitro and in vivo antipicornavirus activity of some P-benzoyl phenoxypyridines. Antiviral Res 6:355–367

18. Otto MJ, Diana GD, McKinlay MA, Felock P, Fancher M (1987) In vitro antipicornavirus activity of WIN 54954, a new analog of disoxaril with improved

spectrum and potency. Twenty-seventh Interscience Conference on Antimicrobial Agents and Chemotherapy, New York, NY, Abstr No 491.

19. Al-Nakib W, Tyrrell DAJ (1987) A "new" generation of more potent synthetic antirhinovirus compounds: Comparison of their MICs and their synergistic interactions. Antiviral Res 8:179–188.

20. Al-Nakib W, Higgins PG, Tyrrell DAJ, Barrow IG, Taylor N, Andries K (1987) Tolerance and prophylactic efficacy of a new antirhinovirus compound, R61837. Seventh International Congress of Virology, Edmonton, Alberta, Canada, Abstr No 32.3

CHAPTER 40
HIV: Antiviral Chemotherapy

SAMUEL BRODER

The term *retrovirus* is used to denote a class of animal and human viruses that employ a special DNA polymerase (reverse transcriptase) to catalyze the conversion of genomic RNA to DNA in the process of viral replication. These viruses are interesting in that a crucial step in their replication is the reverse transcription of viral RNA to produce linear, double-stranded (minus strand and plus strand) proviral DNA. In this process the termini of the RNA genome are duplicated to yield structures in the DNA called *long-terminal repeats* (LTRs). The viral genome is expressed from proviral DNA (the minus strand of DNA), which has enhancer/promoter and polyadenylation signals in the LTRs.

The purpose of this chapter is to summarize possible therapeutic interventions that could be based on the emerging knowledge of the life cycle of the human immunodeficiency virus (HIV), the retrovirus that causes acquired immunodeficiency syndrome (AIDS) and its related disorders [1,2]. The virus has eight known genes. In addition, it has been suggested the DNA plus strand of HIV-1 encodes a protein during viral infection because there is a previously unidentified open reading frame (ORF) in a region complementary to the envelope sequence [3]. In principle, this ORF could encode a 20-kd, highly hydrophobic protein. This HIV replication requires a complex sequence of steps, each one of which might provide an opportunity for therapeutic intervention. It has already been proved that drugs that inhibit reverse transcriptase can suppress HIV replication in vitro and in vivo [4]. Indeed, a broad family of nucleoside analogs, called dideoxynucleosides, exhibits antiretroviral activity [5]. One member of the dideoxynucleoside family, 3'-azido-2', 3'-dideoxythymidine (AZT) [6], has been shown to prolong survival and to improve the clinical status of patients with advanced

AIDS [7,8]. These observations make it likely that intervening in one or more steps in the viral life cycle will have a major impact on the morbidity and morality associated with HIV infection.

Background

AIDS was initially defined as the development of either an opportunistic infection or Kaposi's sarcoma (an unusual neoplasm that had previously been recognized to be associated with certain immunosuppressed states) in a person without a known cause for immunodeficiency [9]. It was also recognized that such patients were at risk for certain lymphomas. Shortly thereafter it became apparent that these patients had a cellular immunodeficiency characterized by an inexorable depletion of helper/inducer (T4$^+$ or CD4$^+$ T cells, and within 3 years it had been shown that HIV has the capacity to replicate within cells of the immune system, leading to the profound destruction of CD4$^+$ T cells, which results in AIDS. In addition to its capacity to damage the immune system, the virus can enter the central nervous system (CNS) and induce a spectrum of neurological disturbances [10,11]. Central nervous system involvement is especially common in children with HIV infection.

Although it is still not known exactly how the virus enters the brain and how it brings about neurological damage, it is a virtual certainty that successful therapeutic strategies must address the consequences of, and probably prevent, viral replication within the CNS. Fortunately, the prevailing assumption that AIDS-virus-induced dementias were uniformly irreversible has been disproved with the advent of antiretroviral chemotherapy [12].

As with any virus, the different stages in the life cycle of HIV present a variety of distinct potential targets for antiviral agents [4]. Reverse transcriptase is one of the most attractive targets, and there have been successes at a clinical level using this as a target for new therapies, notably with AZT [7,8,12]. The testing of new antiretroviral agents has been facilitated by the availability of rapid and sensitive in vitro screening systems that determine whether a putative drug can inhibit the replication and T cell killing activity of HIV [3]. For certain nucleoside analogs, a great deal of knowledge in terms of structure/activity relationships has now emerged. At the outset it is worth stressing that effective therapy of HIV infections may well depend on a combination of therapeutic strategies without relying on any one single agent, in part because the emergence of drug-resistant strains might be less likely. This is because once control of the etiological agent is established, it may be possible to introduce immunopotentiative and adoptive cellular therapy to restore the immune system.

HIV as a Retrovirus

In the era before the etiology of AIDS was recognized, one of the most notable features of retroviruses was their capacity to induce neoplastic transformation in infected target cells; hence, the expressions "RNA tumor virus" or "leukemia virus" appeared in the literature to denote this category of virus. But the virus that causes AIDS has not yet been shown to have a transforming capacity, and it appears to cause certain cancers through a viral derangement of immune surveillance or abnormal tissue growth factor production or both, as already discussed. By definition, retroviruses replicate through a DNA intermediate (i.e., at one step of their cycle of replication, genetic information flows from RNA to DNA, a reverse or "retro" direction) [14,15]. Before human pathogenic retroviruses were recognized, retroviruses were known to contain a standard set of genes called *gag, pol,* and *env* as basic components of a replicating genome. Reverse transcriptase (the viral DNA polymerase) that catalyzes this step is encoded by the *pol* gene of the virus, a gene that is conserved to a considerable extent in its amino-acid and nucleotide sequences among all retroviruses. The reverse transcriptase gene is one component of the *pol* region, and in general, this region is expressed as a large molecular weight species that includes a protease, a reverse transcriptase, RNase H, and an endonuclease (integrase). Posttranslational cleavage is thought to yield appropriate peptide subunits as functioning molecules (vide infra).

In common with other known animal retroviruses, HIV has as its major structural components a core of genomic RNA; group-specific antigen (*gag*) proteins, which play a role both in the structure of the core and the assembly of the virion; a lipid bilayer; and an outside envelope glycoprotein. As the virus is expressed in a given cell, it manufactures a *gag–pol* fusion protein. This retroviral fusion protein (*gag–pol*) can then undergo posttranslational cleavage events to form active *gag* and *pol* products. Various retroviruses, including HIV, use an interesting ribosomal frame shift to translate the starting *gag–pol* polyprotein, and we will return to this point later.

The HIV is the most complex retrovirus yet characterized in that it contains at least eight genes, plus a ninth open reading in a novel location (as discussed earlier). The functions of several of these novel genes are still not known or are poorly understood. One such gene (designated *sor*) has very recently been linked to the ability to replicate by a pathway of cell-free virion infection [16]. The *sor* gene encodes a 23-kd protein that plays a crucial role in the efficient generation of infectious virus. The viruses with proviral genomes that contain mutant *sor* genes were quite limited in their capacity to establish stable infection in vitro. Therefore, in the future, it is conceivable that drugs or biologics could be developed to interfere with *sor* and thereby attenuate the pathogenicity of HIV infection either in individuals who are already infected or as part of an adjunct to an immunization program for individuals who are at risk for acquiring HIV infection. Another as yet poorly characterized gene is 3'-orf. It has recently been shown that the 3'-orf protein is a myristylated, GTP-binding phoshpoprotein with features similar to the cellular *src* and *ras* oncogene products [17]. How the 3'-orf gene influences the clinical features of HIV infection is not known. The functions of two other genes, *tat-III* and *art/trs*, will be discussed later.

Cell Binding and Entry

The first step in the infection of a cell by HIV is its binding to the target cell receptor. In the case of helper/inducer T cells, this receptor is linked to the cell-surface protein that is recognized by T4 or CD4 antibodies. The process of specific binding between the CD4 receptor and parts of the viral envelope glycoprotein may be suceptible to attack by antibodies either to the virus or to the receptor, and theoretically certain chemicals or small peptides could be designed to occupy the receptor and prevent binding. Indeed, it may be possible to use a genetically engineered soluble form of the CD4 protein as a therapeutic agent [18]. Such a construct would be expanded to provide false attachment sites for the virus and thus prevent normal viral attachment.

In addition to a role in the receptor-mediated entry of the virus into target T cells, the CD4 receptor plays a part in the susceptibility of a target T cell, once it begins to produce virus, to be killed by that virus. Precisely how HIV destroys T cells in vivo is not known. The cytopathic effect is thought to be mediated in part by an interaction between the CD4 molecule and the HIV envelope protein (gp120) that elicits lethal cell-to-cell fusions (syncytia) or a surface autofusion phenomenon that destroys the integrity of the cell membrane. (An initial fusion event is absolutely essential to virus infection.) But other factors are also thought to play a role in the cytopathic effects of the virus. Recent results raise the possibility that the carboxyl terminus of the envelope protein, a region different from the portion that directly binds to the CD4 molecule, is important in the destruction of T cells by the virus [19]. Conceivably, therapeutic agents could be designed to alter certain properties (e.g., the lipid or sugar composition) of the viral surface or target cell surface to reduce viral infectivity or cytopathic effects.

An alternative target is the envelope protein itself. Although there can be considerable variation in the protein from one viral isolate to another, the range of alterations in the binding site is most likely constrained by the need to bind to CD4, which is relatively constant in the structure. An antibody directed against this site might bind to (and neutralize) most strains of HIV and perhaps kill infected cells as they begin to express envelope antigens so that spread of virus to uninfected cells is reduced. In this context monoclonal antibodies to the envelope protein could have a therapeutic role in patients with AIDS or related diseases. We have recently been able to produce a complement-fixing human IgG_k monoclonal antibody against the major envelope glycoprotein of the first known pathogenic human T lymphotropic retrovirus (HTLV-1) [20], and similar approaches could be used to develop human monoclonal antibodies against HIV. A potential difficulty of this approach, however, is that virally infected cells could make infectious cell-to-cell contacts. (It is interesting to note that even viral mutants that are defective in *sor,* and are thereby poorly transmitted by cell-free virion infection, may still be transmitted by a process of cell-to-cell spread in vitro.) Antibodies might not gain access to relevant epitopes under certain circumstances. Also, it has been shown that AIDS can occur in the face of what in vitro appears to be neutralizing antibodies to HIV. Whether this occurs because the titers of such antibodies are low or because such antibodies do not block epitopes that mediate in vivo cytopathogenicity is under investigation. One must at least consider that humoral immunity per se does not protect the host and that the protection one sees is mediated by cellular immunity.

After binding to a cell, HIV enters the target cell by an incompletely defined fusion process mediated by specific sites in the transmembrane portion (gp41) of the envelope. It is conceivable that drugs or antibodies could be developed to block this step. Another theoretical target is the stage of "uncoating" of the virus after it enters a target cell. In this stage the virus loses its envelope coat, and RNA (most likely as a ribonucleoprotein) is released into the cytoplasm (each virion is thought to convey a dimer of two identical genomic RNA subunits into the cell). Pharmacological agents that block viral uncoating might eventually be developed to AIDS.

Uncoated viral RNA is used as a template for the production of DNA by reverse transcriptase. As already mentioned, this enzyme has become a prime target for antiretroviral agents both because it should be possible to find inhibitors that will discriminate between this viral DNA polymerase and the DNA polymerases of the host cell (thus lessening side effects) and because a great deal is already known about the *pol* gene and reverse transcriptase [14,15]. The HIV reverse transcriptase uses a lysine transfer RNA primer to make a minus strand DNA copy of the viral RNA as an RNA–DNA hybrid. Retroviral DNA polymerase (reverse transcriptase) possesses an inherent RNase H activity that specifically degrades the RNA of the RNA–DNA hybrid. The C-terminal region of the reverse transcriptase protein is believed to be a domain with RNase H activity. Inhibition of this process would suppress viral replication because an effective and orderly degradation of the viral RNA is a requirement of effective conversion of genomic RNA to proviral DNA. The reverse transcription then catalyzes the production of a plus strand of DNA. The resultant, double-stranded viral DNA can then migrate to the nucleus by an as yet poorly characterized mechanism.

The reverse transcriptase of HIV has been purified and seems to exist as a p51 and a p66 molecule. Large quantities of the HIV reverse transcriptase have become readily available because the enzyme has been expressed in bacteria and yeast using recombinant DNA technology. This should help in the testing of potential reverse transcriptase inhibitors. The use of such inhibiting agents in patients with AIDS is predicated on the assumption that there is continuing viral replication in the disease and that its inhibition will permit some regeneration, or at least prevent further deterioration of the immune system. Two drugs, suramin and HPA-23, that early on were chosen for clinical trials because they inhibit reverse transcriptase (and were already being given to human beings for other purposes), have not appeared to benefit patients clinically. However, only limited ranges of doses and schedules of administration have been tested. Suramin has certain inherent antitumor properties—perhaps in part related to its ability to induce circulating levels of sulfated glycosamino-glycans that are now being adapted to treat malignancies of various types.

Taking all factors into consideration, agents that inhibit reverse transcriptase still have the most likelihood of achieving an immediate clinical impact on the AIDS virus. Several potent agents that inhibit this enzyme are either already available, having been developed for the therapy of conventional viral diseases (e.g., phosphonoformate) or are being developed specifically for AIDS on the basis of in vitro screening systems for activity against HIV. We will return to one class of these inhibitors, the dideoxy-nucleosides, later. As already mentioned, one member of this class of nucleosides (AZT) has already been shown to prolong survival in patients with AIDS and to confer certain other clinical benefits as well.

Integration, Latency, and Reactivation

The proviral DNA apparently can either remain in an unintegrated form or can become integrated into the genome of the host cell. Unfortunately, the capacity of retroviruses to integrate into the genome of host target cells was initially thought to render retroviral diseases inherently untreatable. In the future it is possible that chemicals could be developed to interfere with the viral endonuclease, or "integrase" (thought to be a function of one of the *pol* gene products), that mediates this integration step. Later in the viral life cycle, perhaps after activation of the infected cell by such physiological signals as antigens or regulatory interleukins, the proviral DNA is transcribed to mRNA (and viral genomic RNA) by host cell RNA polymerases, and this RNA is then translated to form viral proteins, again by the biochemical apparatus of the host cell. Certain stimulants of HIV expression (e.g., phorbol esters and polyclonal lectins) appear to operate by NF–KB, a DNA-binding protein that influences transcriptional imitation by binding to sequences in the HIV enhancer [21]. Thus, in theory this provides a target for a future therapeutic strategy. In addition, proteins encoded by certain DNA viruses (e.g., herpes simplex and adenoviruses) can activate HIV expression. Thus, in theory the suppression of certain viruses found in patients with AIDS, but not directly linked to the immediate cause of the disease, might confer a clinical benefit by dampening HIV expression.

Retroviruses use a novel mechanism for the translation of certain genes. For example, they can synthesize a single *gag–pol* polyprotein from two separate reading frames on an RNA template. The coupling of these reading frames requires that the ribosome correctly shift from one reading frame to another, something mammalian cells are not thought to do as part of a physiological genetic translation process. It is theoretically possible that specific chemicals or antibiotic-like agents could interrupt this process of ribosomal frame shifting, leading to impaired viral expression, and thereby, improved clinical status in infected individuals.

It has recently been shown that HIV has a gene, *tat*, coding for a diffusible protein that markedly enhances the expression of other viral genes and viral replication [22]. The *tat* gene, like the *tax* genes of the first two known human pathogenic retroviruses, HTLV-I and HTLV-II (formerly called tat-I and tat-II genes, respectively), is so called because it was originally thought to mediate a *trans*-activation of transcription; that is, it worked through a mechanism that affected the transcription of genes not in direct proximity to itself. The *tat* protein seems to increase the viral products at a transcriptional or posttranscriptional level. Whatever the precise mechanisms, this protein is thought to provide the virus with a positive feedback loop by which a viral product can amplify the production of new virions. The *tat* protein is small (86 amino acids), with a cluster of positively charged amino acids, and is thought to affect the synthesis of viral products by influencing viral LTRs.

It may be possible to find drugs that inhibit the *tat* product itself or a crucial nucleic acid binding site for this protein or both. Using a different reading frame, the *art/trs* gene of HIV produces a different, small (116 amino acids), positively charged protein, which is thought to function as a second essential *trans*-acting factor in viral replication [23–25]. Again, drugs that bind or inactivate this protein would be expected to inhibit viral replication. At present, one can say that *art/trs encodes an essential viral protein, which is localized in the nucleus. The art/trs* gene complements *art/trs*-negative viral mutants in *trans*. The functional domain of this protein has been localized within 80 amino acids of its second exon. A *cis*-acting element required for *art/trs* activity is located in the *env* region of the HIV genome.

The *art/trs* gene regulates HIV at a posttranscriptional level to increase steady-state levels of unspliced viral mRNA in comparison to multiple spliced mRNA. Thus, the net effect of this viral gene is a redirection of unspliced viral mRNA away from splicing and into the cytoplasm; *art/trs* thus augments the stability of unspliced viral mRNA. The *art/trs* gene, in effect, lowers the concentration of multiple spliced viral mRNA. There are three known HIV genes that are expressed by a process of such multiple splicing: *tat, art/trs,* and *3'-orf*. Therefore, *art/trs* (now called *rev*) is an essential viral protein that down-regulates its own expression.

Drugs or chemicals that partially interfere with the function of the *trs* gene might block the formation of full-length, genomic viral RNA (9.2 kb) in infected cells, since these large molecules would be spliced out of existence. The 4-kb *env* gene transcripts might be affected to a lesser extent, whereas the viral genes that are expressed as smaller 2-kb transcripts (e.g., *tat*, 3'-*orf*) would not be affected.

Although the precise regulatory mechanisms are matters of future study, it is clear that this retrovirus has evolved an astonishingly complex system of genetic regulation. Perhaps this is because there is a race between viral replication within a cell and the destruction of the cell that the virus has commandeered, leaving no tolerance for inefficiency or improper timing in the synthesis of viral components. We can expect that the very complexity of the virus will contribute to its defeat.

Protein Production and Assembly

It is conceivable that drugs that interfere with the structure and function of retroviral mRNA transcribed from integrated DNA in infected cells could be of therapeutic value in AIDS. One drug, ribavirin, is believed to act as a guanosine analog that interferes with the 5'-capping of viral mRNA in other viral systems and perhaps could be useful in retrovirally induced disorders [26]. However, to date there are no convincing data that this drug can block HIV replication in vivo, and indeed, several workers have failed to demonstrate a reduction of circulating p24 levels.

Another approach would be the use of "anti-sense" oligodeoxynucleotides, which have already been tested in vitro. Basically this approach employs short sequences of DNA (or DNA that is chemically modified to allow better cell penetration and resistance to enzymatic degradation) whose base pairs are complementary to a vital segment of the viral genome. In theory such anti-sense oligodeoxynucleotides could block expression of the viral genome through a kind of hybridization arrest of translation, or it could possibly interfere with the binding of a regulatory protein such as *tat*. Matsukura et al. [27] discuss the in vitro use of phosphorothioate analogs of various oligodeoxynucleotides, and some of these analogs are exceedingly potent inhibitors of HIV (and other lentiviruses) in vitro. There may be an unexpected level of nonspecificity, however, in terms of the oligonucleotide sequence, and the precise antiviral mechanism is not known. Whether these phosphorothioate analogs work through a process of hybridization arrest or affect earlier stages of viral replication, such as template/primer binding to reverse transcriptase, is now under study.

The final stages in the replicative cycle of HIV involve crucial secondary processing of certain viral proteins by a protease (a function of one of the *pol* gene products) and myristylating and glycosylating enzymes (provided by the host) as a prelude to assembly of infectious virions. Therefore, additional strategies for the treatment of AIDS might conceivably involve certain kinds of protease inhibitors or drugs that dampen or alter myristylation and glycosylation steps in the synthesis of viral components [28]. Finally, retroviruses are released by a process of viral budding, which may be inhibited by interferons or drugs that induce interferon production (e.g., ampligen) [29]. It might be worth stressing that interferons are thought to block several stages in the replication of HIV. In that sense they theoretically can attack during several stages in the life cycle of the virus at one time.

Although the discussion has focused on how to suppress HIV replication per se, it might be worth noting that the virus could set off a chain of secondary events in vivo (autoimmune reaction, toxic lymphokine production, etc.) necessary for the expansion of clinical disease. It is also intriguing to speculate that a combination of antiretroviral therapy coupled with bone marrow transplantation, lymphocyte replacement, or stimulation of bone marrow precursor cells by colony-stimulating factors might be successful in certain subsets of patients with HIV infections. Studies testing whether GM–CSF has a role in therapy of this disease are underway.

Dideoxynucleosides

One can now turn to a discussion of a broad family of 2′, 3′dideoxynucleoside analogs, including certain didehydro congeners, that can be metabolized to become potent, chain-terminating inhibitors of HIV reverse transcriptase [4]. These analogs, even at large viral doses, can completely inhibit in vitro HIV replication and its capacity to destroy T cell cultures at concentrations that are 10- to 20-fold lower than those that impair the proliferation and survival of target cells [5,6]. Several related compounds (e.g., didehydro-dideoxythymidine) have also been shown to have potent activity against HIV in vitro [30]. In several cases these compounds have been studied over the past 20 years or so; in a triphosphate form, they are familiar to every molecular biologist as reagents for the Sanger DNA-sequencing procedure. But their application to human antiretroviral therapy demands a new perspective. The dideoxynucleoside analogs are of special interest because they prove that a simple chemical modification of the sugar moiety can predictably convert a normal substrate for nucleic acid synthesis into a compound with a potent capacity to inhibit the replication and cytopathic effect of HIV, at least in vitro.

Although these drugs may have the same ultimate mechanism of action, they behave as different agents from a clinical and pharmacological point of view. For example, AZT is known to suppress HIV replication in vivo [7,8], and its major toxicity is bone marrow suppression [7,31]. A closely related drug, 2′, 3′-dideoxycytidine, can also suppress HIV replication in vivo, and its major toxicity is a dose-dependent peripheral neuropathy [32]. The latter side effect can be significantly reduced by regimens that have drug-free rest periods [32]. Although many issues related to the antiretroviral effects of 2′, 3′-dideoxynucleosides are as yet unresolved, it would appear that as they are successively phosphorylated in the cytoplasm of a target cell to yield 2′, 3′-dideoxynucleoside-5′-triphosphates, they become analogs of the 2′-deoxynucleoside-5′-triphosphates that are the natural substrates for cellular DNA polymerases and reverse transcriptase. (It is generally thought that nucleoside-5′-triphosphates do not cross cell membranes and are not active as drugs because of their ionic character and their comparatively low lipophilicity.) Such analogs could compete with the binding of normal nucleotides to DNA polymerases (with high relative affinity for reverse transcriptase), or they could be incorporated into DNA and bring about DNA chain termination because normal 5′->3′ phosphodiester linkages cannot be completed [33]. We know that, at concentrations that are achievable in human cells, dideoxynucleotide analogs can serve as substrates for the HIV reverse transcriptase to elongate a DNA chain by one residue, after which the chain is terminated [33].

At doses that are not toxic for mammalian cells, pyrimidine and purine dideoxynucleoside analogs can inhibit the in vitro replication and pathogenic effects of a range of animal and human retroviruses, even when the pathogenic effect being monitored (transformation) requires only a single round of replication; moreover, with certain lentiviruses, these drugs can reduce the in vitro viral infectivity by more than five orders of magnitude [34]. Two dideoxynucleosides, 2′, 3′-dideoxycytidine and 3′-azido-2′, 3′-dideoxythymidine (AZT; vide infra), can block the replication of other human pathogenic retroviruses, HTLV-I and HIV-2, in vitro. However, possible emergence of drug-resistant mutants must always be considered. Indeed, it has recently been shown that the site-specific mutagenesis in the *pol* gene can render the reverse transcriptase less sensitive to AZT-5′-triphosphate [35]. (One might also speculate by analogy to herpes simplex that *pol* mutations might attenuate the pathogenicity of the mutant virus.)

It is important to stress that the phosphorylation reactions crucial to the activation of the nucleoside analogs are catalyzed by host cell kinases, and therefore, extreme caution must be used in extrapolating from cells of one type (or species) to another. Cell lines derived from different species show striking differences in their sensitivity to the cytostatic and antiretroviral activity [34].

If the relevant kinases are lacking in the host cell, the retrovirus will appear resistant to the nucleoside analogs, but if the retrovirus replicates in a different target cell that has the appropriate enzymes for anabolic phosphorylation, it will appear sensitive again. Similarly, one dideoxynucleoside, 2′, 3′-dideoxythymidine, behaves as a relatively poor substrate for human thymidine kinase. The substitution of an azido group at the 3′-carbon of this analog, yielding 3′-azido-2′, 3′dideoxythymidine (AZT), produces a compound that is an excellent substrate for thymidine kinase ($K_m = 3\mu M$) and is a very potent inhibitor of HIV replication. The substitution of a cyano moiety at the 3' carbon, however, does not produce a a good antiretroviral agent. At present, one cannot predict in advance which substitutions will work and which will fail.

Some data suggest that the HIV reverse transcriptase is much more susceptible to the inhibitory effects of these drugs as triphosphates than is mammalian DNA polymerase alpha, an enzyme that drives key DNA synthetic and repair functions in cells. Indeed, the reverse transcriptase has a higher affinity for dideoxynucleotides than normal substrates. This parallels what had been learned in animal retroviral systems. Although most 2', 3'-dideoxynucleosides-5'-triphosphates can inhibit mammalian DNA polymerase beta (a repair enzyme), and DNA polymerase gamma (a mitochondrial enzyme), we have observed that dideoxynucleosides can suppress HIV replication and protect sensitive helper/inducer target T cells in vitro for long periods of time without interfering with the function and survival of T cells [5,33]. It is interesting to note that AZT (as a triphosphate) is a comparatively poor inhibitor of DNA polymerase alpha and beta, compared to its inhibitory capacity for reverse transcriptase. As discussed above, our working explanation for the activity of these drugs against pathogenic retroviruses is that, following anabolism to nucleoside-5'-triphosphates, they bind to the viral DNA polymerase and/or bring about a selective retroviral DNA chain termination as the RNA form of the virus attempts to make DNA copies of itself [4,33].

AZT

Synthesized about 25 years ago by Horwitz et al. and shown to inhibit C type murine retrovirus replication in vitro by Ostertag et al. more than 13 years ago [37], no medical application of 3'-azido-2', 3'-dideoxythymidine (AZT) had emerged prior to our studies. This drug is a very potent in vitro inhibitor of HIV replication, and it has a cytopathic effect in susceptible target T cells and an antiretroviral effect against widely divergent strains of HIV [6]. As discussed above, the drug undergoes anabolic phosphorylation in human T cells to a nucleoside-5'-triphosphate, which can compete with thymidine-5'-triphosphate (TTP) and serve as a chain-terminating inhibitor of HIV reverse transcriptase. In that sense, AZT parallels the other dideoxynucleosides. It is interesting to note that AZT is rather poorly phosphorylated in monocytes/macrophages. Since the competing physiological nucleotide, thymidine triphosphate, is very low in such cells, the drug was found to be an excellent antiretroviral agent in vitro when a monocytotropic strain of HIV was studied (C-F Perno and S Broder, unpublished). From a clinical point of view, AZT can suppress bone marrow significantly (e.g., to cause megaloblastic anemia), which can be a serious side effect in patients with advanced AIDS [31], although it seems to be less of a problem in patients with earlier disease; it should be used with caution in patients with preexisting marrow dysfunction. This feature of the drug might lend itself to regimens that combine AZT with agents that do not have the same marrow-suppressing capacity, such as dideoxycytidine (ddC) [32]. A pilot protocol involving a weekly regimen of AZT alternating with ddC has been initiated at the National Cancer Institute to see if the nonoverlapping toxicities of these drugs could be put to clinical advantage. In the case of AZT, the key toxicity one would try to reduce would be bone marrow suppression [7,30]; in the case of ddC, it would be peripheral neuropathy [31]. Successful antiretroviral chemotherapy may depend on the use of combinations of drugs that compensate for each other's side effects.

What one can say at this time is that AZT, a drug chosen on the basis of its selective in vitro antiviral effect against HIV [5,6, has been shown, as a single agent, to clinically benefit patients with advanced disease in a double-blind placebo trial [7,8]. Although the drug does have serious side effects in some patients [7,31], it lacks cardiac, renal, and hepatic toxicity. It is worth stressing that the drug may also at least partially improve HIV-related dementias [12]. Such neurological improvements may be particularly evident in children.

Future Strategies

As a possible cure for AIDS, AZT represents no more than a first step in developing practical chemotherapy against such pathogenic human retroviruses as HIV. In the long term, its main value may be to validate the key assumptions underlying antiviral strategies for intervention against established AIDS. Retroviral infections were, until recently, viewed as inherently untreatable. We now have proof that HIV replication can be suppressed in vivo and that such suppression can lead to prolonged survival and improved quality of life for certain patients with HIV infections.

Among the most important questions now facing clinicians is whether an early antiretroviral therapy can block or substantially delay the onset of AIDS in individuals who carry the virus but are still well. In the future it may be possible to devise relatively tolerable regimens to inhibit HIV replication and thereby prevent the perturbation of the immune system that leads to AIDS [4,38]. This is an important direction for research because a very high proportion of individuals who are seropositive for HIV will develop AIDS or a related condition [39]. It is of interest that elevated β2-microglobulin is a powerful predictive factor.

No technology now available, or likely to become available soon, can get rid of the latent form of the virus in an infected person. In this sense there is no known cure for HIV infection. However, if one defines cure as the restoration of actuarial survival to that of uninfected age-matched individuals, it is entirely possible that many of the therapies now under discussion can be curative, at least at early stages of the infection.

References

1. Barre-Sinoussi F, Chermann JC, Rey F, Nugeyre MT, Chamaret S, Gruest J, Daughet C, Axler-Blin C, Vezinet-Brun F, Rouzioux C, Rozenbaum W, Montagnier L (1983) Isolation of a T cell lymphotropic virus from a patient at risk for acquired immunodeficiency syndrome (AIDS). Science 220:868–871

2. Popovic M, Sarngadharan MG, Read E, Gallo RC 1984) Detection, isolation, and continuous production of cytopathic retrovirus (HTLV-III) from patients with AIDS and pre-AIDS. Science 224:497–500

3. Miller R (1988) Human immunodeficiency virus may encode a novel protein on the genomic DNA plus strand. Science 239:1420–1422

4. Mitsuya H, Broder S (1987) Strategies of antiviral therapy in AIDS. Nature 325:773–778

5. Mitsuya H, Broder S (1986) Inhibition of the *in vitro* infectivity and cytopathic effect of human T-lymphotropic virus type III/lymphadenopathy associated virus (HTLV-III/LAV) by 2′,3′-dideoxynucleosides. Proc Natl Acad Sci USA 83:1911–1915

6. Mitsuya H, Weinhold KJ, Furman PA, St Clair MH, Lehrman SN, Gallo RC, Bolognesi D, Barry DW, Broder S (1985) 3′-Azido-3′-deoxythymidine (BW A509U): An antiviral agent that inhibits the infectivity and cytopathic effect of human T-lymphotropic virus type III/lymphadenopathy associated virus *in vitro*. Proc Natl Acad Sci USA 82:7096–7100

7. Yarchoan R, Klecker RW, Weinhold KJ, Markham PD, Lyerly HK, Durack DT, Gelmann E, Lehrman SN, Blum RM, Barry DW, Shearer GM, Fischl MA, Mitsuya H, Gallo RC, Collins JM, Bolognesi DP, Myers CE, Broder S (1986) Administration of 3′-azido-3′-deoxythymidine, an inhibitor of HTLV-III/LAV replication, to patients with AIDS or AIDS-related complex. Lancet i:575–580

8. Fischl M, Richman DD, Grieco MH, Gottlieb MS, Volberding PA, Laskin OL, Leedom JM, Allan JD, Mildvan D, Schooley RT, Jackson GG, Durack DT, King D, AZT Collaborative Working Group (1987) The efficacy of 3′-azido-3′-deoxythymidine (azidothymidine) in the treatment of patients with AIDS and AIDS-related complex: A double-blind placebo-controlled trial. New Engl J Med 317:185–191

9. Gottlieb MS, Schroff R, Schanker HM, Weisman JD, Fan PT, Wolf RA, Saxon A (1981) *Pneumocystis carinii* pneumonia and mucosal candidiasis in previously healthy homosexual men. Evidence of a new acquired cellular immunodeficiency. New Engl J Med 305:1425–1431

10. Snider WD, Simpson DM, Nielsen S, Gold JWM, Metroka CE, Posner JB (1983) Neurological complications of acquired immune deficiency syndrome: Analysis of 50 patients. Ann Neurol 14:403–418

11. Shaw GM, Harper ME, Epstein LG, Gajdusek DC, Price RW, Navia BA, Petito CK, O'Hara CJ, Groopman JE, Cho ES, Oleske JM, Wong-Staal F, Gallo RC (1985) HTLV-III infection in brains of children and adults with AIDS encephalopathy. Science 227:177–182

12. Yarchoan R, Berg G, Brouwers P, Spitzer AR, Fischl MA, Thomas RV, Schmidt P, Safai B, Perno CF, Myers CE, Broder S (1987) Preliminary observation of the response of HTLV-III/LAV-associated neurological disease to the administration of 3′-azido-3′-deoxythymidine. Lancet i:132–135

13. Mitsuya H, Matsukura M, Broder S (1987) Rapid *in vitro* systems for assessing

activity of agents against HTLV-III/LAV. *In* Broder S (ed) AIDS: Modern Concepts and Therapeutic Challenges. Marcel Dekker, New York, pp 303–333

14. Baltimore D (1970) Viral RNA-dependent DNA polymerase. Nature 226:1209–1211

15. Temin HM, Mizutani S (1970) RNA-dependent DNA polymerase in virions of Rous sarcoma virus. Nature 226:1211–1213

16. Strebel K, Daugherty D, Clouse K, Cohen D, Folks T, Martin MA (1987) The HIV 'A' (sor) gene product is essential for virus infectivity. Nature 328:728–730

17. Guy B, Kieny MP, Riviere Y, Peuch CL, Dott K, Girard M, Montagnier L, Lecocq JP (1987) HIV F/3′-*orf* encodes a phosphorylated GTP-binding protein resembling an oncogene product. Nature 330:266–269

18. Smith DH, Byrn RA, Marsters SA, Gregory T, Groopman JE, Capon DJ (1987) Blocking of HIV-1 infectivity by a soluble, secreted form of the CD4 antigen. Science 228:1704–1707

19. Fisher AG, Ratner L, Mitsuya H, Marselle LM, Harper ME, Broder S, Gallo RC, Wong-Staal F (1986) Infectious mutants of HTLV-III with changes in the 3′-*orf* region and markedly reduced cytopathic effects. Science 233:655–659

20. Matsushita S, Robert-Guroff M, Trepel J, Cossman J, Mitsuya H, Broder S (1986) Human monoclonal antibody directed against an envelope glycoprotein of human T-cell leukemia virus type I. Proc Natl Acad Sci USA 93:2672–2676

21. Nabel G, Rice S, Knipe D, Baltimore D (1988) Alternative mechanisms for activation of human immunodeficiency virus enhancer in T cells. Science 239:1299–1301

22. Felber B, Pavlakis G (1988) A quantitative bioassay for HIV-I based on transactivation. Science 238:184–187

23. Sodroski J, Rosen C, Wong-Staal F, Salahuddin SZ, Popovic M, Arya S, Gallo RC (1985) Transacting transcriptional regulation of human T-cell leukemia virus type III long terminal repeat. Science 227:171–173

24. Sodroski J, Goh WC, Rosen C, Dayton A, Terwilliger E, Haseltine WA (1986) A second post-transcriptional transactivator gene required for HTLV-III replication. Nature 321:412–417

25. Feinberg MB, Jarrett RF, Aldovini A, Gallo RC, Wong-Staal F (1986) HTLV-III expression and production involve complex regulation at the level of splicing and translation of viral RNA. Cell 46:807–817

26. McCormick JB, Getchell JP, Mitchell SW, Hicks DR (1984) Ribavirin suppresses replication of lymphadenopathy-associated virus in culture of human adult lymphocytes. Lancet ii:1367–1369

27. Matsukura M, Shinozuka K, Zon G, Mitsuya H, Reitz M, Cohen JS, Broder S (1987) Phosphorothioate analogs of ologodeoxynucleotides: Inhibition of replication and cytopathic effects of human immunodeficiency virus. Proc Natl Acad Sci USA 84:7706–7710

28. Gruters RA, Neefjes JJ, Tersmette M, Goede RY, Tulp A, Huisman HG, Miedema F, Ploegh HL (1987) Interference with HIV-induced syncytium formation and viral infectivity by inhibitors of trimming glucosidase. Nature 330:74–77

29. Montefiori DC, Mitchell WM (1987) Antiviral activity of mismatched double-stranded RNA against human immunodeficiency virus *in vitro*. Proc Natl Acad Sci USA 84:2985–2989

30. Baba M, Pauwels R, Herdewijn P, De Clercq E, Desmyter J, Vandeputte M (1987) Both 2′-3′-dideoxythymidine and its 2′-3′-unsaturated derivative (2′-3′-

dideoxythumidinene) are potent and selective inhibitors of human immunodeficiency virus replication *in vitro*. Biochem Biophys Res Commun 142:128–134

31. Richman DD, Fischl MA, Grieco MH, Gottlieb MS, Volberding PA, Laskin OL, Leedom JM, Groompan JE, Mildvan D, Hirsch MS, Jackson GG, Durack DT, Lehrman SN, AZT Collaborative Working Group (1987) The toxicity of azidothymidine (AZT) in the treatment of patients with AIDS and AIDS-related complex. A double-blind, placebo-controlled trial. New Engl J Med 317:192–197

32. Yarchoan R, Perno CF, Thomas RV, Klecker RW, Allain JP, Willis RJ, McAtee N, Fischl MA, Dubinsky R, McNeely MC, Mitsuya H, Pluda JM, Lawley TJ, Leuther M, Safai B, Collins JM, Meyers CE, Broder S (1988) Phase I studies of 2′-3′-dideoxycytidine in severe human immunodeficiency virus infection as a single agent and alternating with Zidovudine (AZT). Lancet i:76–81

33. Mitsuya H, Jarrett RF, Matsukura M, DiMarzo Veronese F, DeVico AL, Sarngadharan MG, Johns DG, Reitz MS, Broder S (1987) Long-term inhibition of human T-lymphotropic virus type III/lymphadenopathy-associated virus (human immunodeficiency virus) DNA synthesis and RNA expression in T cells protected by 2′-3′-dideoxynucleosides *in vitro*. Proc Natl Acad Sci USA 84:2033–2037

34. Dahlberg JE, Mitsuya H, Broder S, Blam SB, Aaronson SA (1987) Broad spectrum anti-retroviral activity of 2′-3′-dideoxynucleosides. Proc Natl Acad Sci USA 84:2469–2473

35. Larder BA, Purifoy JM, Powell KL, Darby G (1987) Site-specific mutagenesis of AIDS virus reverse trnascriptase. Nature 327:716–717

36. Balzarini J, Pauwells R, Baba M, Herdewijn P, de Clercq E, Broder S, Johns DG (1988) The *in vitro* and *in vivo* anti-retrovirus activity, and intracellular metabolism of 3′-azido-2′-3′-dideoxythymidine and 2′-3′-dideoxycytidine are highly dependent on the cell species. Biochem Pharm 37:897–903

37. Ostertag W, Roesler G, Krieg CJ, Kind J, Cole T, Crozier T, Gaedicke G, Steinheider G, Kluge N, Dube S (1974) Induction of endogenous virus and of thymidine kinase by bromodeoxyuridine in cell cultures transformed by Friend virus. Proc Natl Acad Sci USA 71:4980–4985

38. De Wolfe F, Lange JMA, Goudsmit J, Cload P, De Gans J, Schellekens PTH (1988) Effect of Zidovudine on serum human immunodeficiency virus antigen levels in symptom-free subjects. Lancet i:373–376

39. Moss A, Bacchetti P, Osmond D, Krampf W, Chaisson R, Stites D, Wilber J, Allain J, Carlson J (1988) Seropositivity for HIV and development of AIDS or AIDS-related condition: Three year follow-up of the San Francisco General Hospital cohort. Br Med J, 745–750

CHAPTER 41
Treatment of Cytomegalovirus Infections

CLYDE S. CRUMPACKER

The goal of effective treatment of serious infections with human cytomegalovirus (CMV) has been difficult to achieve. Human cytomegalovirus is the most significant infection complicating renal transplantation [1] and the major cause of death in patients who receive bone marrow transplants [2]. In these patients, pneumonia due to CMV is the most severe consequence of the infection. Attempts at therapy for pneumonia due to CMV in immuno-compromised patients have been unsuccessful, in spite of regimens that decrease immunosuppressive therapy, use of such antiviral agents, either alone or in combination, as acyclovir [3], adenine arabinoside [4], and leukocyte interferon [5], or employment of transplant nephrectomy.

In patients with AIDS, CMV is a serious cause of sight-threatening retinitis or life-threatening colitis, meningoencephalitis, and pneumonitis. In patients with AIDS, CMV is frequently present in the lung with other pathogens; it is rarely present as a single agent [6]. Recently, however, ganciclovir, 9-(1,3-,dihydroxy-2-propoxy) methyl guanine (DHPG), a derivative of acyclovir and a new antiviral agent, has been used successfully to treat clinically important CMV infections [7]. This review summarizes important features of this landmark antiviral drug, its mechanism of action and clinical use, and an alternative antiviral drug, foscarnet.

Mechanisms of Action of Ganciclovir

Ganciclovir, independently synthesized in four laboratories in 1982, is called by four different names: DHPG (Syntex) [8], 2′NDG (Merck) [9], BIOLF-62 (Bio Logicals Inc.) [10], and BW-B759U (Burroughs-Wellcome) [11]. The initial report on BIOLF-62 showed anti-HSV activity but did not show CMV inhibition [12]. Other reports clearly showed that laboratory strains of CMV were inhibited [13–16]. The first report of inhibition of CMV clinical strains and CMV DNA synthesis directly in infected cells [17] was confirmed and extended [18]. To actively inhibit CMV replication, DHPG needs to be phosphorylated to DHPG-triphosphate, a selective inhibitor of CMV DNA polymerase [19].

In herpes simplex virus-infected cells, DHPG is phosphorylated by the herpes simplex virus (HSV) thymidine kinase, and HSV mutants that are thymidine kinase deficient are resistant to DHPG. The use of recombinant HSV-1/HSV-2 viruses showed that resistance to DHPG could also be mapped to the DNA polymerase gene of HSV and that resistance was mediated by a different region of the polymerase gene than was resistance to acyclovir and phosphonoacetic acid (PAA) [20]. In studies on mutant polymerase enzymes, the determinants of nucleotide binding specificity of DHPG-TP and ACV-TP were different, which suggests that DHPG TP and ACV-TP interact with different regions of the HSV polymerase enzyme [21]. The studies employing recombinant viruses and mutant polymerase enzymes clearly establish that DHPG-TP interacts with the viral polymerase enzyme to inhibit viral DNA synthesis.

Since CMV does not encode a viral thymidine kinase enzyme in its genome, activation by viral thymidine kinase cannot be used to account for the high levels of DHPG-TP in CMV-infected cells in which DHPG-TP is 10-fold higher than the levels found in uninfected cells [22]. The levels of DHPG-TP are also 10-fold higher than the levels of ACV-TP when compared under similar conditions of CMV infection. The levels of DHPG-TP also persist in infected cells long after the drug is removed from the culture medium [23]. The mechanisms by which high levels of DHPG-TP are achieved and maintained has been intensely studied, but the specific mechanism remains unknown. Three possible explanations have been considered:

1. Activation by a cellular enzyme, a "ganciclovir kinase," following infection of cells by CMV. Mitochondrial deoxyguanosine kinase has been suggested as a candidate for such an enzyme [24].
2. Phosphorylation by a CMV-encoded enzyme yet to be identified, a CMV ganciclovir kinase. A mutant of CMV that induces greatly decreased amounts of DHPG-TP provides support for the existence of such a viral encoded enzyme [25].
3. Prevention of dephosphorylation of DHPG-TP by cellular enzymes in CMV-infected cells allowing for a buildup and persistence of high levels of DHPG-TP.

The mechanism of interaction of the DHPG-TP and CMV viral DNA replication has been explored using partially purified DNA polymerase preparations that show that DHPG-TP is a competitive inhibitor with deoxyguanosine triphosphate (dGTP) [26]. DHPG-TP can substitute for dGTP, can act as a substrate for DNA polymerase of CMV, and can be incorporated into the DNA of CMV to decrease DNA chain elongation dramatically [19]. When DHPG is removed, chain elongation proceeds. The substitution of DHPG-TP for dGTP in an in vitro system of CMV DNA replication shows that CMV DNA terminated by DHPG-MP is an inefficient substrate for chain elongation. It is likely that DHPG-MP is excised from chain termini before elongation resumes and that long chain CMV DNA contains minimal amounts of DHPG [27]. Inhibition of viral DNA synthesis probably accounts for all the antiviral effects of DHPG. The inhibitory concentration for 90% of viral protein synthesis is greater than 10 μM, whereas 90% of viral DNA synthesis occurs at concentrations less than 1 μM, which suggests that input CMV DNA may transcribe and translate viral protein in the presence of DHPG [28]. The evidence, however, does not suggest that DHPG has any direct effect on viral production after DNA replication.

Treatment of CMV Retinitis

Retinitis caused by CMV is by far the most serious sight-threatening infection in patients with AIDS. Ganciclovir has had its greatest success in treating this infection. Results must be interpreted with some caution, however, because the data obtained are not based on placebo/control double-blind studies, but rather almost exclusively on uncontrolled, compassionate-use studies of ganciclovir. In 108 immunocompromised patients, most of whom had AIDS, 84% had a favorable clinical response with ganciclovir [29]. The diagnosis of CMV retinitis is usually based on clinical criteria of retinal hemorrhage, white perivascular lesions, retinal edema, and necrosis [30]. With ganciclovir therapy at a dose of 5 mg/kg intravenously twice a day for 14 days, a healing of retinal lesions and an improvement in vision commonly occur. The virological response to ganciclovir in patients who had positive baseline cultures for CMV in the blood or urine was a complete clearing of CMV from the blood in 83% of the patients and complete clearing from the urine in 87% of the patients. The median time for a virological response was 8 days [29].

A high rate of relapse of CMV retinitis occurs in patients when ganciclovir is discontinued. In a group of 61 patients who responded initially to ganciclovir induction therapy, time to relapse was evaluated in three subgroups: 20 received no maintenance therapy, 9 received low dose maintenance with ganciclovir (10–15 mg/kg per week), and 32 received high-dose maintenance therapy with ganciclovir (25–35 mg/kg per week). The time to relapse of retinitis following completion of initial induction was evaluated with the use of survival analysis (Kaplan–Meir). The three groups of patients were comparable with regard to age, cumulative induction dose, and concurrent disease except that more patients with Kaposi's sarcoma were present among those who did not receive maintenance therapy. The results showed that patients who received high-dose maintenance had a longer median time (105 days) to relapse of CMV retinitis compared to the relapse time (47 days) for those who received no maintenance or low-dose maintenance ($P < .0002$). These results indiciate that ganciloVir is an effective treatment for CMV retinitis, but it is not a cure. Relapse is common, but can be delayed by maintenance therapy with ganciclovir.

Treatment of CMV Colitis

Disseminated CMV infection is common in patients with AIDS. A recent series of 97 autopsies in patients with AIDS found disseminated CMV infection in 68% of cases; over half of these patients had evidence of gastrointestinal CMV disease [31]. Diarrhea is a common feature in patients with AIDS, and CMV ranks as a common cause of this diarrhea. In an uncontrolled study of gastrointestinal CMV infection in patients, treatment with ganciclovir produced a positive clinical response in 75% of the patients [32]. The sites of infection included colon (67%), esophagus and stomach (22%), rectum (77%), liver (3%), and small bowel (1.4%). Of these patients, 39% responded completely with resolution of all symptoms of fever, dysphagia, and diarrhea. This group of patients were well enough to return to work. Endoscopic evaluation usually confirmed healing of lesions. Virological response was more difficult to evaluate because the study lacked a control group and because the natural history of CMV colitis in patients with AIDS is unknown. Placebo controlled, blind studies with ganciclovir will be important to evaluate fully its role on clinical and virological parameters in well-defined AIDS patients with CMV colitis.

Treatment of CMV Pneumonia

The most difficult life-threatening CMV infection to treat is pneumonia. A careful, well-done initial study in 10 patients with CMV pneumonia following bone marrow transplantation reported an impressive antiviral effect on clearing of CMV, but only 1 patient survived with ganciclovir therapy [33].

In a multicenter uncontrolled study of ganciclovir treatment for well-documented CMV pneumonia in 21 bone marrow transplant patients, 8 patients survived; 2 other patients survived for 60 and 79 days before relapsing with CMV pneumonia or dying of disseminated aspergillosis [34]. This survival of 38% is still small, but it may represent the beginning of successful treatment for CMV pneumonia in bone marrow transplant recipients. The use of CMV immune globulin with ganciclovir may enhance the benefit of treatment of CMV pneumonia, since 3 of the 8 survivors received CMV immune globulin and ganciclovir. More extensive controlled studies of ganciclovir plus CMV immune globulin to treat CMV pneumonia in bone marrow transplant patients are warranted.

The most striking success of treatment for CMV pneumonia has been observed in renal transplant patients treated with ganciclovir. A study of life-threatening primary CMV pneumonia in four renal transplant recipients revealed that 2 patients recovered completely and survived CMV pneumonitis, even though 1 patient required assistance with a respirator at the start of treatment [35]. This was extended in a study of 11 patients with CMV pneumonia, in which 4 patients were cured with ganciclovir therapy (36%); 2 other patients showed clinical improvement [36]. It is remarkable that 3 of 10 ventilator-dependent patients survived their episodes of pneumonia. This decrease in mortality is encouraging, although not statistically significant. This represents the best success with any therapy for any group of patients with CMV pneumonia.

Treatment of CMV pneumonia with ganciclovir in patients with AIDS is still controversial. The entity of CMV pulmonary infection in patients with AIDS needs to be more precisely defined, since other pathogens are almost always present with CMV in the lung. In a large study of 441 patients who had pulmonary complication of AIDS, CMV was identified by either culture or characteristic cytopathic changes in 17% of the patients. In two-thirds of the patients, *Pneumocystis carinii* pneumonia was also present; CMV was the only pathogen in only 4% of the patients [36]. In the protocol for compassionate use of ganciclovir, 46 patients with AIDS were enrolled who had CMV documented by culture or histologically from bronchial alveolar lavage or lung biopsy specimens. The condition of 28 (60%) of these patients improved with ganciclovir therapy; the condition of 16 others (34%) deteriorated, and in 2 other patients (4%), it stabilized. Thirty-nine of these patients were treated with ganciclovir for more than 10 days, and 72% of these patients improved. These results suggest that ganciclovir treatment was associated with a favorable outcome in a majority of AIDS patients. Unfortunately, these apparently favorable results need to be interpreted with caution, as patients were not adequately analyzed for presence of other pathogens or severity of disease when treatment was begun [35].

The major toxicity associated with ganciclovir treatment is neutropenia. In one study, the absolute neutrophil count decreased to less than 1,000 cells/dl in 42% of the patients. Thrombocytopenia occurred in 19% of patients, and nausea was observed in 6%. In 314 immunocompromised patients treated with ganciclovir, 5 neutropenic patients died of bacterial or fungal sepsis. In 127 patients who received maintenance treatment, 17% had treatment interrupted due to an adverse effect, most commonly neutropenia [29]. Since neutropenia is the main adverse effect limiting the use of ganciclovir, a combination of ganciclovir with a growth stimulating factor, such as granulocyte/macrophage colony-stimulating factor (GMC-SF), may permit the use of ganciclovir in neutropenic patients.

Foscarnet for CMV

Foscarnet (phosphonoformic acid, PFA), a pyrophosphate analog with broad antiviral activity against human herpesviruses, blocks the pyrophosphate receptor site on viral DNA polymerase [37]. Replication of human CMV in tissue culture is completely inhibited at a PFA concentration of 500 μM (ID_{50} = $130_{\mu M}$) [38]. In a preliminary study, PFA at a concentration of 3.37 to 7 mg/kg an hour was employed for the treatment of severe CMV infection in 3 renal transplant and three bone marrow transplant recipients. A favorable clinical response was seen in 5 patients, and 2 patients were doing well 5 and 8 months after treatment [39]. Foscarnet has been used to treat CMV infection in patients with AIDS, and its clinical efficacy was evaluated in 44 patients with CMV retinitis and AIDS. A good response was seen in 41% of 35 patients being treated for the first episode of retinitis, and a favorable response was seen in 89% of 9 patients undergoing retreatment. Relapse was common, and virological data were minimal. In 2 additional patients, however, shedding of CMV in the urine and blood ceased with PFA therapy [40].

The main toxicity seen with PFA is renal impairment with increased serum creatinine, hyperphosphatemia, and hypocalcemia. Oberg has published a significant review on the effects of PFA [41].

Further evaluation of PFA for CMV infection will require placebo-controlled trials and an assessment of the dose needed to prevent relapse. The chief role of PFA may be as an alternative therapy for patients who cannot tolerate ganciclovir.

Recently, the development of progressive disease due to ganciclovir-resistant CMV has been reported in three immunocompromised patients with AIDS or leukemia [42]. By the use of restriction endonuclease analysis, it was established that one patient became infected with a resistant virus, another patient was infected with a susceptible virus that became resistant, and a third patient was infected by two strains of CMV, an initial sensitive one and a subsequent genetically distinct resistant strain. This report points out that resistant CMV strains causing significant disease will become an increasing clinical problem. This will make the development of alternative treatments for CMV, such as foscarnet, an important medical goal.

References

1. Peterson PK, Balfour HH, Marker SC, Fryd DS, Howard RJ, Simmons RL (1980) Cytomegalovirus disease in renal allograft recipients: A prospective study of the clinical features, risk factors and impact on renal transplantation. Medicine 57:283–300
2. Myers JD, Flournoy N, Thomas ED (1982) Nonbacterial pneumonia after allogeneic marrow transplantation; a review of ten years experience. Rev Inf Dis 4:1119–1132

3. Wade JC, Hintz M, McGriffin RW, Springmeyer SC, Connor JD, Meyers JD (1982) Treatment of cytomegalovirus pneumonia with high-dose acyclovir. Am J Med 73:249–256

4. Meyers JD, McGriffin RW, Bryson YJ, Cantell K, Thomas ED (1982) Treatment of cytomegalovirus pneumonia after marrow transplantation with combined vidarabine and human leukocyte interferon. J Infect Dis 146:80–84

5. Shepp DH, Newton BA, Meyers JD (1984) Intravenous lymphoblastoid interferon and acyclovir for treatment of cytomegalovirus pneumonia. J Infect Dis 150:776–777

6. Murray JF, Felton CP, Garay SM, Gottlieb MS, Hopewell PC, Stover DE, Tierstein AS (1985) Pulmonary complications of the acquired immunodeficiency syndrome. Report of a National Heart, Lung, and Blood Institute Workshop. New Engl J Med 310:1682–1688

7. Collaborative DHPG Treatment Study Group (1986) Treatment of serious cytomegalovirus infections with 9-(1,3-dihydroxy-2-propoxymethyl)guanine in patients with AIDS and other immunodeficiencies. New Engl J Med 314:801–805

8. Martin JC, Dvorak CA, Smee DF, Matthews TR, Verheyden JPH (1983) 9-(1,3-Dihydroxy-2-propoxymethyl)guanine: A new potent and selective antiherpes agent. J Med Chem 26 759–761

9. Ashton WT, Karkas JD, Field AK, Tolman RL (1982) Activation by thymidine kinase and potent antiherpetic activity of 2'-nor-2'-deoxyguanosine (2'NDG). Biochem Biophys Res Comm 108:1716–1721

10. Ogilvie KK, Cheriyan UO, Radatus BK, Smith KO, Gallaway KS, Kennell WL (1982) Biologically active acyclonucleoside analogues. II. The synthesis of 9-[[2-Hydroxy-1(hydroxymethyl)ethoxy]methyl]-guanine (BIOLF-62). Canadian J Chem 60:3005–3010

11. Schaeffer HJ (1982) Nucleosides with antiviral activity. In Rideout JL, Henry DW, Beacham LM III (eds) Nucleosides, Nucleotides and Their Biological Applications. New York, Academic Press, pp 1–7

12. Smith KO, Galloway KS, Kendell WI, Ogilvie KK, Radatus BK (1982) A new nucleoside analog 9-[2-hydroxy-1-(hydroxymethyl)ethoxymethyl]guanine, highly active in vitro against herpes simplex virus types 1 and 2. Antimicrob Agents Chemother 22:55–61

13. Smee DF, Martin TR (1983) Antiherpes virus activity of the acyclic nucleoside 9-(1,3,-dihydroxy-2-propoxymethyl)guanine. Antimicrob Agents Chemother 23:676–682

14. Mar EC, Cheng YC, Huang ES (1983) Effects of 9-(1-3-dihydroxy-2-propoxymethyl)guanosine on human cytomegalovirus replication in vitro. Antimicrob Agents Chemother 24:518–521

15. Cheng YC, Huang ES, Lin JC, Mar EC, Pagano JS, Dutschman GE, Grill SP (1983) Unique spectrum of activity of 9-[(1,3-dihydroxy-2-propoxy)methyl]guanine against herpes viruses in vitro and its mode of action against herpes simplex virus type 1. Proc Natl Acad Sci USA 80:2767–2770

16. Field AK, Davies ME, Dewitt C, Perry HC, Liou R, Germershausen J, Karkas JD, Ashton WT, Johnston DBR, Tolman RC (1983) 9-[(2-Hydroxy-1-(hydroxymethyl)-ethoxy] methyl guanine: A selective inhibitor of herpes group virus replication. Proc Natl Acad Sci USA 80:4139–4143

17. Tocci MJ, Livelli TJ, Perry HC, Crumpacker CS, Field AK (1984) Effects of nucleoside analogue 2' nor 2'deoxyguanosine on human cytomegalovirus replication. Antimicrob Agents Chemother 25:247–252

18. Plotkin SA, Drew WL, Felsenstein D, Hirsch MS (1985) Sensitivity of clinical isolates of human cytomegalovirus to 9-(1,3-dihydroxy-2-propoxymethyl)-guanine. J Infect Dis 152:833–834

19. Mar EC, Chiou JF, Cheng YC, Huang ES (1985) Inhibition of cellular DNA polymerase and human cytomegalovirus-induced DNA polymerase by the triphosphates of 9-(2-hydroxyethoxymethyl)guanine and 9-(1,3-dihydroxy-2-propoxymethyl)guanine. J Virol 53:776–780

20. Crumpacker CS, Kowalsky PN, Oliver SA (1984) Resistance of herpes simplex virus to 9-(2-hydroxy-1-(hydroxymethyl)ethoxy)methyl guanine (2′NDG)—Physical mapping of drug synergism within the viral DNA polymerase locus. Proc Natl Acad Sci USA 81:1556–1560

21. Frank KB, Chiou JF, Cheng YC (1984) Interaction of herpes simplex virus induced DNA polymerase with 9-(1,3-dihydroxy-2-propoxymethyl)guanine, triphosphate. J Biol Chem 259:1566–1569

22. Smee DF, Boehme R, Chernow M, Binko BP, Matthews TR (1985) Intracellular metabolism and enzymatic phosphorylation of 9-(1,3-dihydroxy-2-propoxymethyl)guanine in herpes simplex virus infected and uninfected cells. Biochem Pharmacol 34:1049–1056

23. Biron KK, Stanat SC, Sorrell JB, Fyfe JA, Keller PM, Lambe CV, Nelson DJ (1985) Metabolic activation of the nucleoside analog 9-[2-hydroxy-1-hydroxymethyl)ethoxy]methyl, Proc Natl Acad Sci 82:2473–2477

24. Smee DF (1985) Interaction of 9-(1,3-dihydroxy-2-propoxymethyl)guanine with cytosol and mitochondrial deoxyguanosine kinases: Possible role in anti-cytomegalovirus activity. Molec Cell Biochem 69:75–81

25. Biron KK, Fyfe JA, Stanat SC, Leslie LK, Sorrell JB, Lambe CV, Coen DM (1986) A human cytomegalovirus mutant resistant to the nucleoside analog 9-[2-hydroxy-1-(hydroxymethyl)ethoxy]methylguanine (BW7590). Proc Natl Acad Sci 83:8769–8773.

26. Freitas VR, Smee DF, Chernow M, Boehme R, Matthews TR (1985) Activity of 9-(1,3-dihydroxy-2-propoxymethyl)guanine compared with that of acyclovir against human, monkey and rodent cytomegaloviruses. Antimicrob Agents Chemother 28:240–245

27. Matthews T, Boehme R (1988) Antiviral activity and mechanism of action of ganciclovir. Rev Inf Dis 10:S490–S494

28. Rasmussen LE, Chen PT, Mullenax JG, Merigan TC (1984) Inhibition of human cytomegalovirus by 9-(1,3-dihydroxy-2-propoxymethyl)guanine alone and in combination with human interferons. Antimicrob Agents Chemother 26:441–445

29. Buhles WC, Mastre BJ, Tinker AJ, Strand V, Koretz SH, and the Syntex Collaborative Ganciclovir Treatment Group (1988) Ganciclovir treatment of life- or sight-threatening cytomegalovirus infection: Experience in 314 immunocompromised patients. Rev Inf Dis 10:S495–S506

30. Palestine AG (1988) Clinical aspects of cytomegalovirus retinitis. Rev Inf Dis 10:S515–S519

31. Dannenberg AJ, Margulis SJ (1987) Cytomegalovirus infection of the gastrointestinal tract in AIDS. Gastroenterology 92:1362

32. Dieterrich DT, Chachoua A, Lafleur F, Worrell C (1988) Ganciclovir treatment of gastrointestinal infections caused by cytomegalovirus in patients with AIDS. Rev Inf Dis 10:S532–S537

33. Shepp DH, Dandliker PS, de Miranda P, Burnett TC, Cederberg DM, Kirk LE, Meyers JD (1985) Activity of 9-[2-hydroxy-1-(hydroxymethyl)ethoxymethyl]gua-

nine in the treatment of cytomegalovirus pneumonia. Ann Int Med 103:368–373

34. Crumpacker CS, Marlowe S, Zhang JL, Abrams S, Watkins P, and the Ganciclovir Bone Marrow Transplant Treatment Group (1988) Treatment of cytomegalovirus pneumonia. Rev Inf Dis 10:S538–S546

35. Hecht DW, Syndman DR, Crumpacker CS, Werner BG, Heinze-Lacey B, and the Boston Renal Transplant CMV Study Group (1988) Treatment of renal transplant associated primary cytomegalovirus pneumonia with ganciclovir (DHPG). J Infect Dis 157:187–191

36. Syndman DR (1988) Ganciclovir therapy for cytomegalovirus disease associated with renal transplants. Rev Inf Dis 10:S554–S560

37. Verheyden JP (1988) Evolution of therapy for cytomegalovirus infection. Rev Inf Dis 10:S477–S489

38. Wahren B, Oberg B (1980) Inhibition of cytomegalovirus late antigens by phosphonoformate. Intervirology 12:335–339

39. Klintmalm G, Lonnqvist B, Oberg B, Gahrton G, Lernestedt J-O, Lundgren G, Ringden O, Robert K-H, Wahren B, Groth C-G (1985) Intravenous foscarnet for the treatment of severe cytomegalovirus infection in allograft recipients. Scand J Infect Dis 17:157–163

40. Mills J, Jacobson MA, O'Donnell JJ, Cederberg D, Holland GN (1988) Treatment of cytomegalovirus retinitis in patients with AIDS. Rev Inf Dis 10:S522–S527

41. Oberg B (1983) Antiviral effects of phosphonoformate (PFA, foscarnet sodium). Pharmacol Ther 19:387–415

42. Erice A, Chou S, Biron KK, Stanat SC, Balfour HH, Jordan MC (1989) Progressive disease due to ganciclovir-resistant cytomegalovirus in immunocompromised patients. New Engl J Med 320:289–293

CHAPTER 42

The Mechanism of Action of Amantadine and Rimantadine Against Influenza Viruses

ALAN J. HAY

Amantadine (1-aminoadamantane hydrochloride) and rimantadine (1-methyl-1-adamantane methylamine hydrochloride), currently the only effective antiviral drugs for use against influenza, are effective both in the prophlyaxis and treatment of infection [1–4]. The selectivity of their action against influenza A virus, however, is a major drawback, since they provide no protection against influenza B virus infections that account for a significant proportion of disease in humans. This feature, in particular, has restricted their application, even though amantadine was initially licensed for use in the United States as long ago as 1966. To date the use of rimantadine has been largely confined to Russia.

Replication of the majority of influenza A isolates, whether from humans, mammals, or birds, is specifically inhibited in cell culture by amantadine and rimantadine. Infections in horses, mice, and various domestic fowl are also effectively reduced by these agents. However, drug-resistant viruses do arise readily in cell culture as a result of passage in the presence of the drug and have been isolated during the treatment of children as well as of mice and chickens [4–6]. The basis of this drug resistance, similar both in vivo and in vitro, is central to our current understanding of the mechanism of the anti-influenza action.

Genetic Basis of Drug Resistance

Evidence compiled to date from studies of drug action, comparisons of the phenotypic and genetic properties of recombinant (reassortment) viruses, and the identification of mutations in drug-resistant variants has implicated the products of only two genes, those encoding the hemagglutinin (HA) and

M proteins, in the actions of amantadine and rimantadine. It may be of significance in a discussion of the actions of these drugs that analyses of reassortments derived from a combination of sensitive and drug-resistant human isolates point to the M gene as the sole determinant of their sensitivity, whether the resistant viruses were isolated from cell culture or from children undergoing treatment with rimantadine [6–8]. Equivalent studies of certain avian (H7 or H5) viruses, however, clearly showed that their susceptibility to amantadine or rimantadine is also governed by properties of their HAs [9,10].

In every instance, nucleotide sequence analysis has identified a mutation in the M gene as responsible for conferring the drug resistance phenotype [6,11,12]. These occur within the region that codes only for the M2 membrane protein, the 97 amino-acid product of a spliced mRNA [13]. The resulting single amino-acid substitutions of residues 27, 30, 31, or 34 all fall within the hydrophobic membrane-spanning domain of the molecule, comprising residues 25–43 (Figure 42.1). The highly specific nature of these changes is further emphasized by the selection of mutants containing the same amino-acid substitution (e.g., serine 31 by asparagine), both in cell cultures from a variety of human, avian, and equine viruses and from children and chickens during treatment with either amantadine or rimantadine. The occurrence and predominance of certain other substitutions are clearly strain dependent; thus, substitution of glycine 34 by glutamic acid has only been observed in drug-resistant mutants of one avian virus, whereas isolates from another strain are altered only in amino acid 27. Whether this reflects features of M2 itself or the properties of other virus components is not known. Also the significance of amino-acid substitutions in the HAs of approximately one in four drug-resistant isolates of certain avian viruses has yet to be determined. In this regard it should be noted that the mutations in M2 completely abolish drug susceptibility and therefore may be selected subsequent to a mutation in HA that simply reduces drug sensitivity.

Genetic evidence therefore points to the transmembrane domain of the M2 protein as the prime target of amantadine and rimantadine action. Although it is clear that HA is also involved, there is as yet no evidence to indicate whether the matrix (M1) protein is also implicated in the inhibition of virus replication.

Mechanism of Action

The drugs are not virucidal and do not prevent adsorption of virus to cells or virus penetration. The block to virus replication can occur at two stages: (1) early during infection per se and prior to the onset of primary transcription by inhibiting an as yet undefined feature of virus uncoating or (2) late

Asp 44

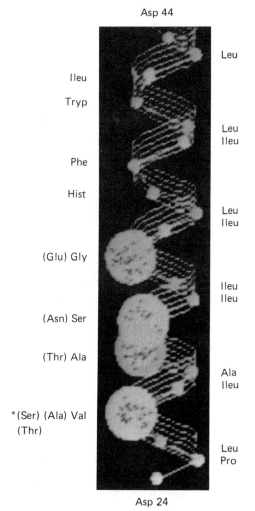

Asp 24

Figure 42.1. α-Helical representation of the hydrophobic transmembrane domain of the M2 protein of A/Chicken/Germany/27 (H7N7). The sequence spans asportic acid 24 to 44. The N terminus (bottom) is disposed toward the exterior of the cell or virus or the interior of vesicles; and the C terminus (top), toward the cytoplasm [13]. The large dots indicate amino acids substituted in drug-resistant mutants; the parentheses, prominant amino acid substitutions in the drug-resistant mutants of A/Chicken/Germany/27; and the asterisk, predominant substitutions in the mutants of A/Chicken/Germany/34.

within the cycle by preventing virus maturation [10]. In cell culture susceptibility to the early action of these drugs is a feature common to most viruses; only certain avian strains (H7 and H5) that undergo permissive infections of, for example, chick embryo cells and MDCK cells, have so far been shown to succumb to the late effect. The clearest indication as to the mechanism of drug action and the function of the M2 protein has come from studies of the latter phenomenon.

Inhibition occurs at the final stage in virus assembly by blocking the release of virus particles without interfering with bud formation [14]. This appears to be the direct consequence of a specific alteration of the HA structure during its transport to the surface of drug-treated infected cells. Addition of drug after the initial infection period does not significantly affect the synthesis of other virus components or the amount of HA synthesized; its posttranslational modification including glycosylation, acylation, and proteolytic cleavage; or its transfer to the plasma membrane. However, the conformation of the molecule exposed on the cell surface is indistinguishable from that acquired following exposure of mature HA to a pH below about 5.5, as shown by several criteria, in particular susceptibility to proteolytic digestion and reducing agent and recognition by monoclonal antibodies specific for the low pH form of HA [15]. Once inserted into the plasma membrane, the HA is refractory to drug action. Pulse-chase experiments have shown that the sensitive stage occurs some 20 minutes after synthesis of HA, following the 20°-sensitive step, and coincides with the proteolytic cleavage of HA_o to HA_1 and HA_2. This points to a compartment late in the transport pathway, such as the trans-Golgi or post-Golgi vesicles. The coincidence with protelytic cleavage is not particularly surprising in view of the nature of the phenomenon, since only the cleaved form of HA undergoes an irreversible conformational change upon exposure to low pH [16]. Furthermore, it also explains why the effect has only been observed with viruses possessing HA, which is cleaved intracellularly. Whether this action occurs more generally in natural permissive infections of the human upper respiratory tract has not been ascertained.

That alterations in the primary sequence of M2 alone can abolish the effect of amantadine argues against a direct interaction with HA and points to interference with an interaction between HA and M2 proteins—either a direct structural interaction or some indirect interrelationship between the two proteins. Evidence favors the latter, and there are indications that the acidity of the cellular compartment relative to the pH at which the conformational transition occurs is important in eliciting the effect. Certain viruses are insensitive to the action in MDCK cells, in contrast to their susceptibility in chick embryo fibroblasts; whereas mutants possessing HAs that undergo the conformational transition at higher pH values of 6.0 to 6.4 (wild type, 5.6) are sensitive in both cell types. Of interest in this regard are reports of differences in the acidities of the cisternae and vesicles of the Golgi complex of different cells and, in particular, the lower pH in corresponding compartments of fibroblasts as compared to epithelial cells [17]. Furthermore, the specific action of amantadine is counteracted by agents that elevate the pH of intracellular compartments involved in endocytosis and exocytosis. These include, for example, an extracellular pH greater than 8.5, millimolar concentrations of acidotropic amines, for example, methylamine and NH_4Cl, and low concentrations of such ionophores as monensin and nigericin. By analogy, the concentration optimum of the specific action of amantadine and rimantadine, in the range of 0.01 to 10 μM, can be explained by the antagonistic action of higher drug concentrations. The corollary of this, since M2 is instrumental in drug action, is that this protein effects a similar change in pH, which is blocked by amantadine or rimantadine. This role of M2 is supported by observations that, in the absence of drug, cells infected with certain viruses, including HA mutants (as above) and reassortants containing a heterologous M gene, may express elevated levels of low pH HA that display a cell dependence and reversibility similar to that just discussed. This further emphasizes the important relationship between HA and an M gene product, presumably M2, in preserving the integrity of the HA, apparently by influencing the pH of certain vesicular compartments during its transport to the plasma membrane.

The conclusion from these observations is that the M2 protein forms a transmembrane channel involved in reducing the pH gradient across the membrane of normally acidic vesicles and that amantadine and rimantadine interfere by blocking ion flux. This is therefore analogous to the anticholinergic action of amantadine and related compounds [18] and is similar to that of other noncompetitive antagonists that block the ion channel of the nicotinic acetylcholine receptor. A further interesting parallel concerns the possible sites of amantadine interaction. Affinity labeling of the nicotinic acetylcholine receptor with certain noncompetitive anatagonists indicated that their interaction with serine residues occurred toward the inner surface of the membrane, in homologous positions of the MII transmembrane segments of the four different polypeptide subunits of the pentamer [19]. The interaction of amantadine with an equivalent region of the virus M2 protein is indicated by the locations of the amino-acid substitutions in drug-resistant mutants. The positions of these altered amino acids and the polar and charged amino acids in wild-type and mutant viruses on the same face of an α-helical configuration of the transmembrane domain of M2 (Figure 42.1) are consistent with its proposed amphiphilic nature. Recent evidence indicating that M2 polypeptides form disulfide-linked pentamers (R. Sugrue, unpublished) is consistent with the ability of the protein to form a transmembrane channel. The consequences of the amino-acid substitutions are not known—whether, for example, they simply alter the interaction of amantadine with M2 or more drastically alter the protein structure. Studies of amphiphilic peptides have shown that similar changes in sequence can dramatically alter the conductance and cation selectivity of the ion channel [20]. The reduced "background" expression of low pH HA in cells infected with drug-resistant mutants, as opposed to wild-type virus, supports this possibility.

What are the implications for the mechanism of drug action early in infection? Since the same mutations in M2 abolish susceptibility to both early and late actions of amantadine and rimantadine, it is evident that the two actions have the same basis and are directed against a related function of M2. Concentrations of amantadine and rimantadine comparable to those effective in cell culture specifically reduce the rate of in vitro membrane fusion between virus and liposomes [21]. This is not accompanied by any reduction in the extent of fusion or the rate of the conformational transition in the HA. It may be, therefore, that the effect is exerted upon some other feature of the uncoating process, for example, disassembly of the matrix protein shell [22]. The observations that monensin increases the rate of virus membrane fusion in vitro (S. Wharton, personal communication) and that matrix protein dissociates at acidic pH lend support to the idea that in endosomes the M2 protein is involved in the transfer of protons across the viral membrane that destabilizes the virus structure.

Conclusion

In summary, therefore, current evidence regarding the specific anti-influenza A action of amantadine and rimantadine supports the notion that they act as ion channel blockers and interfere with a function of the virus M2 protein involving the transport of hydrogen ion or some counterion. This role of M2 is important both in uncoating virus particles and in preserving the integrity of HA during its transport through acidic compartments to the cell surface. Although drug-resistant mutants exhibit cross-resistance to most compounds with an activity similar to that of amantadine and rimantadine, in view of the diverse nature of "channel blockers," it is conceivable that other molecules may inhibit replication by interacting with different regions of the M2 protein.

Acknowledgments

I thank Dr. C. Orengo and Dr. S. Tendler for the computer display of the M2 sequence.

References

1. Dolin R, Reichman RC, Madore HP, Maynard R, Linton PN, Webber-Jones J (1982) A controlled trial of amantadine and rimantadine in the prophylaxis of influenza A infection. N Engl J Med 307:580–584
2. Hayden FG, Monto AS (1986) Oral rimantadine hydrochloride therapy of influenza A virus H3N2 subtype infection in adults. Antimicrob Agents Chemother 29:339–341
3. Sears SD, Clements ML (1987) Protective efficacy of low-dose amantadine in adults challenged with wild-type influenza A virus. Antimicrob Agents Chemother 10:1470–1473

4. Hall CB, Dolin R, Gala CL, Markovitz DM, Zang YQ, Madore PH, Disney FA, Tapley WB, Green JL, Francis AB, Pichichero ME (1987) Children with influenza A infection: Treatment with rimantadine. Pediatrics 80:275–282

5. Webster RG, Kawaoka Y, Bean WJ, Beard CW, Brugh M (1985) Chemotherapy and vaccination: A possible strategy for the control of highly virulent influenza virus. J Virol 55:173–176

6. Belshe RB, Hall-Smith M, Hall CB, Betts R, Hay AJ (1988) Genetic basis of resistance to rimantadine emerging during treatment of influenza virus infection. J Virol 5:1508–1512

7. Lubeck MD, Schulman JL, Palese P (1978) Susceptibility of influenza A viruses to amantadine is influenced by the gene coding for M protein. J Virol 28:710–716

8. Hay AJ, Kennedy NTC, Skehel JJ, Appleyard G (1979) The matrix protein gene determines amatadine-sensitivity of influenza viruses. J Gen Virol 42:189–191

9. Scholtissek C, Faulkner GP (1979) Amantadine-resistant and sensitive influenza A strains and recombinants. J Gen Virol 44:807–815

10. Hay AJ, Zambon MC, Wolstenholme AJ, Skehel JJ, Smith MH (1986) Molecular basis of resistance of influenza A viruses to amantadine. J Antimicrob Chemother 18(suppl b):19–29

11. Hay AJ, Wolstenholme AJ, Skehel JJ, Smith MH (1985) The molecular basis of the specific anti-influenza action of amantadine. EMBO J 4:3021–3024

12. Bean WJ, Threlkeld SC, Webster RG (1989) Biological potential of amantadine-resistant influenza A virus. J Inf Dis (in press)

13. Lamb RA, Zebedee SL, Choppin PW (1985) Influenza virus M2 protein is an integral membrane protein expressed on the infected cell surface. Cell 40:627–633

14. Ruigrok RWH, Hirst L, Hay AJ (1989) Inhibition of influenza virus maturation by amantadine—An electron microscopic examination. Virology (submitted)

15. Sugrue R, Bahadur G, Zambon MC, Hay AJ (1989) Altered expression of the influenza haemagglutinin by amantadine (submitted)

16. Wiley DC, Skehel JJ (1987) The structure and function of the haemagglutinin membrane glycoprotein of influenza virus. Ann Rev Biochem 56:365–394

17. Anderson RGW, Orci L (1988) A view of acidic intracellular compartments. J Cell Biol 106:539–543

18. Warnick JE, Maleque MA, Bakry N, Eldefrawi AT, Albuquerque EX (1982) Structure–activity relationships of amantadine: 1. Interaction of the N-alkyl analogues with the ionic channels of the nicotinic acetylcholine receptor and electrically exictable membrane. Molec Pharmacol 22:82–93

19. Changeaux JP, Giraudat J, Dennis M (1987) The nicotinic acetylcholine receptor: Molecular architecture of a ligand-regulated ion channel. Trends Pharmacol Sci 8:459–465

20. Lear JD, Wasserman ZR, Degrado WF (1988) Synthetic amphiphilic peptide models for protein ion channels. Science 240:1177–1181

21. Wharton SA, Belshe R, Hall-Smith M, Hay AJ (1989) Inhibition of membrane fusion by amantadine (submitted)

22. Bukrinskaya AG, Vorkunova NK, Pushkarskaya NL (1982) Uncoating of a rimantadine-resistant variant of influenza virus in the presence of rimantadine. J Gen Virol 60:61–66

CHAPTER 43
Control of Insect Pests by Baculoviruses

DAVID H.L. BISHOP

Two families of viruses (the Baculoviridae and the Nudaurelia β virus group) are infectious and specific in their host range to arthropod species [1]. Of these, certain baculoviruses are pathogenic for insects, and since the last century, they have been used to control pests [2–4]. Granulosis viruses (GV), the nuclear polyhedrosis viruses (NPV), and the nonoccluded *Oryctes rhinoceros* (rhinoceros beetle) viruses are among the baculoviruses that have been used commercially as insecticides [5]. Their use has been primarily focused on the control of lepidopteran (moth), hymenopteran (sawfly), and coleopteran (beetle) pests.

Viruses that infect and replicate within an arthropod host include members of approximately 12 dozen virus families [1]. Some of these viruses also infect vertebrates (the so-called arboviruses). Arboviruses (yellow fever virus, the dengue and tick-borne encephalitis viruses, etc.) appear to have little effect on their arthropod hosts, even though the same viruses may cause diseases in vertebrates. In view of such properties, arboviruses have not been used to control arthropod pests (mosquitoes, ticks, etc.). Many sap-feeding and stem-boring insects (aphids, leafhoppers, etc.) both acquire and transmit plant viruses. The relationship of such viruses with insects appears to be passive. Viruses that cannot replicate in insects are unsuitable as control agents.

Usually caterpillars that are permissive to a particular virus are susceptible to infection throughout their larval life. The dose of virus required to kill a species varies according to the permissiveness of the species and in proportion to the instar and size of the caterpillar. For highly permissive species, a single polyhedral inclusion body is sufficient to establish a lethal infection in early instar larvae; for later stages, lethal infections may require tens, hundreds, or thousands of virus inclusion bodies [6]. The probability of a pest being infected by a virus therefore depends on the particular virus–host system and the distribution of virus on the plant on which the pest feeds. In general, it is the larger, later stages of the pest that do the most crop damage. However, for crops in which cosmetic appearance is important (e.g., fruit), minor surface damage may not be commercially acceptable. Control of the pest (by any procedure) has to take this into account. In some cases it is not possible to infect later stages of larvae with virus because of the feeding habits of the pest (e.g., *Cydia pomonella*, codling moth larvae, that infest apples, pears, quinces, peaches, and walnuts), since, after initially feeding on the outer plant surface, the pest rapidly bores into internal regions of the plant (fruit or buds). The time of application of virus and many chemical insecticides is often critical to their success and to crop protection. Overlapping generations of a pest (as in the tropics), resulting in the coexistence of all its stages in a crop, and the presence of other pests complicate this problem.

A major environment advantage of viral insecticides over chemical insecticides (e.g., the organochorines, organophosphates, and the recently developed synthetic pyrethroids), or the alternative juvenile hormone analogs and chitin synthesis inhibitors, is their host specificity. From the commercial view, host specificity may be a disadvantage when there are other pest species to control. In natural ecosystems and in monoculture crops, there will be both pest and beneficial insect species, as well as insects that may be of conservation and aesthetic value (e.g., butterflies). Application of a broadly acting chemical insecticide to a crop may dramatically reduce many components of the desirable resident insect fauna (e.g., natural predators and pollinators). In addition, there can be other negative attributes of chemicals (pollution of the environment, phytopathology, and food-chain effects following absorption by other animal species such as birds and fish and other invertebrates). In some circumstances these effects can be reduced by restricting the frequency of application and the amount of chemical insecticide applied. When insect populations are severely depleted by the use of conventional chemical insecticides, recolonization of the site may occur in an unbalanced fashion, since the various insect species will recolonize at different times and at different rates. For example, in temperate regions, insects with one replication cycle per year may not recolonize a site until the following year, by which time rapidly multiplying and/or migrating insect species will have reestablished themselves. There is, therefore, an increasing and warranted concern about the immediate and long-term effects on the environment of chemical pest-control agents.

The specificy of a virus insecticide means that nonpermissive and nontarget insects, or other species, are unaffected by virus treatment of the pest. Food-chain and other species are also unharmed, since the virus cannot infect them. Thus from an environmental viewpoint, virus insecticides with a narrow host range form ideal control agents for particular pests, since they are nonpolluting and environmentally less damaging or disruptive than chemical insecticides.

After the death of the host, baculoviruses may persist in the environment (e.g., on foliage and in soil) and may be passively distributed by other insects or vertebrates that feed on the remains of the infected insect (e.g., birds, rodents). As a consequence, control may be amplified and extended to other sites and to late-developing generations of the pest species. Virus persistence depends on a variety of factors (the particular virus–host interactions; the amount of derived progeny viruses; attributes of the site; climatic considerations [sunlight, etc.]; the plants, predators, and the frequency of new pest incursions into a site; etc.). Virus persistence has been demonstrated to occur and in some cases shown to contribute to the control of subsequent generations of a pest, or its attempts to recolonize a treated site [7].

The host specificity of a virus has a negative side, since such specificity reduces the market potential of the agent. Some viruses (e.g., that of *Mamestra brassicae,* cabbage moth, NPV) exhibit wider host ranges than others (e.g., that of *Euproctis chrysorrhoea,* brown-tail moth, NPV). Because a virus insecticide may only be effective against a single pest insect, it will not eliminate coexisting, nonsusceptible pests. In rich insect faunas that include multiple pest species, the commercial use of virus insecticides has been limited, although a recent report describes the use of mixtures of virus insecticides to control two pests [8]. Viruses can also be employed in conjunction with other control procedures (i.e., in integrated pest management programs) and may be employed to a greater extent in the future when resistance of a target species develops against chemical insecticides (e.g., in certain *Heliothis* spp.), or when important environmental issues (e.g., conservation) must be considered.

A second negative side to the more general use of viral control of pests is the lack of a well-developed technology for very large-scale or continuous virus production. Since viruses are biological entities, they cannot be synthesized chemically. They have to be produced biologically. Production is usually restricted by the difficulties of working with living biological systems (e.g., the availability and culture of the host to produce the insecticide). Although for small infestations (e.g., the control of *Neodiprion sertifer,* pine sawfly, or *Panolis flammea,* pine beauty moth, in a few thousand hectares of forests in the UK) [9], adequate production of the appropriate NPVs or GVs can be achieved at minimum cost. The technology for large-scale production of baculoviruses (to apply to tens of thousands of hectares of forest or other crops) has not been developed, except in a few cases (e.g., for the control of *Lymantria dispar,* gypsy moth [10].

Despite their negative side, virus insecticides have found to be of use in various contexts. In environmentally sensitive areas (e.g., forests in parts of the United States, the United Kingdom, and Western Europe), viruses and other types of biological control (e.g., *Bacillus thuringiensis* toxins) [11,12] have been preferred to chemical insecticides. In developing countries where the cost of importing chemicals is prohibitive, virus insecticides have found particular favor.

A synopsis of the infection cycle of baculoviruses is provided below, together with some information on the viruses that have been used for pest control. In addition, new studies on the development of baculovirus insecticides by genetic engineering are discussed.

The Baculovirus Infection Cycle

Baculoviruses have a genome of double-stranded, circular supercoiled DNA (*ca* 10^5 base pairs). The viral DNA is infectious per se. In nonoccluded virions, the viral DNA and associated structural proteins (viz., the nucleocapsid) are enveloped in lipid and coated with glycoprotein. The nucleocapsids of baculoviruses may also be singly or multiply enveloped and occluded into a crystalline protein structure. The viral glycoprotein found on free virions is not present in occluded structures. The principal occluding protein is the polyhedrin protein (for the NPVs) or granulin protein (for the GVs). The two proteins have related sequences [13]. Three major subgroups of baculoviruses are recognized. Viruses of subgroup A may be singly enveloped (Figure 43.1) or multiply enveloped (Figure 43.2), occluded, and polyhedral-shaped NPVs (type species: *Autographa californica*, NPV, AcNPV). Viruses of subgroup B (Figure 43.3) are the occluded and ovicylindrical, but usually only singly enveloped GVs (type species: *Trichoplusia ni* GV). Viruses of subgroup C are nonoccluded, rod-shaped virions (type species: *Oryctes rhinoceros* virus). Apart from DNA homology studies, the genetic relationships of the different virus types have not been studied in detail [14].

An infection cycle of an HPV (e.g., AcNPV) is illustrated diagrammatically in Figure 43.4. The NPV is ingested with food (foliage) by the larval form of the host. In the alkaline midgut of the animal, the polyhedral body is digested, releasing virions to infect the columnar and other cells of the midgut. Within such cells the viral DNA is introduced into the cell nucleus, and virus replication ensues. Progeny nonoccluded virions are produced that may progressively spread the infection to other cells and tissues. Four phases of mRNA synthesis and corresponding synthesis of viral gene products are recognized (variously termed, e.g., "very early or α, early or β, late or γ, and very late or σ [15]. The very early phase involves transcription of certain viral genes by the cellular, DNA-directed RNA polymerase. Some of these mRNA species may be spliced by the cellular machinery [16]. Following their translation, other mRNA species are made. What regulates the synthesis of the viral genes that are expressed afterwards is not known. Viral DNA replication occurs, apparently involving a de novo synthesized viral DNA polymerase [17]. It is not known whether the phases that follow the very early phase mRNA transcription are controlled by viral or cellular proteins (e.g., polymerases), or which other viral gene products are involved. During the late phase, infectious nonoccluded virions are

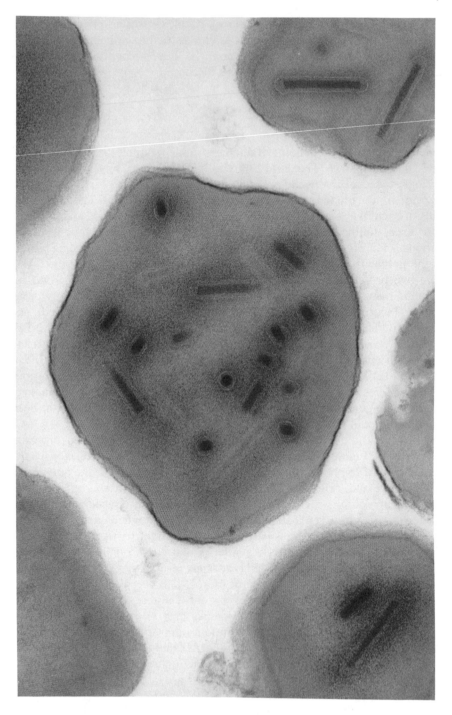

Figure 43.1. Sectioned polyhedron of *Heliothis zea* NPV, showing singly enveloped virus nucleocapsids (magnification × 99,000).

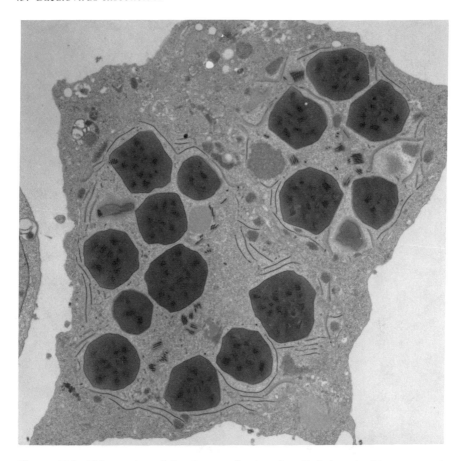

Figure 43.2. Thin section of *Spodoptera frugiperda* cells infected with *Autographa californica* NPV, showing singly and multiply enveloped viral nucleocapsids free in the cell nucleus and packaged in polyhedra (magnification × 18,750).

budded from the cell plasma membrane (thereby acquiring viral glycoprotein). By budding from the non-gut side of the midgut cells, these extracellular virions may spread the infection to other (internal) cells and tissues of the host. During the very late phase of infection, which is particularly evident in the insects' fat bodies, occlusion of viral nucleocapsids occurs, producing polyhedra. Upon death of the host, the occluded viruses are released into the environment. Different courses of infection have been recognized with other baculoviruses, depending on the virus and its host species (e.g., infections limited to midgut tissues).

Pertinent to their role as insecticides is the ability of NPVs and GVs to persist in the environment. Persistence is primarily due to the presence of an occlusion body. The importance of the polyhedrin protein has been demonstrated by removal of the polyhedrin gene from AcNPV and the use of the derived nonoccluded virus as a nonpersistent (i.e., "self-destructive")

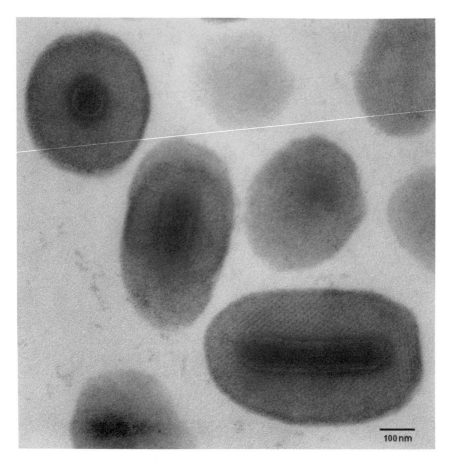

Figure 43.3. Thin section of *Darna trima* GV (magnification × 180,000).

insecticide [18]. Such nonoccluded viruses do not survive on plants, in soil, or in the remains of caterpillars in the environment and are of low intrinsic infectivity to larvae [18].

Baculovirus Insecticides

When baculovirus insecticides are used, one has to consider the characteristics of the viruses, the target pests, other pests, the crops and other plants that the pests invade, as well as attributes of the sites. For example, certain plants produce exudates that inhibit virus infectivity. *Heliothis* NPV is less infectious on cotton than on sorghum because of the cotton exudates. In part it is possible to overcome this by incorporating chelating or other agents in the formulation of the virus insecticide. Like all biological material, viruses are subject to degradation and inactivation in the environment (albeit for

BACULOVIRUSES

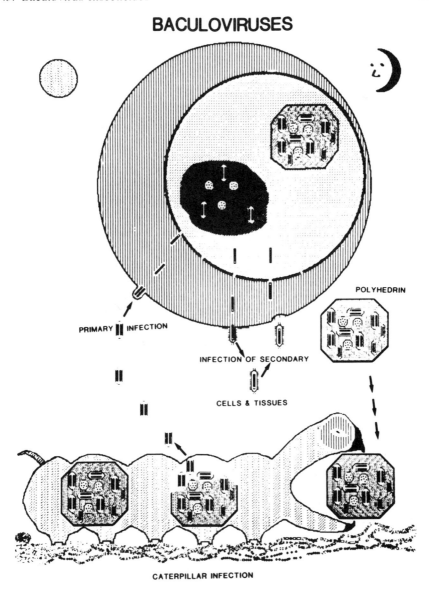

Figure 43.4. Schematic infection cycle of a lepidopteran larva by an NPV.

NPVs or GVs at reduced rates because of the occluding proteins). Protection against inactivation by sunlight may be afforded by the site (e.g., shelter by plants) and by incorporating in the insecticide preparations protectants against ultraviolet light inactivation.

The majority of pests for which there are virus control agents are members of the Lepidoptera; these include pests of agriculture (horticulture), amenity grasses, orchards, forests, and other plantation and stored crops. Some virus insecticides are applicable to hymenopteran or coleopteran pests. Space constraints do not permit a comprehensive discussion of the use of the dozens of virus insecticides that have been employed for pest control. Instead, a brief survey of the viruses that have been used to control pests of forests, orchards, and plantations will be provided, in addition to a discussion of the use of virus control in agriculture (e.g., for control of the important *Heliothis* spp. pests).

Forest, Orchard, and Plantation Pests

Some 30% of the landmass of the world is covered by forests (*ca* 4000 million hectares). In addition, there are large parts of the world with some scrub or other tree cover that is not considered to be forested. Forests are important in water conservation, the oxygen cycle, and in the support of wildlife. They are natural, renewable resources that are used in the pulp, building, and energy industries.

There are many serious insect and other pests of forests and orchards; pests that cause defoliation, bark loss, disease, and/or death of the plants. Bark beetles are considered the most serious pests of forests. Other important pests include *Hylobius abietus*, pine weevil; *Lambdina fiscellaria*, hemlock looper; *Adelges piceae*, balsam woolly aphid; *Bupalus piniaria*, pine looper; *Leucoma salicis*, white satin moth; *Tortrix viridana*, oak leaf roller and other leaf rollers; *Operophtera brumata*, winter moth; *P. flammea; N. sertifer; Orgyia pseudotsugata*, Douglas fir tussock moth; *Thaumetopoea pityocampa*, pine processionary moth; *Lymantria monacha*, nun moth; *L. dispar*, gypsy moth; *Erannis defoliaria*, mottled umber—also known as winter moth; *Malkacosoma disstria*, forest tent caterpillar; *Choristoneura fumiferana*, eastern spruce budworm; and other budworms, such as *C. occidentalis*, western spruce budworm; *Bucculatrix ainsliella*, oak leaf skeletonizer; and various aphids. Many of these pests are restricted in distribution to particular parts of the world (temperate regions, tropics, Old World, the Americas, etc.) The annual wastage due to pests is difficult to assess because of the distribution and amount of natural forest and the difficulties of survey; however, it is estimated to involve millions of hectares of forest. Reduced timber growth, the exacerbation of the situation by disease and secondary results of pest infestations (e.g., opportunisitic diseases involving fungi), pollution, acid rain, etc., complicate the issue of damage assessment.

There is increasing concern about the use of chemical insecticides to control forest pests. In view of this, viruses and other forms of biological control (e.g., *B. thuringiensis* toxin) have found favor as control agents against certain forest pests.

In view of its feeding preferences (oak, fruit, shade, ornamental, and other cultivated and wild trees), *L. dispar* is an important pest of forests and urban areas in Europe, the northeastern USA, Canada, Japan, and elsewhere. *Lymantria dispar* NPV has been used successfully in Europe, Russia (product name, Virin-ENSh), and the United States (product name, Gyp-chek). In Russia some 50,000 hectares of forest were treated with virus in 1978 alone. In Europe the virus has been used experimentally on some 200 hectares. Over the last 10 years in the United States and Canada, approximately 5,000 hectares of forest have been treated with virus. However, this amounts to only a small proportion of the total area treated during that time for infestation by *L. dispar* (viz., some 1 million hectares). For a variety of reasons, including the particular habits of the pest, *B. thuringiensis* toxin and to a lesser extent alternative chemical insecticides have been more widely used. The NPV of *L. salicis* (a pest of sallows, poplars, and willows) is cross-infective against a variety of lymantrids including *E. chrysorrhoea* (on fruit trees, hawthorn, blackthorn, brambles), *O. antiqua* (the vaporer moth on deciduous trees, conifers), *L. monacha* (oak and conifers), and *L. dispar* (broad-leaf deciduous trees such as oak) [19].

The NPV of *N. sertifer* (a pest of most *Pinus* species in Europe, Scandinavia, and the northeastern United States) has been found to be a very effective control agent. To date, over 10,000 hectares (30 million trees) of forest in Europe and Scandinavia have been successfully protected against infestations. The virus is sold by Kemira Oy of Finland and Oxford Virology in the United Kingdom. Other *Neodiprion* spp. can also be controlled by NPVs, although they are relatively minor pests. *Orgyia pseudotsugata* (Douglas fir and other coniferous trees in Canada, the western United States, and Mexico) is another serious pest that can be controlled by NPVs. The NPV of *P. flammea* (a pest of various pine species in Europe) has been used cost-effectively by staff of the NERC Institute of Virology together with the UK Forestry Commission to control pine beauty moth infestations in some 1,000 hectares of forest in the United Kingdom. *Mamestra brassicae* NPV is closely related [20] and cross-infective to *P. flammea* and has also been used for its control.

Viruses for other forest pests have been identified, but due to low efficiencies and other reasons (e.g., sporadic pest outbreaks), or effective control by *B. thuringiensis* toxins, they have not been widely employed. Among others, these viruses include the NPVs and GVs of *C. fumiferana* and *C. occidentalis* budworms of spruces in the United States and Canada and an NPV for *T. pityocampa,* a pest of pines and cedars in southern Europe.

Insect and fungal pests of plantation crops (coffee, tea, cocoa, coconut, oil palm, rubber, banana, citrus, dates, and nuts) are usually controlled by the application of chemical insecticides. Concerning coconut palm plantations, control of rhinoceros beetle infestations has been achieved by staff of the NERC Institute of Virology in conjunction with local government in the Seychelles by infecting adult beetles and releasing them to infect the sequestered *Oryctes* larvae that are found in decaying palm logs and other habitats [21]. Prior studies against *Oryctes* infestations of oil palm in southeast Asia also demonstrated effective beetle control by virus.

Codling moth (*C. pomonella*) and the related oriental fruit moth (*C. molesta,* another peach pest) are important orchard pests. In apple orchards in which no control measures are taken, infestations involving 80% or more of the crop can occur. The world annual apple and pear production is of the order of 40 and 10 million tons, respectively. Control of apple and pear pests is usually afforded by the application of a variety of chemical insecticides. A GV for codling moth affords effective control, particularly where there is a need to maintain populations of predatory insects to control other members of an orchard pest complex (e.g., red spider mite, *Panonychus ulmi*) [22]. However, because the timing required for virus application is critical (just after the larvae emerge and before they bore into the fruit) and because of the need to control other associated pests, codling moth virus insecticides have found only limited application.

Agricultural (Horticultural) Pests

Other than a brief discussion on *Heliothis* species, space precludes an in-depth discussion of the variety of viruses that have been used to control the various insect pests of agriculture and horticulture. Readers are referred to other articles for further information [4,23,24].

Larvae of the *Heliothis* species are among the most serious pests of agricultural crops. In recent years control of *Heliothis* pests on cotton is estimated to account for some 30% of the world annual production of chemical insecticides. The various *Heliothis* species are distributed widely in the world (between 40° north and south of the equator). *Heliothis zea* (cotton bollworm, corn earworm, tomato fruitworm) and *H. virescens* (tobacco budworm, tomato budworm) are pests of several important agricultural crops, particularly cotton, maize, tobacco, beans and other vegetable crops, clover, sesame, sunflowers, and other farm products). Depending on the crop (site, etc.), they can be effectively controlled using *Heliothis* spp. NPV. *Heliothis zea* and *H. virescens* are restricted in their distribution to the New World (i.e., the Americas and the West Indies). *Heliothis armigera* (Old World bollworm; African cotton bollworm; a pest of cotton, maize, sorghum, millet, tobacco, Sudan grass, legumes such as chick-peas, pigeon pests, citrus fruits, tomatoes, soya beans, and sunflower) can also be effectively controlled by virus. *Heliothis armigera* is a major pest in Southern Europe, Africa, and the Middle East as well as most of Asia and Australia. It is not found in the Americas. Other *Heliothis* spp. that are important agricultural pests include *H. assulta* (a pest of tobacco, maize, tomato, and other crops in Africa, Australia, and Asia), *H. geotoporon* (cotton, sorghum, flax in the Americas), *H. punctigera* (tobacco, cotton, lucerne in Australia), and *H. viriplaca* (cotton and various legumes in Russia and Asia). Many of these and other *Heliothis spp.* are regional pests, ofen occurring together and in association with other pests (pest complexes) [14,19,23].

NPVs isolated from various *Heliothis* spp. and other Lepidoptera have been shown to be effective control agents of *Heliothis, Spodoptera,* and *Anticarsia* on a variety of crops (cotton, soybeans, tobacco, maize, sorghum, etc.) [4,23,24]. Based on *H. zea* NPV, a commercial product (Elcar) was produced by Sandoz Chemical Company. However, production of this insecticide lapsed with the development of the new synthetic pyrethroids. Although, at present, chemicals are preferred control agents for the *Heliothis* and many other agricultural pests, with the emergence of chemical insecticide-, including synthetic pyrethroid-resistant species, virus insecticides may again be used as control agents.

Integrated Pest Management

In addition to viruses and the traditional forms of broad-acting chemical insecticides (the organophosphorous and organochlorine compounds), a variety of new insecticides and other agents have been developed or are under development for the control of arthropod pests (e.g., Lepidoptera, Hymenoptera, Coleoptera, ticks, and mites), as have compounds that may be used in combination with other chemicals or in conjunction with viruses (i.e., in integrated pest-management programs). The new forms of insecticide that have been developed are more selective than the earlier chemical insecticides and have low mammalian toxicity. Some of these compounds, such as the insect growth regulators, may have a reduced effect on beneficial predators or parasitoids. New insect growth regulators include the benzyolureas (teflubenzuron, chlorofluazuron, diflubenzuron-Dimilin) and the acylureas. Some of the new candidate pesticides are ovicidal; others, acaricidal. Other compounds that may be used to control pests include the benzaphenone hydrazones and thiazoles, both of which have different modes of action on growth regulators. Antibiotics (e.g., avermectin) and *B. thuringiensis* toxins are control agents that are suitable for integrated pest-management programs. All these compounds have to be ingested to exert an effect and can remain active after application for several weeks. Finally, pheromones have been used in integrated pest-management programs both to bait insects and to disrupt insect breeding habits.

Genetically Engineered Virus Insecticides

There is considerable room to improve the known virus insecticides by genetic engineering. Issues to be addressed include their virulence, speed of action, host range, and prevention of resistance. In the future, insecticides may be developed to address alternative arthropod pests (mosquitos, ticks, mites, locusts, flies, cockroaches, etc.).

The speed of action of a baculovirus insecticide may be improved by including "foreign" genes into the viral DNA. The foreign genes under investigation include those that should affect the physiology of the host species and inhibit its eating habits or development (e.g., insect hormone genes) and genes that are lethal in their own right (e.g., toxins).

Baculoviruses are specific in their host range to arthropods. In view of the complexities of virus–host relationships at the molecular level, it is inconceivable that they could be engineered to be otherwise. Since baculovirus insecticides have track records of safety and suitability, they are appropriate subjects for genetic engineering. Some baculoviruses are more amenable to genetic engineering than others because of the availability of tissue culture and cloning systems. Most of the genetic engineering work has focused on AcNPV, which can be engineered to express high levels of many foreign genes [25]. The level of expression varies according to the viral promoter used as well as the particular attributes of the foreign gene product [26].

In addition to commercial issues, any genetically engineered virus insecticide that is developed for field use will have to satisfy both the general public and local and national (and international) regulatory authorities with regard to suitability and safety. In some countries the use of genetically engineered organisms in the environment is an emotional issue that is under review with regard to regulation and the appropriate legislation.

In the United Kingdom, four field releases of genetically engineered baculoviruses have been undertaken with the agreement of the necessary regulatory authorities [27]. The first release was undertaken in 1986; the second, in 1987 [18]. Two more releases occurred in 1988. In the first release, a genetically marked AcNPV was used to follow the fate (persistence, distribution) of the virus in the environment. The second release involved a genetically marked AcNPV from which the polyhedrin gene had been removed. The polyhedrin-negative virus was developed as a "self-destructive" viral insecticide. The experimental details of the studies have been published [18]. The studies undertaken in 1988 also involved the polyhedrin-negative virus and another AcNPV derivative in which a marker "junk" gene has been incorporated. Based on these and future risk-assessment studies, as well as the required laboratory tests (see ref. 18), it is anticipated that it will be possible to develop comprehensive environmental impact analyses to determine if there are any risks associated with the release of genetically engineered baculoviruses into the environment.

References

1. Matthews REF (1982) Classification and nomenclature of viruses. Intervirology 17:1–199
2. Cunningham JC (1982) Field trials with baculovirues: Control of forest insect pests. In Kurstak E (ed) Microbial and Viral Pesticides. Marcel Dekker, New York, pp 335–386
3. Evans HF, Entwistle PF (1987) Viral diseases. In Fuxa JR, Tanada Y (eds) Epizootiology of Insect Diseases. John Wiley and Sons, New York, pp 257–322

4. Huber J (1986) Use of baculoviruses in pest management programs. *In* Granados RR, Federici BA (eds) The Biology of Baculoviruses. II Practical Application for Insect Control. CRC Press, Boca Raton, Florida, pp 181–202

5. Entwistle PF, Evans HF (1985) Viral control. *In* Gilbert LI, Kerkut GA (eds) Comprehensive Insect Physiology, Biochemistry and Pharmacology 12:347–367

6. Evans HF (1981) Quantitative assessment of the relationships between dosage and response of the nuclear polyhedrosis virus of *Mamestra brassicae*. J Invert Pathol 37:101–109

7. Carruthers WR, Cory JS, Entwistle PF (1988) Recovery of pine beauty moth (*Panolis flammea*) nuclear polyhedrosis virus from pine foliage. J Invert Pathol 52:27–32

8. Doyle CJ, Entwistle PF (1988) Aerial application of mixed virus formulations to control joint infestations of *Panolis flammea* and *Neodiprion sertifer* on lodgepole pine. Ann Appl Biol 113:119–127

9. Kelly PM, Entwistle PF (1988) *In vivo* mass production in the cabbage moth (*Maestra brassicae*) of a heterologous (*Panolis*) and a homologous (*Mamestra*) nuclear polyhedrosis virus. J Virol Methods 19:249–256

10. Shapiro M (1982) *In vivo* mass production of insect viruses for use as pesticides. *In* Kurstak E (ed) Microbial and Viral Pesticides. Marcel Dekker, New York, pp 463–492

11. Cunningham JC (1988) Baculoviruses: Their status compared to *Bacillus thuringiensis* as microbial insecticides. Outlook on Agriculture 17:10–17

12. Luthy P. Cordier J-L, Fischer H-M (1982) *Bacillus thuringiensis* as a bacterial insecticide: Basic considerations and application. *In* Kurstak E (ed) Microbial and Viral Pesticides. Marcel Dekker, New York, pp 35–74

13. Rohrmann GF (1986) Polyhedrin structure. J Gen Virol 67:1499–1573

14. Smith GE, Summers MD (1982) DNA homology among subgroup A, B and C Baculoviruses. Virology 123:393–406

15. Kelly DC, Lescott T (1981) Baculovirus replication: Protein synthesis in *Spodoptera frugiperda* cells infected with *T. ni* NPV. Microbiologica 4:35–57

16. Chisholm GE, Henner DJ (1988) Multiple early transcripts and splicing of the *Autographa californica* nuclear polyhedrosis virus IE-1 gene. J Virol 62:3193–3200

17. Wang X, Kelly DC (1983) Baculovirus replication: Purification and identification of the *Trichoplusia ni* nuclear polyhedrosis virus-induced DNA polymerase. J Gen Virol 64:2229–2236

18. Bishop DHL, Entwistle PF, Cameron IR, Allen CJ, Possee RD (1988) Field trials of genetically engineered baculovirus insecticides. *In* Sussman M, Collins CH, Skinner FA, Stewart-Tull DE (eds) The Release of Genetically Engineered Micro-Organisms. Academic Press, London, pp 143–179

19. Skatulla Von U (1985) Untersuchunges zur Wirkung eines Kernopolyeder-virus aus *Leucoma salicis* L. (Lep., Lymantriidae) auf einige Lymantriiden-Arten. Anzeiger fur Schadlingskunde Pflanzenschutz, Umweltschultz 58:41–47

20. Possee RD, Kelly DC (1988). Physical maps and comparative DNA hybridization of *Mamestra brassicae* and *Panolis flammea* NPV genomes. J Gen Virol 69:1285–1298

21. Lomer CJ (1986) Release of *Baculovirus oryctes* into *Oryctes monoceros* populations in the Seychelles. J Invert Pathol 47:237–246

22. Glen DM, Wiltshire CW, Milsom NF, Brain P (1984) Codling moth granulosis

virus: Effects of its use on some other orchard arthropods. Ann Appl Biol 104:99–106

23. Ignoffo CM, Couch TL (1981) The nucleopolyhedrosis virus of *Heliothis* species as a microbial insecticide. *In* Burges HD (ed) Microbial Control of Pests and Plants Diseases 1970–1980. Academic Press, London, pp 329–362

24. Yearian WC, Young SY (1982) Control of insect pests of agricultural importance by viral insecticides. *In* Kurstak E (ed) Microbial and Viral Pesticides. Marcel Dekker, New York, pp 387–423

25. Luckow VA, Summers MD (1988) Trends in the development of baculovirus expression vectors. Bio/Technology 6:47–55

26. Vlak JM, Klinkenberg FA, Zaal KJM, Usmany M, Klingeroode EC, Geervliet JBF, Roosien J, Van Lent JWM (1988) Functional studies on the p10 gene of *Autographa californica* nuclear polyhedrosis virus using a recombinant expressing a p10-beta-galactosidase fusion gene. J Gen Virol 69:765–776

27. Bishop DHL (1986) UK release of genetically marked virus. Nature 323:496

Index